Mechanochromic Fluorescent Materials
Phenomena, Materials and Applications

RSC Smart Materials

Series Editors:
Hans-Jörg Schneider, *Saarland University, Germany*
Mohsen Shahinpoor, *University of Maine, USA*

Titles in this Series:
1: Janus Particle Synthesis, Self-Assembly and Applications
2: Smart Materials for Drug Delivery: Volume 1
3: Smart Materials for Drug Delivery: Volume 2
4: Materials Design Inspired by Nature
5: Responsive Photonic Nanostructures: Smart Nanoscale Optical Materials
6: Magnetorheology: Advances and Applications
7: Functional Nanometer-Sized Clusters of Transition Metals: Synthesis, Properties and Applications
8: Mechanochromic Fluorescent Materials: Phenomena, Materials and Applications

How to obtain future titles on publication:
A standing order plan is available for this series. A standing order will bring delivery of each new volume immediately on publication.

For further information please contact:
Book Sales Department, Royal Society of Chemistry, Thomas Graham House, Science Park, Milton Road, Cambridge, CB4 0WF, UK
Telephone: +44 (0)1223 420066, Fax: +44 (0)1223 420247
Email: booksales@rsc.org
Visit our website at: www.rsc.org/books

Mechanochromic Fluorescent Materials
Phenomena, Materials and Applications

Edited by

Jiarui Xu
Material Sciences Institute, School of Chemistry and Chemical Engineering, Sun Yat-sen University, P. R. China
Email: xjr@mail.sysu.edu.cn

and

Zhenguo Chi
Material Sciences Institute, School of Chemistry and Chemical Engineering, Sun Yat-sen University, P. R. China
Email: chizhg@mail.sysu.edu.cn

ROYAL SOCIETY OF CHEMISTRY

THE QUEEN'S AWARDS FOR ENTERPRISE: INTERNATIONAL TRADE 2013

RSC Smart Materials No. 8

Print ISBN: 978-1-84973-821-7
PDF eISBN: 978-1-78262-322-9

ISSN: 2046-0066

A catalogue record for this book is available from the British Library

Published by The Royal Society of Chemistry,
Thomas Graham House, Science Park, Milton Road,
Cambridge CB4 0WF, UK

Registered Charity Number 207890

For further information see our web site at www.rsc.org

Foreword

The recent, rapid expansion in our understanding of the structure-properties relationship of many chemical compositions at the molecular level enables us design new functional materials useful for practical applications. One such group of materials is the so-called "smart materials", which are also called "stimuli-responsive" or "intelligent" materials. Some scientists and engineers prefer to describe them as "designed materials". "Smart materials", responding to external stimuli, undergo alteration in their functional properties of mechanical, optical, including spectroscopic or electromagnetic, behavior due to changes occurring in their molecular structure, composition or morphology. I believe that there would be no objection to calling this century "the age of smart materials". These materials are leading the rapid innovative developments that we are presently experiencing in many technological fronts, responding to pressing global demands.

This book, *Mechanochromic Fluorescent Materials: Phenomena, Materials and Applications*, is a valuable addition to the limited list of monographs published on the topic of smart materials. This book is unique in that its content covers a complete spectrum of the science and engineering of mechanofluorochromic compounds and materials: low molar mass compounds, polymers and their composites, organometallic complexes and molecular aggregates or assemblies. This book not only guides us to novel exciting new discoveries about how some of compounds and polymers reveal optical and spectroscopic responses to external stimuli, but also teaches us molecular-level understanding of the changes, both in the chemical structural and micro- or macro-morphological levels. There are many expectations that those materials will soon find a wide variety of advanced technical applications. The topics of each chapter are superbly connected and the overall contents are very well balanced. This book is expected to be

RSC Smart Materials No. 8
Mechanochromic Fluorescent Materials: Phenomena, Materials and Applications
Edited by Jiarui Xu and Zhenguo Chi

Published by the Royal Society of Chemistry, www.rsc.org

very helpful for advanced undergraduates, graduate students and those who would like to enter this particular area of research to attain an overall, but relatively comprehensive view of the subject of this monograph.

I would like to extend a warm compliment to Prof. Jiarui Xu and Prof. Zhenguo Chi, editors of this book, and all the contributors, for bringing us the science of the exciting new developments on the phenomena and materials of mechanofluorochromism. It is my sincere hope that this book decorates brilliantly the desks and bookshelves of concerned students, scientists and engineers. On behalf of the authors, I would like to thank the Royal Society of Chemistry for having selected such a wonderful topic and helping to publish this book in a very timely fashion.

Jung-Il Jin
Former President of IUPAC
Distinguished Professor of Korea University

Contents

RSC Smart Materials No. 8
Mechanochromic Fluorescent Materials: Phenomena, Materials and Applications
Edited by Jiarui Xu and Zhenguo Chi

Published by the Royal Society of Chemistry, www.rsc.org

CHAPTER 1

Mechanofluorochromism: An Overview

JIARUI XU* AND ZHENGUO CHI

PCFM Lab, State Key Laboratory of Optoelectronic Materials and Technologies, School of Chemistry and Chemical Engineering, Sun Yat-sen University, Guangzhou, China
*Email: xjr@mail.sysu.edu.cn

Smart materials are designed materials possessing one or more properties that can be significantly changed in a controlled fashion by external stimuli, such as stress, temperature, moisture, pH and electric or magnetic fields. Thus, smart materials are also called stimuli-responsive materials. Smart materials have been considered to be the fourth generation of materials after natural materials, synthetic polymer materials and artificial design materials, which is one of the important developing trends in high-tech new materials and will support the development of modern high technology, resulting in the disappearance of the boundaries of the traditional functional and structural materials to realize the functionalization of structural materials and the diversification of functional materials. It is believed that the development and large-scale application of smart materials will lead to a great revolution in the development of material sciences. Smart materials and systems have been widely investigated as physical atomic-level switches, memories to chemical sensing systems and bio-related controlled release systems. In particular, organic and biological soft materials, which are generally not restricted by crystal lattices or inflexible bonding, are anticipated to find various applications as functional materials because of their flexible responses to the applied stimuli. Some prototypical smart materials could be operated through rather ambiguous mechanisms, where precise

RSC Smart Materials No. 8
Mechanochromic Fluorescent Materials: Phenomena, Materials and Applications
Edited by Jiarui Xu and Zhenguo Chi

Published by the Royal Society of Chemistry, www.rsc.org

control of the component molecules was often neglected. Because of this, these materials might be used in a variety of applications but are usually inferior in their specific response to a selected stimulus. To develop more specifically responsive materials, control over particular molecular activity is critical. The stimulus selectivity through molecular recognition or related interactions, together with the specific changes of molecular properties, defined by the component molecular structures would lead to materials with appropriate outputs upon application of selected inputs. Such materials are often referred to as molecular smart materials.[1]

Generally, the external stimuli mainly fall into one of two categories: one is physical stimulus, which is used to promote alteration of molecular properties, including optical, electrical, magnetic and thermal inputs, and the other is chemical/biological stimulus, such as the addition of chemicals and biomolecules. Research into the molecular responses to these various stimuli has been extensive, and the design of molecular-responsive smart materials have been well explored. Most of these stimuli are suitable for direct interaction with the molecules dispersed in appropriate media. Also, several of these stimuli, including magnetic and thermal effects, can propagate through space and interact at a distance. Under these circumstances, the behaviors of the molecules upon application of these stimuli can be easily observed using conventional analytical tools.

A simple but important stimulus for controlling the properties of smart materials is mechanical stimulus. In fact, smart materials can response to external mechanical stimuli such as shearing, grinding, rubbing, stretching and bending, and provide controlled functions.

Mechanochromism is an overarching term that describes the phenomenon of color change caused by mechanical grinding, crushing or rubbing, which has also been known as tribochromism or pressing, and also termed *piezochromism*,[2] of the solid sample and reversion to the original color by, for example, heating or recrystallization. The grinding process usually generates a metastable form. One example of a typical piezochromic organic compounds is diphenylflavylene.[3] Here, piezochromism means the change of sample colors under mechanical force, not the change of the emission colors of the sample.

On the other hand, mechanochromic fluorescent (or *mechanofluorochromic*) materials change their emission colors (spectra) when an appropriate external mechanical force stimulus is applied. As a class of "smart" materials, they possess mechanical responsiveness that provides a fundamental basis for fluorescence switches, mechanosensors, indicators of mechanohistory, security papers, optoelectronic devices and data storage in various fields.[4–7]

Mechanofluorochromic behavior can be generally achieved by either chemical or physical structural changes to the materials. Although the modification of molecular structures containing open/closed cyclic forms[8] and double-bond E/Z configurations[3] is the most common approach for tuning the emissions of fluorescent compounds, limited success has also been achieved in switching the fluorescence of the solid-state materials with high efficiency and reproducibility.[9]

Chemical structural change is implemented using chemical reactions, such as bond breaking or forming at the molecular level. In these cases, a relatively high pressure or rigorous conditions would be necessary to promote the chemical reactions.[10] Moreover, insufficient conversion, irreversible reactions, or loss of the fluorescence capability of the compound may frequently occur during the solid-state chemical reactions, which have been considered a drawback in such systems. Thus, the successful example reported for mechanofluorochromism based on the chemical structural change is very limited (Figure 1.1A).[8]

Control of the molecular phenomena by macroscopic mechanical stimuli is not well appreciated. For example, it may be claimed that elongation, deformation, disentanglements and ruptures of molecules may occur when mechanical stress is applied, but what is really occurring at the molecular level is not well understood. Although the activation of covalent bonds has primarily been demonstrated in solution, activation in the solid state has not been thoroughly investigated. To date, there are still many experimental and theoretical challenges to be overcome for a better comprehension of molecular behaviors in the solid state under mechanical stimuli.

By contrast, the fluorescent properties of molecules in the solid state depend on the molecular arrangement, conformational flexibility and intermolecular interactions in the materials. Any modification or alteration of the molecular packing and conformation of the fluorophore would affect the HOMO-LUMO energy levels and alter the fluorescent properties (Figure 1.1B). Thus, controlling the mode of molecular packing (aggregation states) to achieve the dynamic control of highly efficient and reversible solid-state fluorescence is more attractive for both fundamental research and practical applications because of low pressure demand and good reversibility for these compounds. In this book, recent progress in the development of mechanofluorochromic materials is reviewed and discussed.

By now, compounds reported to have mechanofluorochromic properties mainly include organic compounds, organometallic complexes, polymers and dye-doped polymer composites.

Dye-doped polymer composites are a representative family of mechanofluorochromic materials. The typical characteristic of these systems is that the fluorescence colors may change drastically along with the tensile deformation of matrix polymers. On the other hand, a great deal of attention has been paid to the mechanochromic properties of single-component small organic dye molecules. In comparison with the packing changes in the organic compounds, metal complexes can utilize metal-metal interactions to adjust their optoelectronic properties.

However, the mechanochromic fluorescent materials that are dependent on changes in the physical molecular packing modes are still extremely rare to this day. This rarity may be attributed to two major issues.[5] Firstly, predicting and designing for such materials exhibiting polymorphism with different luminescent properties are difficult. Each identified compound reported in the literature seems an isolated event, which makes

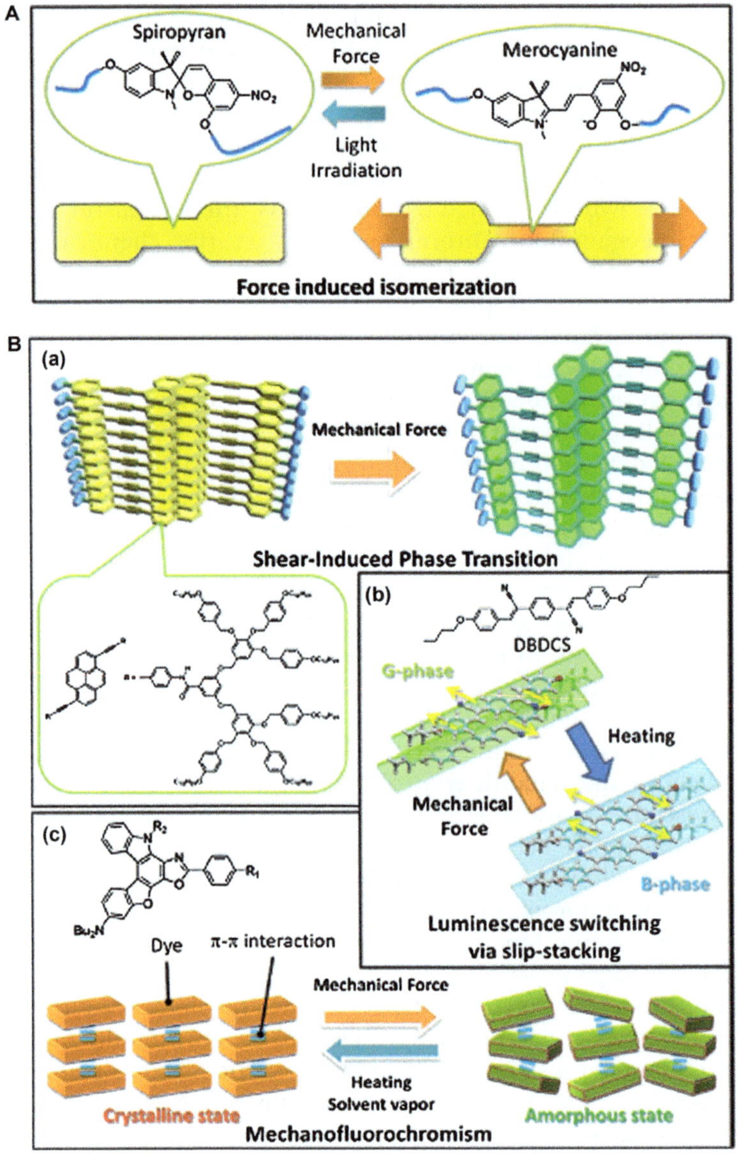

Figure 1.1 (A) Mechanoresponsive polymer: control by mechanical stretching of isomerization between spiropyran and merocyanine attached to polymer chain. (B) Mechanochromic luminescence by varying intermolecular interactions: (a) control of stacking of liquid crystalline molecules through mechanical shearing; (b) shear induced luminescence color switching *via* slip-stacking of molecular sheets; (c) mechanofluorochromism between crystalline and amorphous states with changes of molecular arrangement and intermolecular π-π interactions.
Adapted with permission from reference 7; Copyright 2012 Wiley-VCH.

identification of a general characteristic and regularity difficult. Secondly, the luminescent efficiency of organic fluorescent materials often becomes very weak when they are in the solid state because of the so-called aggregation-caused quenching (ACQ) effect. Consequently, observation of the mechanochromic fluorescent phenomenon in the solid samples becomes difficult.

The discovery of so-called aggregation-induced emission (AIE) materials, first reported by Tang *et al.* in 2001,[11] opened up an important opportunity for the exploitation of new mechanochromic fluorescent materials. AIE materials are a class of anti-ACQ materials that emit more efficiently when they are in the aggregated state than when they are in the dissolved form, and have attracted considerable research attention for their potential application in various fields, such as organic light-emitting devices and chemosensors.[12–17] Meanwhile, a number of AIE compounds with different AIE moieties have been found to possess mechanochromic emission properties. Hence, the use of AIE becomes important in the synthesis of various mechanochromic fluorescent materials.

The miraculous properties and potential applications of mechanochromic fluorescent materials comprise a new branch of modern smart materials, and more mechanochromic luminescent systems are expected to be discovered.

Thus, seven chapters are included in this book. Chapter 1 serves as an introduction. Chapter 2 will describe organic small molecule mechanofluorochromic materials, which are considered to be a major component of the mechanochromic fluorescent materials reported in the literature. In Chapter 3, mechanochromic luminescent metal complexes are described. Although the number of mechanochromic luminescent metal complexes is still limited to date, the unique properties of organometallic or coordination compounds should make these materials a novel source for mechanochromic fluorescent materials. Mechanofluorochromic polymers and dye-doped polymer composites are described in Chapter 4 and Chapter 5, respectively, some of the systems are genetic examples of mechanochromic fluorescent materials. Chapter 6 will deal with the relatively new types of mechanofluorochromic AIE compounds. Many of the compounds have been prepared by our group, and this chapter can be considered a comprehensive summary by the authors in this field. In the last chapter, Chapter 7, the major mechanofluorochromic mechanisms are summarized, which is vital to understanding how mechanofluorochromism happens.

Readers are also encouraged to refer to other published valuable reference reviews about mechanochromic luminescent materials,[1,4,5,18–31] which are extremely helpful for understanding the concepts, mechanism and applications of these new materials.

References

1. K. Ariga, T. Mori and J. P. Hill, *Adv. Mater.*, 2012, **24**, 158–176.
2. H. Bouas–Laurent and H. Dürr, *Pure Appl. Chem.*, 2001, **73**, 639–665.

3. A. Schönberg, M. Elkaschef, M. Nosseir and M. M. Sidky, *J. Am. Chem. Soc.*, 1958, **80**, 6312–6315.
4. Y. Sagara and T. Kato, *Nat. Chem.*, 2009, **1**, 605–610.
5. Z. G. Chi, X. Q. Zhang, B. J. Xu, X. Zhou, C. P. Ma, Y. Zhang, S. W. Liu and J. R. Xu, *Chem. Soc. Rev.*, 2012, **41**, 3878–3896.
6. A. Pucci and G. Ruggeri, *J. Mater. Chem.*, 2011, **21**, 8282–8291.
7. K. Ariga, T. Mori and J. P. Hill, *Adv. Mater.*, 2012, **24**, 158–176.
8. D. A. Davis, A. Hamilton, J. Yang, L. D. Cremar, D. V. Gough, S. L. Potisek, M. T. Ong, P. V. Braun, T. J. Martínez, S. R. White, J. S. Moore and N. R. Sottos, *Nature*, 2009, **459**, 68–72.
9. J. W. Chung, Y. You, H. S. Huh, B. K. An, S. J. Yoon, S. H. Kim, S. W. Lee and S. Y. Park, *J. Am. Chem. Soc.*, 2009, **131**, 8163–8172.
10. D. L. Fanselow and H. G. Drickamer, *J. Chem. Phys.*, 1974, **61**, 4567–4574.
11. J. Luo, Z. Xie, J. W. Y. Lam, L. Cheng, H. Chen, C. Qiu, H. S. Kwok, X. Zhan, Y. Liu, D. Zhu and B. Z. Tang, *Chem. Commun.*, 2001, 1740–1741.
12. Y. Hong, J. W. Y. Lam and B. Z. Tang, *Chem. Commun.*, 2009, 4332–4353.
13. Y. Hong, J. W. Y. Lam and B. Z. Tang, *Chem. Soc. Rev.*, 2011, **40**, 5361–5388.
14. B. K. An, D. S. Lee, J. S. Lee, Y. S. Park, H. S. Song and S. Y. Park, *J. Am. Chem. Soc.*, 2004, **126**, 10232–10233.
15. Z. Y. Yang, Z. G. Chi, T. Yu, X. Q. Zhang, M. N. Chen, B. J. Xu, S. W. Liu, Y. Zhang and J. R. Xu, *J. Mater. Chem.*, 2009, **19**, 5541–5546.
16. H. Y. Li, Z. G. Chi, X. Q. Zhang, B. J. Xu, S. W. Liu, Y. Zhang and J. R. Xu, *Chem. Commun.*, 2011, **47**, 11273–11275.
17. B. Xu, J. B. Zhang, S. Q. Ma, J. L. Chen, Y. J. Dong and W. J. Tian, *Prog. In Chem.* (Chinese), 2013, **25**, 1079–1089.
18. M. M. Caruso, D. A. Davis, Q. L. Shen, S. A. Odom, N. R. Sottos, S. R. White and J. S. Moore, *Chem. Rev.*, 2009, **109**, 5755–5798.
19. C. Weder, *Chimia*, 2009, **63**, 758–763.
20. S. Varghese and S. Das, *J. Phys. Chem. Lett.*, 2011, **2**, 863–873.
21. J. M. Hu and S. Y. Liu, *Macromolecules*, 2010, **43**, 8315–8330.
22. A. Pucci, R. Bizzarri and G. Ruggeri, *Soft Matter*, 2011, **7**, 3689–3700.
23. K. Ariga, T. Mori and J. P. Hill, *Chem. Sci.*, 2011, **2**, 195–203.
24. D. R. T. Roberts and S. J. Holder, *J. Mater. Chem.*, 2011, **21**, 8256–8268.
25. A. Pucci and G. Ruggeri, *J. Mater. Chem.*, 2011, **21**, 8282–8291.
26. A. L. Black, J. M. Lenhardt and S. L. Craig, *J. Mater. Chem.*, 2011, **21**, 1655–1663.
27. A. L. Balch, *Angew. Chem. Int. Ed.*, 2009, **48**, 2641–2644.
28. F. Ciardelli, G. Ruggeri and A. Pucci, *Chem. Soc. Rev.*, 2013, **42**, 857–870.
29. J. Gierschner and S. Y. Park, *J. Mater. Chem. C*, 2013, **1**, 5818–5832.
30. Y. Q. Dong, C. Y. Li, W. J. Zhao, Y. P. Dong and B. Z. Tang, *J. Mol. Eng. Mater.*, 2013, **1**, 1340010(1)–1340010(13).
31. X. Q. Zhang, Z. G. Chi, Y. Zhang, S. W. Liu and J. R. Xu, *J. Mater. Chem. C*, 2013, **1**, 3376–3390.

CHAPTER 2

Organic Small-molecule Mechanofluorochromic Materials

MIAO LUO[a] AND XIE ZHOU*[b]

[a] College of Basic Education, Zhanjiang Normal University, Zhanjiang, Guangdong, China; [b] School of Pharmaceutical Sciences, Sun Yat-sen University, Guangzhou, China
*Email: zhouxie@mail.sysu.edu.cn

2.1 Introduction

In the last two decades, organic π-conjugated materials have attracted immense interest due to their versatility in tuning optical and electronic properties, as promising candidates for various opto-electronic applications, such as organic light emitting diodes (OLEDs), field effect transistors, solar cells and photodetectors.[1–4] While the majority of such materials have initially been based on the polymers due to their easy processability and affordable mechanical properties, their ill-defined and complicated structures and morphologies have seriously limited a clear understanding of their inherent properties, which has hindered exploitation of the full potential of π-conjugated materials by demanding tedious protocols for device optimization. Thus, the focus has been gradually shifted to small organic molecules and oligomers, which benefit from a well-defined chemical structure and precisely tunable intra- and intermolecular interactions, toward active

RSC Smart Materials No. 8
Mechanochromic Fluorescent Materials: Phenomena, Materials and Applications
Edited by Jiarui Xu and Zhenguo Chi

Published by the Royal Society of Chemistry, www.rsc.org

components for organic sensors, field effect transistors, solar cells, light-emitting diodes and lasers.[5]

Thus, although polymers and dye-doped polymers are a representative family of mechanofluorochromic materials, reports on the successful synthesis of these polymers are very limited. A great deal of attention has thus been given to the mechanofluorochromic properties of single-component small dye molecules.

The mechanofluorochromic compounds so far reported are attributed to a reversible switching between crystalline and amorphous states, between stable and metastable liquid crystalline phases, or between the two different crystalline phases with changes of intermolecular interactions, such as π–π interaction, dipole-dipole interaction, and hydrogen bonding before and after grinding of solids.

Because of such technological evolution, it is therefore urgent to establish the design concepts of solid state properties, which remains a challenging issue because of the subtle interplay of intra- and intermolecular interactions which strongly alter their opto-electronic properties.[6] To disentangle these different contributions, however, and to drive molecules into a specific stacking geometry which induces the desired electronic and/or optical properties, a detailed and systematic understanding of the structure-property relationships is required. This can only be achieved by combining (i) complete structural (X-ray) characterization by using high-quality crystals, (ii) accurate and thorough spectroscopic investigation, (iii) an appropriate theoretical model and, most importantly, (iv) modular molecular systems which enable systematic variation of the relevant opto-electronic properties by varying the intermolecular interactions.[7]

2.2 Organic Small-molecular Compounds

2.2.1 Tetraphenylpyrene Carboxylic Acid Derivatives

Araki *et al.* (2007) reported the use of tetraphenylpyrene(TPPy)-based dye **1** that contains four hexyl amide groups at the *p*-position of phenyl units.[8] Their study pioneered the research on pure organic mechanofluorochromic materials. The addition of methanol to a chloroform solution of dye **1** results in the precipitation of a white powder ("B-form") that exhibits strong blue fluorescence (Figure 2.1a). When pressed with a spatula, the B-form transforms into a yellowish solid ("G-form") with a strong greenish fluorescence (Figure 2.1b). After grinding, the G-form solid readily dissolves in chloroform and re-precipitates from the solution to produce the B-form powder again. The B-form solid does not re-dissolve in chloroform or in other common organic solvents. The G-form powder can be converted back into the B-form upon heating at above 112 °C. The B-form is stable with ordered hydrogen-bonding arrays (Figure 2.1c, left), whereas the G-form is in a metastable state in which the hydrogen bonding was more random (Figure 2.1c, right). When exposed to a pressure stimulus, the tight hydrogen bonding interactions are

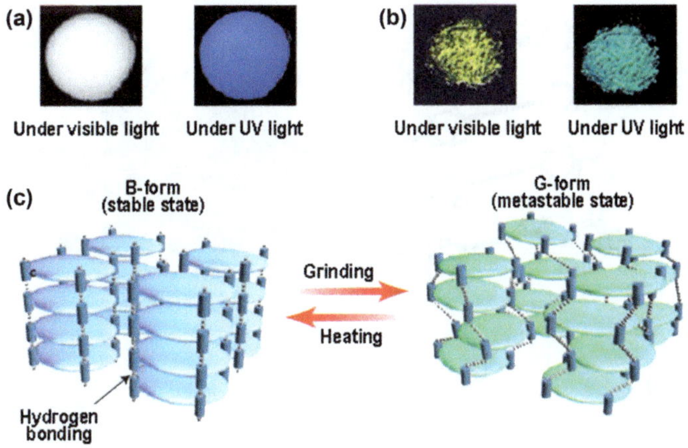

Figure 2.1 Images of dye **1** in B-form (a) and G-form (b) under visible light and UV light (365 nm). (c) Schematic representation of the assembled structures of dye **1** in B-form and G-form.
Reproduced with permission from reference 9. Copyright 2009, Nature Publishing Group.

destroyed and the pyrene packings become mismatched, which inhibits the formation of H-aggregates in the columnar-assembled structures. Thus, the emission color changes.

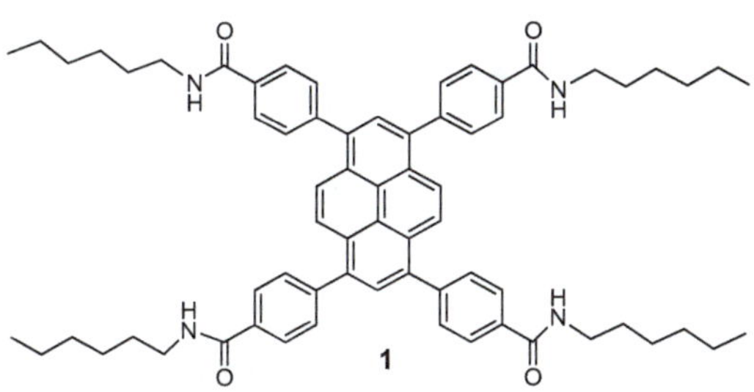

1

Araki *et al.* (2011)[10] continued to design and synthesize a series of TPPy-based dyes (**2a** to **2e**) as analogs of dye **1** to study further the role of amide hydrogen bonds and side chains in the molecular self-assembly. Dyes **2a** and **2d** have no mechanofluorochromism. However, dyes **2b**, **2c** and **2e** exhibit relatively evident mechanofluorochromic responses similar to dye **1**. After grinding, their emission wavelengths red-shift 15, 20 and 13 nm, respectively. These results indicate that the formation of molecular packing requires appropriately sized side chains. The mode of molecular packing is different, although the molecular size and shape of the amide dye **1** and the ester dye **2e** are quite similar (Figure 2.2). The tetraphenylpyrene cores of dye **1** are firmly connected by four amide hydrogen bonds to form the H-type columnar aggregate, whereas the ester derivative **2e** does not form this type of molecular assembly.[10] After increasing the pressure from 2.0 MPa to 9.8 MPa, dye **1** fully returns to its initial emission after heating above 112 °C.

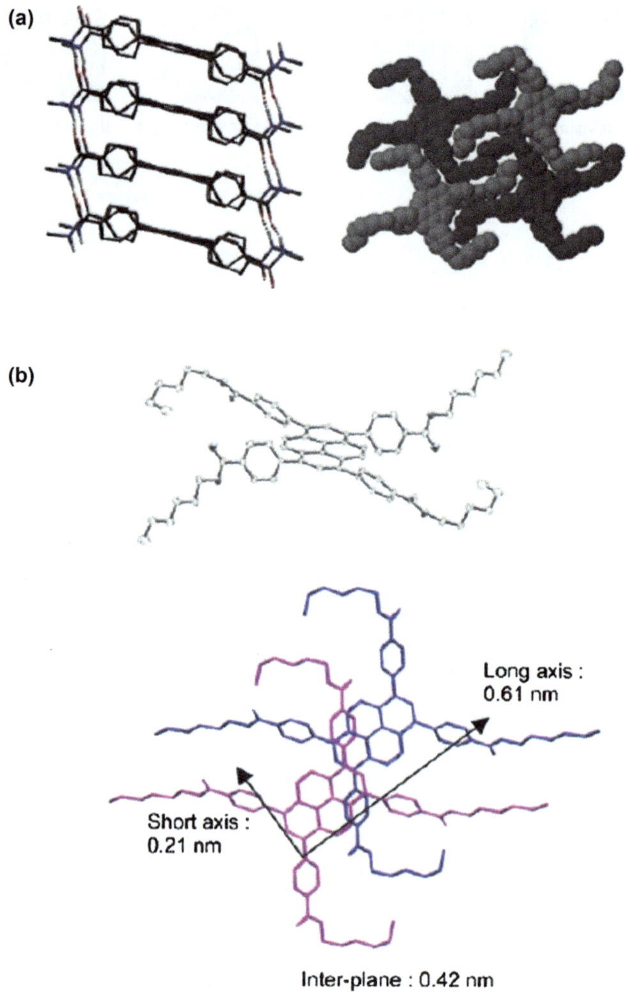

Figure 2.2 (a) Packing structure of 1b along c-axis (left) and normal to c-axis (right). (b) An ORTEP drawing of 2b and a view of two adjacent molecules normal to pyrene plane.
Reproduced with permission from reference 10. Copyright 2011, The Royal Society of Chemistry.

However, at a pressure of 9.8 MPa, the emission of the sample is further red shift, and the fluorescence cannot be restored to the initial emission. Application of pressure between 2.0 and 9.8 MPa is believed to have partially disrupted the hydrogen bonds, and the corresponding directed molecular packing can be restored by heating. Moreover, application of pressure exceeding 9.8 MPa can result in the excessive destruction of hydrogen bonds, which makes the full recovery of hydrogen bond-directed molecular packing difficult in the heating conditions. Compared with dye **1**, the ester derivative **2e** requires a higher pressure (greater than 9.8 MPa) to induce a noticeable color change in its fluorescence from blue-green to green. The blue-green fluorescence of the pressed sample of dye **2e** can be fully recovered by heating at around 40 °C, which indicates that the green-emitting pressed sample is unstable at the atmospheric pressure. Thus, the presence of amide hydrogen bonds is essential in eliciting a sensitive and repeatable mechanofluorochromic response, and that the excessive destruction of hydrogen bonds restricts the re-formation of the hydrogen bond-directed structure.[10]

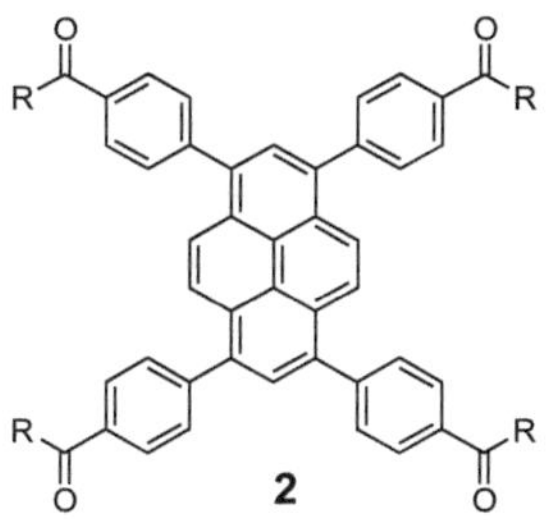

2

2a: R = -NH-C_3H_7
2b: R = -NH-$C_{14}H_{29}$
2c: R = -NH-$CH_2CH(C_2H_5)C_4H_9$

2d: R = -O-CH_3
2e: R = -O-C_6H_{13}

Araki *et al.* (2012)[11] studied the effect of mechanical stress on the solid-state luminescence of dye **3**, a carboxylic acid derivative of TPPy. Grinding the powdery solid of dye **3** in a mortar results in a blue shift of luminescence (from yellow to green) instead of the red shift observed for amide **1** and ester **2e** derivatives of TPPy. The initial yellow luminescence is recovered by exposure to solvent vapor. In contrast, application of compressive stress using an oil press causes practically no change of the initial yellow luminescence of dye **3**. The yellow luminescence is ascribed to the dimeric form of dye **3**, while the green luminescence is ascribed to the monomeric state. The macroscale shear stress applied by grinding induces the dimer-to-monomer transition at the molecular level, which causes the blue shift of the luminescence. Because of the tremendous size gap between the materials and molecules, it is difficult to understand the effect that mechanical force applied to the macroscale materials has on nanoscale molecular packing. In the case of amide and ester derivatives **1** and **2e**, both compression and grinding are effective at inducing red shift of the luminescence to green. Since compressive stress is effective alone, the green-emitting compressed state should be more closely packed, not only on the macroscopic scale, but also on the nanoscopic-level molecular packing as compared to the blue-emitting H-type columnar state of dye **1**, or the blue-green-emitting state of dye **2e** in the slip-stacked packing before compression. Increased inter-luminophore interaction by closer packing due to higher compressive stress might cause a further red shift of the luminescence of dyes **1** and **2e** to green. Longer distances of the four hydrogen bonds at the amide units of dye **1** compared to the stacking distance of the planar polycyclic aromatics realized the hydrogen-bond-directed H-type columns with lower-density molecular packing.[10] On the other hand, the shear stress (rather than the compressive stress) effectively causes a blue shift of the luminescence of dye **3** to green. Therefore, compression of the molecular packing is not sufficiently effective

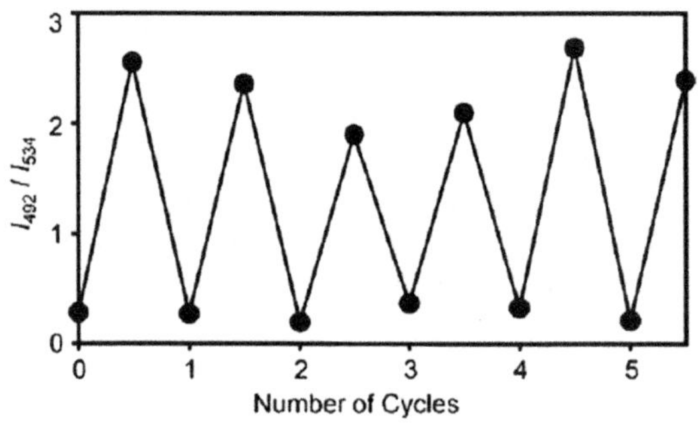

Figure 2.3 Relative intensities of the emission at 492 and 532 nm as a function of the repeated cycle of grinding followed by exposure to acetone vapor. Reproduced with permission from reference 11. Copyright 2012, The Royal Society of Chemistry.

in this case. Since the macroscale shear stress applied to the solid of dye **3** induced alteration of the molecular packing from the dimeric form to the monomeric form, the applied shear stress might cause slipping of the stacked dimer of dye **3** having a relatively flat molecular shape. In order to restore the initial state from the green-emitting state, the thermal process that has been effective in the case of dye **1** and dye **2e** is ineffective, and exposure to solvent vapors is required (Figure 2.3).

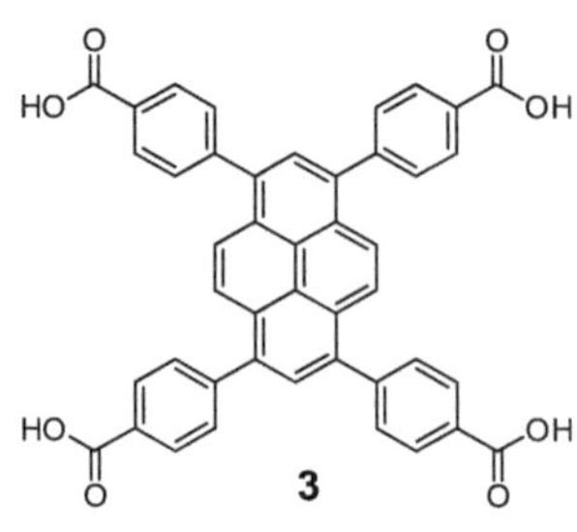

3

2.2.2 Cyano-substituted Oligo(*p*-phenylene vinylene) Derivatives

Weder *et al.* have done several studies on the mechanofluorochromism of the dye-doped polymer composites since 2002.[12] However, they did not report on pure organic mechanofluorochromic compounds until 2008, *i.e.* dye **4** series (**4a–c**).[13] The latter comprises analogs of dyes **4d** and **4e**, whose methoxyl group is replaced with 12- and 18-carbon long alkoxy chains, respectively. In the solid states, dyes **4d, 4e** and **4c** exhibit comparably small bathochromic shifts in their emission spectra after pressing, which indicates that the dyes present mechanofluorochromic properties only when they are doped in a polymer matrix.[12,14] However, when compressed, the blue-emission crystalline powders of dyes **4a** and **4b** emit a yellow color, which shows significant mechanofluorochromism. Heating the compressed samples of dyes **4a** and **4b** facilitates the recovery of the original fluorescent

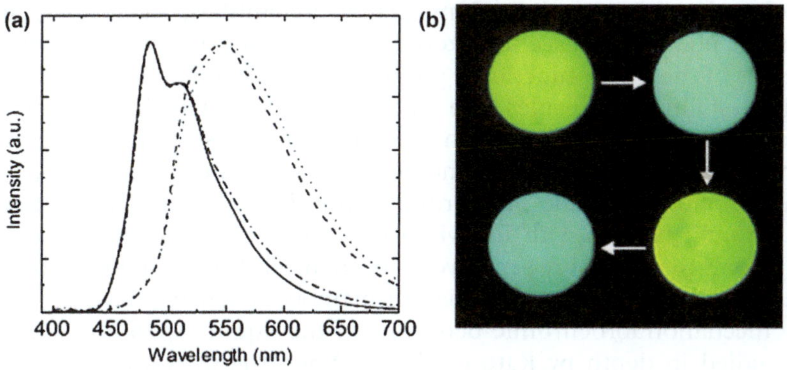

Figure 2.4 (a) Normalized fluorescence spectra of **4b** as-prepared (solid), pressed for 1 min at 1500 psi (dashed), pressed for 19.5 h at 1500 psi (dotted), and pressed for 1 min at 1500 psi/heated to 130 °C for 5 min, and quenched to room temperature by placing on a cold surface (dashdotted). (b) Pictures of a **4b**/KBr mixture pressed for 1 min at 1500 psi (top left), the same sample subsequently heated to 130 °C and quenched to room temperature by placing on a cold surface (top right), the same sample pressed again as before (bottom right), and the same sample heated and cooled as before (bottom left). The pictures were taken under illumination with UV light (365 nm) and the sample diameter is 13 mm. Reproduced with permission from reference 13. Copyright 2008, Wiley-VCH.

color (Figure 2.4). The authors attributed the observed yellow emission to the excimer formation of fluorescent π-conjugated moieties, and they conjectured that balancing the π-π and aliphatic interactions in the dyes was made possible by varying the electron density in the central ring and in the length of peripheral aliphatic tails. Dye **4b** undergoes smectic and nematic liquid crystalline phase transitions during the heating process.

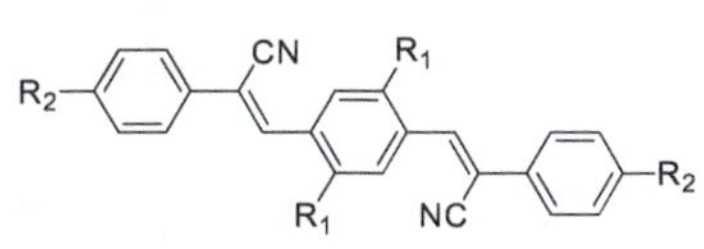

4

Code	R_1	R_2
4a	H	$OC_{12}H_{25}$
4b	H	$OC_{18}H_{37}$
4c	OMe	$OC_{18}H_{37}$
4d	OMe	OMe
4e	H	OMe

2.2.3 Dumbbell-shaped Liquid Crystalline Dendritic Compounds

As mentioned above, dye **4b** exhibits liquid crystalline properties. In fact, liquid crystals are functional, soft materials with ordered, dynamic and metastable properties. The dynamic properties of the liquid crystalline materials have attracted much attention because of their great potential as functional materials. Recently, photo-functional liquid crystalline materials that show photoluminescence, electroluminescence, photochromism and acidichromism have been intensively developed.[15–31] These photo-functional properties of molecular assemblies, such as stimuli-responsive photoluminescent materials, strongly depend on the assembled structures of the photo-functional groups. If the assembled structures of the liquid crystalline molecules are changed by external factors such as mechanical or thermal stimulation, then photoluminescent properties of the materials can be tuned by these methods. These phenomena are referred to as mechanofluorochromic or thermochromic luminescence. If nematic or smectic liquid crystals with low viscosities show mechanofluorochromism, the materials that are more sensitive to mechanical stimuli could be obtained. These materials have great potential for memory or sensing devices.[32]

The mechanofluorochromic behaviors of the liquid crystalline dyes **5a–c** were studied in depth by Kato *et al.*[33–35] When the dyes are mechanically sheared, the yellow emission of the liquid crystalline states switches to blue-green for the pyrene derivative **5a** and to light blue for the anthracene derivative **5b**, and the light blue emission to blue for the naphthalene derivative **5c** (Figure 2.5). The structures are converted from a cubic metastable liquid crystalline phase into a stable columnar form. The metastable cubic phases are formed by hydrogen bonding and π-π interactions. After phase transitions, the metastable cubic phases can again be obtained when the samples are annealed to their isotropic states.[9] According to the proposed mechanism, the pyrene, anthracene and naphthalene groups are arranged in a π-stacked structure in the metastable cubic phases, which enables the formation of excimers and subsequently facilitates the yellow (pyrene and anthracene) or light blue (naphthalene) emission. However, the fluorescent cores are fixed in a molecular arrangement in the columnar phases wherein no excimer formation occurs. Thus, the emission changes to a shorter wavelength.

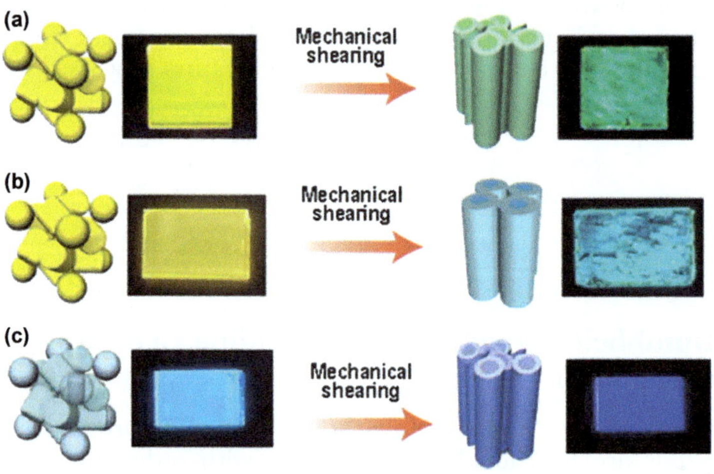

Figure 2.5 The change of fluorescent color and molecular assemblies for the dyes (a) **5a**, (b) **5b** and (c) **5c**.
Adapted with permission from reference 9. Copyright 2009, Nature Publishing Group and reference 35. Copyright 2011, Taylor & Francis.

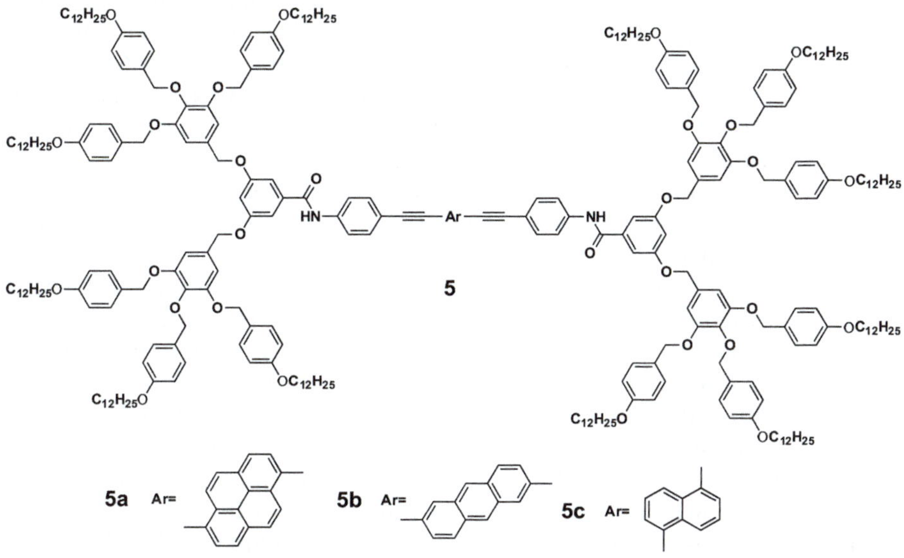

The proposed structures for the self-assembled cubic and shear-induced columnar phases of **5a** are shown in Figure 2.6.[33] Compound **5a** forms micellar structures in the P_{m3n} cubic phase (Figure 2.6a, left). Each micelle in the micellar cubic phase of **5a** consists of a segmented column formed by approximately 20 individual molecules. The number of molecules forming each micelle is calculated based on the results of X-ray measurements.

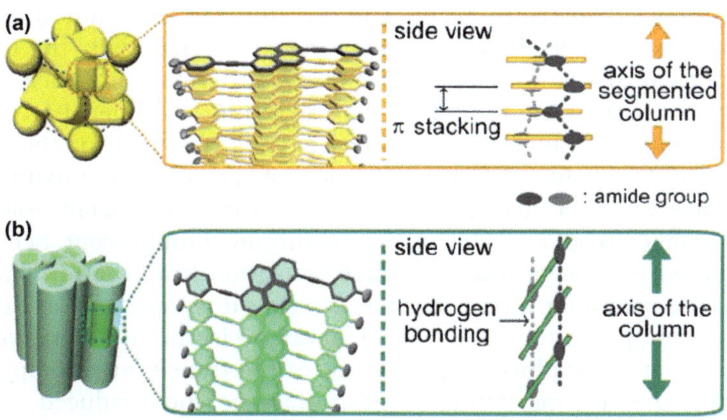

Figure 2.6 Illustrations of the assembly of **5a** in the cubic phase (a) and in the shear-induced columnar phase (b). The dendritic side-chains of **5a** are omitted in the middle and right-hand images of the assembled luminescent cores.
Reproduced with permission from reference 33. Copyright 2008, Wiley-VCH.

The segmented columnar structures in the cubic phase of **5a** result from hydrogen bonding between the amide groups of adjacent molecules and the formation of π-stacked structures of the emission cores (Figure 2.6a, middle and right). This type of self-assembly for **5a** leads to yellow excimer emission. Compound **5a** should be arranged in disordered stacks in these segmented columns (Figure 2.6a, middle and right). This proposed self-assembled structure is supported by the fact that the length between the H-bonded amide groups is generally longer than the distance between the adjacent arenes that form π-stacked structures. The shear-induced cubic–columnar phase transition of **5a** at 160 °C results in the formation of non-segmented columnar structures (Figure 2.6b, left). The stronger H-bonded states that are not observed in the cubic phase are formed in these columnar structures (Figure 2.6b, middle and right). Each molecule in these structures is expected to form hydrogen bonds to the adjacent molecules with a distance of around 5 Å. In light of these crystalline structures, compound **5a** with a similar structure may assemble into the linear H-bonded structures shown in Figure 2.6b, which should result in interference with the excimer formation of the 1,6-diethynylpyrene moieties of compound **5a**. The blue-green photoluminescence for the shear-induced columnar phase is observed because of the self-assembled structures.[33]

Kate and Sagara (2011)[36] reported a stimuli-responsive luminescent liquid crystal that exhibits three luminescent colors, which can be switched by mechanical and thermal stimuli (Figure 2.7A). The liquid crystal is composed of equimolar amounts of the dumbbell-shaped compound **6** and compound 7. The three luminescent colors observed are reddish-orange, yellow and green, which are easily distinguished by the naked eye. Moreover, the liquid crystal mixture of **6** and 7 contains only one type of luminophore, 9,10-bis(phenylethynyl)anthracene, and no additives are required to induce the luminescent color changes. Unlike crystalline materials, the thin-film states are easily prepared. The mixture forms a thermotropic micellar cubic phase upon heating from room temperature to 146 °C. Under UV irradiation (365 nm), reddish-orange photoluminescence is observed for the micellar cubic phase. It was found that mechanical shearing to the mixture in the cubic phase at 90 °C triggers a change in the luminescent color from reddish-orange to green. The mechanochromic luminescent behavior is accompanied by a shear-induced phase transition from the micellar cubic phase to the columnar phase. The shear-induced columnar phase is stable from room temperature to 146 °C and shows no isothermal transition back to the micellar cubic phase. The increase of the transition enthalpy corresponding to the clearing points suggests that the shear-induced columnar phase ($\Delta H = 89$ kJ mol^{-1}) is more thermodynamically stable than the cubic phase ($\Delta H = 23$ kJ mol^{-1}). However, once the mixture of compounds **6** and 7 forms the isotropic phase, the columnar phase does not appear without mechanical shearing. It was also found that the mixture exhibits another change in luminescent color. When the mixtures in the micellar cubic or columnar phases are mechanically sheared at room temperature, the mixtures show an unidentified mesomorphic phase exhibiting yellow emission. Upon heating the mixture in the mesomorphic phase, it shows a transition to the isotropic phase at 145 °C. On subsequent cooling, reddish-orange emission is observed again from the mixture in the micellar cubic phase. And on heating the mesomorphic phase from −50 °C, a few exothermic peaks were observed on the differential scanning calorimetry trace. The results imply that the mesomorphic phase is a thermodynamically metastable phase.

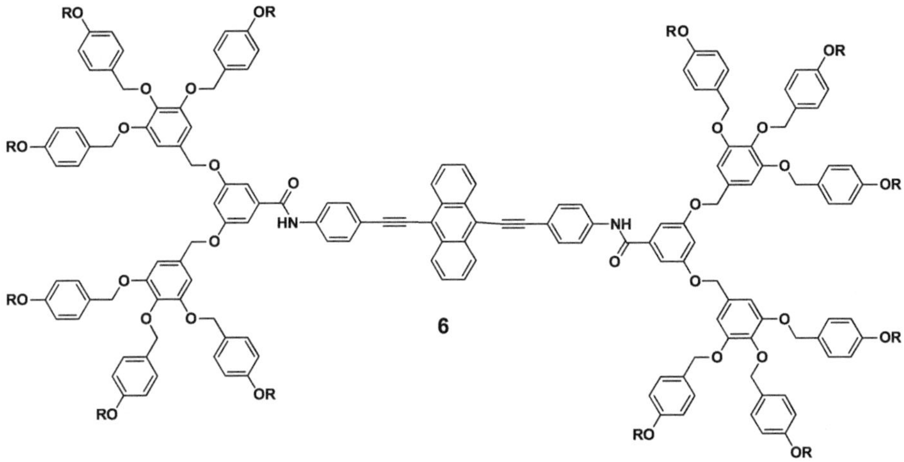

6

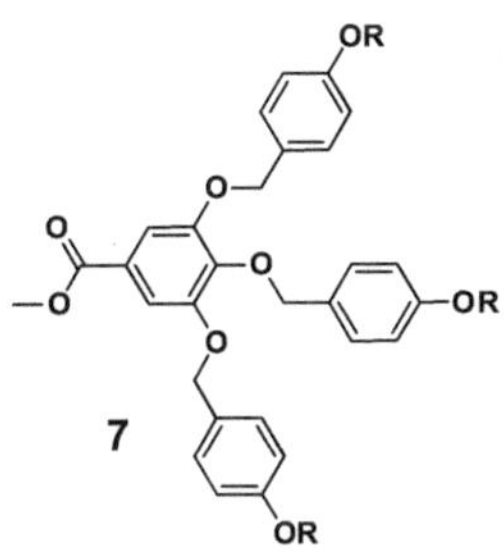

7

R=

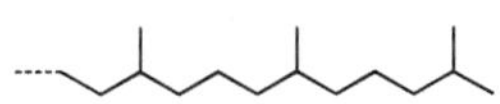

The self-assembled structures of compounds **6** and **7** in the liquid crystalline phases and the mesomorphic phase are proposed in Figure 2.7B.[36] In the cubic phase, compounds **6** and **7** form micellar structures. Each micelle contains an equal number of compounds **6** and **7**. The total number of molecules forming each micelle is approximately 32, which is calculated based on the results of the X-ray measurements. In each micelle, approximately 16 molecules of compound **6** form the segmented column. The segmented columnar structure is built through both the formation of π–π stacked structures of the emission cores and the formation of hydrogen bonding between the amide groups of adjacent molecules. The emission cores are arranged in a disordered stacking arrangement in the segmented columns. The combination of the dumbbell-shaped compound **6** with compound 7 leads to the formation of stable segmented columnar assembly. As a consequence, reddish-orange excimer emission occurs in the cubic

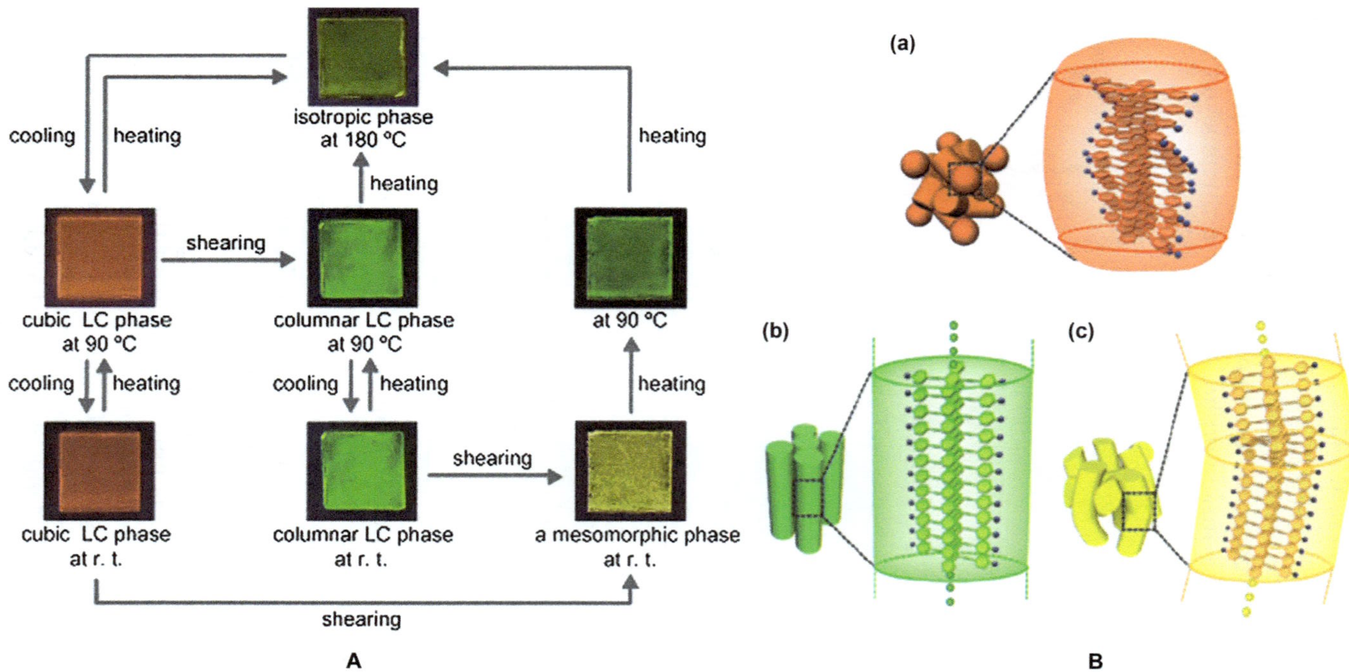

Figure 2.7 (A) Change in photoluminescent color of the liquid-crystalline mixture of compounds **6** and **7**. Photoluminescent images were taken for the mixtures between quartz substrates under UV irradiation at 365 nm. (B) Schematic illustration of the assembled structures of compounds **6** and **7**. (a) Cubic phase, (b) columnar phase and (c) mesomorphic phase. Amide groups of compound **6**: blue spheres. Compound **7** and dendritic moieties of compound **6** are omitted in the detailed illustration to the right of each structure.
Reproduced with permission from reference 36. Copyright 2011, Wiley-VCH.

phase. Such assembled structures are supported also by their previous results on the assembled structures of pyrene, anthracene and naphthalene derivatives having similar molecular structures to compound **6**.[33–35] The molecularly assembled structures in the columnar phase are depicted in Figure 2.7B. In each column, compound **6** forms columnar structures through the formation of a linear hydrogen-bonding array that is not observed in the cubic phase. The luminescent cores are spaced by ~5 Å intervals, thus leading to the interference with the excimer formation of 9,10-bis(phenylethynyl)anthracene moieties. The green photoluminescence of the mixtures in the columnar phase (Figure 2.7A) is attributed to these depicted assembled structures. Upon formation of the shear-induced columnar phase, compound **7** may exist among the hydrogen-bonded columns formed by the dumbbell-shaped compound **6**. The existence of **7** may stabilize the columnar structures consisting of compound **6**. Without compound **7**, compound **6** does not form a columnar phase. In the mesomorphic phase, it is assumed that compound **6** forms less-ordered columnar structures, though no clear peaks appear in the X-ray diffraction pattern. This assumption is based on the fact that almost all of the amide groups of compound **6** are involved in the formation of linear hydrogen bonds similar to those in the columnar phase. These results suggest that the luminescent cores in the mesomorphic phase are partially arranged in a linear arrangement similar to that in the columnar phase. In addition, some of the cores partially overlap, thus leading to the partial-overlap excimer emission (Figure 2.7B). Energy migration and energy transfer may occur from non-overlapped luminescent cores to the partial-overlap excimer sites. Therefore, the mixture exhibits yellow emission in the mesomorphic phase (Figure 2.7A).

Kato *et al.* (2013)[32] prepared a series of smectic liquid crystals (**8a–c**) consisting of an aromatic moiety based on 10,10′-bis(phenylethynyl)-9,9′-bianthryl at the central part and one and two mesogenic moieties on each side connected through oxyethylene spacers as stimuli-responsive materials. The compound (**8b**) having two mesogenic moieties on each side forms an ordered smectic phase. Mechanochromic luminescence is observed for the smectic liquid crystal (Figure 2.8A). Compound **8b** displays blue-green photoluminescence in the isotropic phase and the luminescent color does not change at room temperature on cooling at a scanning rate of 5 K min^{-1}. On the other hand, the photoluminescent color of **8b** changes to green after annealing at 115 °C (State B) and the emission color turns more greenish upon cooling to room temperature (State C). The green photoluminescence changes to blue-green by mechanical shearing (State D). In addition, State B can be obtained again from State D by heating to 115 °C without heating to the isotropic phase. These changes were considered the results of the disturbance of intermolecular π–π interactions. The self-assembled structures of **8b** before and after the mechanical shearing were proposed by the authors and shown in Figure 2.8B. Compound **8b** may form the tilted and interdigitated layer structure in State C. In this state, π–π interactions among the adjacent cores of **8b** may induce the green emission. After mechanical

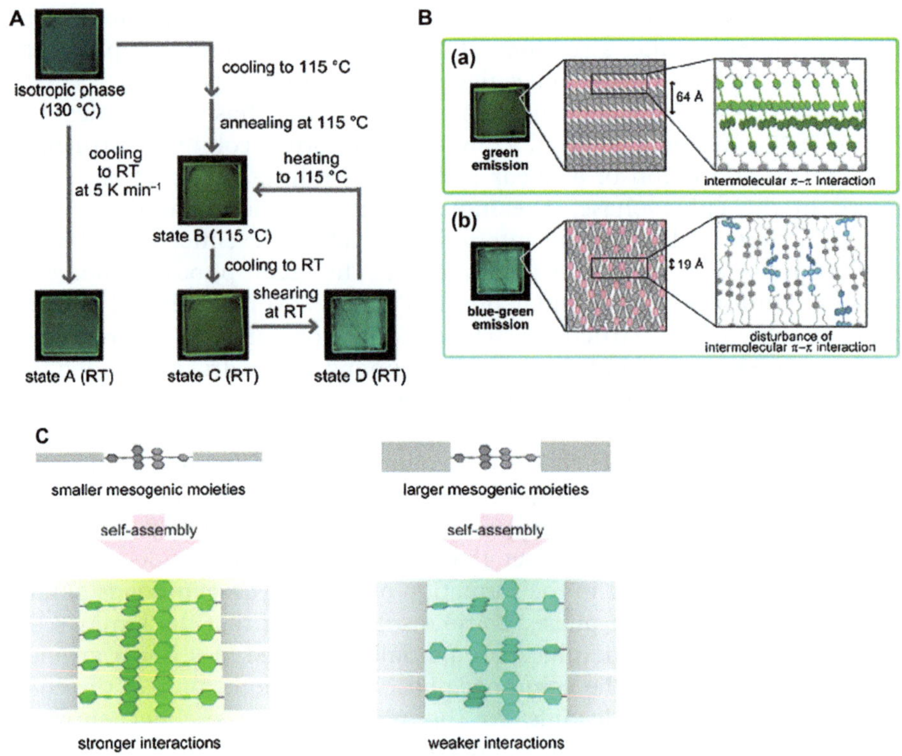

Figure 2.8 (A) Photoluminescent images of compound **8b** taken between quartz plates under UV irradiation at 365 nm. (B) Possible schematic illustrations of the assembled structures of **8b** (a) in State C and (b) in State D. (C) Schematic illustrations of the assembled structures of **8a** in the ordered smectic phase formed by thermal annealing (left) and **8c** in the monolayer smectic phase (right).
Reproduced with permission from reference 32. Copyright 2013, The Royal Society of Chemistry.

shearing, **8b** forms a disordered layer structure (State D), in which the luminescent cores of **8b** are surrounded by the mesogenic cyclohexyl phenyl moieties of adjacent molecules and the intermolecular π–π interactions of the bianthryl rings are disrupted, leading to blue-green photoluminescence. The emission spectrum measured after cooling to room temperature from isotropic phase is almost the same as that measured before mechanical shearing. Mechanochromic luminescent properties are also observed for compound **8a**. However, compound **8c** does not exhibit mechanochromic luminescence, although **8c** also shows the monolayer smectic phase. The volume of terminal mesogens affects the assembled structure and stimuli-responsive properties. The possible assembled structures of **8a** in the ordered smectic phase formed by thermal annealing and **8c** in the monolayer smectic phase are shown in Figure 2.8B. It was considered that the formation of the proper assembled structures with intermolecular π–π

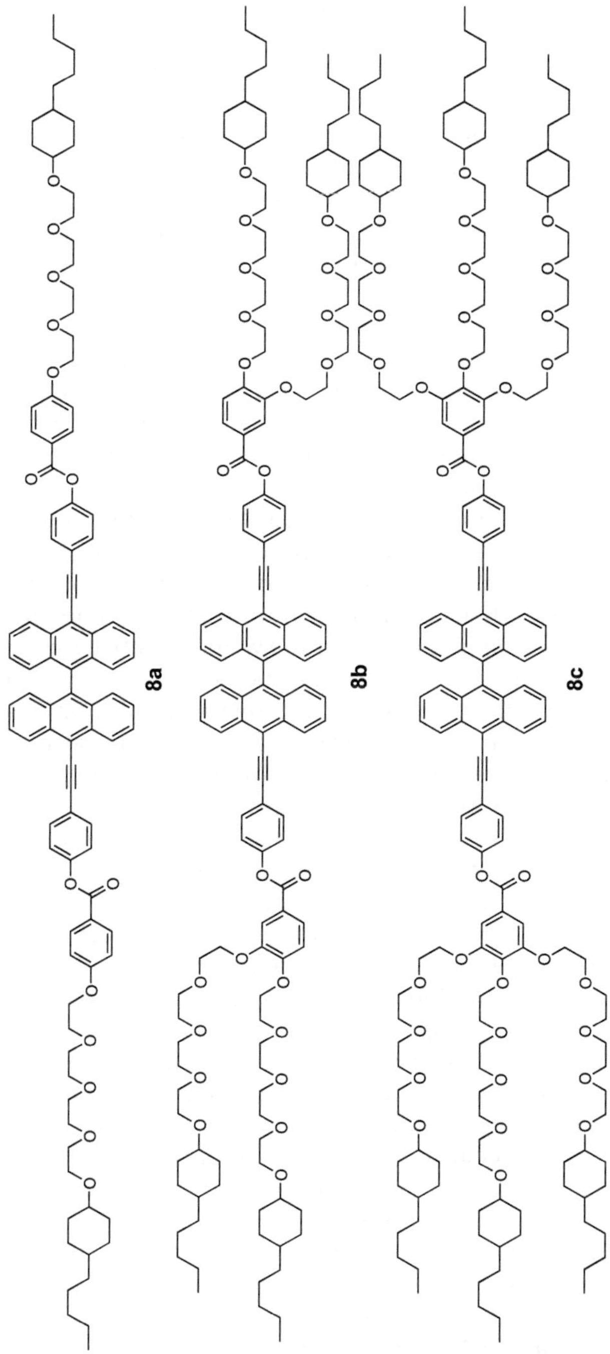

8a

8b

8c

8

interactions of the bianthryl rings is a necessary prerequisite for mechanochromic luminescence for this type of molecular structure because mechanical shearing disrupts the intermolecular π–π interactions. In the cases of **8a** and **8b**, thermal annealing is used to induce π–π interactions of the bianthryl rings. Compound **8c** forms a monolayer smectic phase without intermolecular π–π interactions of the bianthryl rings due to the bulky terminal groups, thus, no mechanochromic luminescence is observed in this case. It was suggested that emission properties and mechanochromic properties of the 9,9′-bianthryl derivatives can be controlled by changing the volume of the mesogenic moieties: smaller mesogenic moieties would lead to stronger ground state interactions between the adjacent luminescent molecules and red-shifted emission, and in contrast, the large mesogenic moieties disturb intermolecular interactions, resulting in no detectable change in the assembled structure of luminescent moieties upon mechanical stimulation (Figure 2.8C). It is perhaps the first report on the mechanochromic luminescent liquid crystalline compounds based on a smectic–smectic phase transition.

For some of the mechanochromic luminescent compounds, the initial photoluminescence color can be recovered through subsequent exposure to vapor of organic solvents such as methylene chloride and tetrahydrofuran. However, few organic and/or organometallic mechanochromic luminescent materials that recover their initial photoluminescence colors upon exposure to water vapor have been prepared. But the water-sensitive mechanochromic luminescent material would potentially be useful for humidity sensors. Nagano *et al.*[37] reported a water-soluble mechanochromic luminescent pyrene derivative **9**, which possesses a pyrene group as the luminescent core and two dendritic groups that are attached to the luminescent core through amide groups. To increase the water-solubility, thirty-two hydroxyl groups were incorporated into the dendrons at the peripheral positions. It was also expected that compound **9** would form molecular assemblies in water, because of its amphiphilic character. The dumbbell-shaped compound was obtained as a cottony yellow solid (Y-form) after purification by gel permeation chromatography and lyophilization of the aqueous solution. Under UV irradiation (365 nm), yellow photoluminescence was observed for compound **9** in the Y-form. Mechanical grinding caused a luminescence color change from yellow to green (Figure 2.9A). This is a typical mechanochromic luminescent behavior. It was also found that the initial yellow photoluminescence was recovered when the green luminescence-emitting state (G-form) was exposed to water vapor. No previously reported luminescent organic and/or organometallic material shows such reversible changes in photoluminescence color in response to mechanical stimulation and exposure to water vapor. As expected, compound **9** is soluble in polar solvents such as methanol, ethanol, *N*,*N*-dimethylformamide, dimethylsulfoxide, pyridine and water at room temperature, but not in hexane, ethyl acetate, methylene chloride, chloroform, toluene, tetrahydrofuran or acetonitrile at room temperature.

Grinding of this sample under ambient conditions induced green photoluminescence under UV irradiation. This cycle between the two

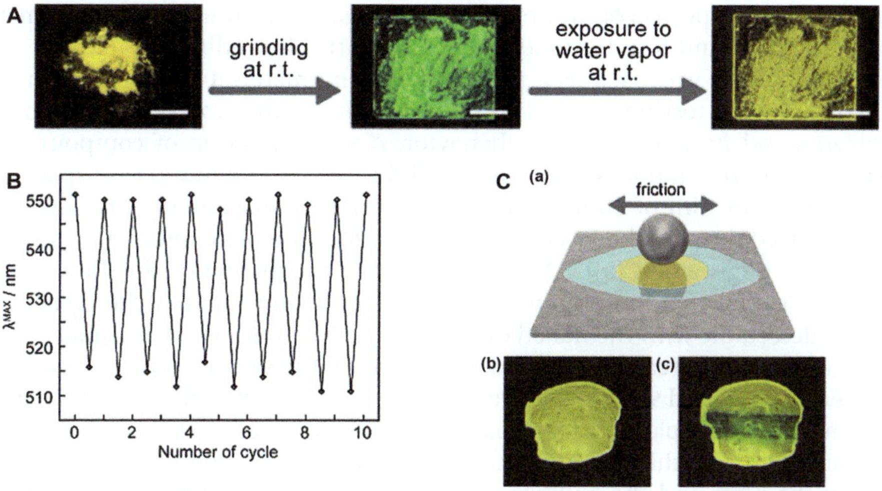

Figure 2.9 (A) Change in the luminescence color of compound **9** in the solid state. The photoluminescence images were taken on quartz substrates under UV irradiation at 365 nm. Scale bar: 5 mm. (B) Changes in the photoluminescence color of compound **9** upon repeated cycles of grinding followed by exposure to high humidity (RH = 100%). $\lambda_{ex} = 400$ nm. (C) (a) Schematic illustration of frictional wear test for grease. The yellow part represents the thin film of compound **9**. The blue part represents the thin film of grease. (b) Photoluminescence image of compound **9** on the metallic plate under UV irradiation (365 nm). (c) Photoluminescence image of compound **9** after the ball had been rolled to-and-fro.
Reproduced with permission from reference 37. Copyright 2011, Wiley-VCH.

different photoluminescent colors can be repeated at least ten times (Figure 2.9B). Therefore, the pyrene derivative **9** appears to be a good candidate for both mechano-sensors and humidity sensors.

The mechanism for the mechanofluorochromism was proposed that mechanical grinding induced changes in the arrangement of the luminescent cores, resulting in interference with normal excimer formation and leading to the appearance of green photoluminescence derived from partial-overlap excimers of the luminescent cores in the G-form. Because hydrogen bonds between the amide groups are retained after grinding, the luminescent cores are not separated sufficiently to show monomer emission. It is speculated that exposure to RH = 100% may lead to an increase in the mobility of the peripheral hydrophilic moieties of the dendritic groups, allowing recovery of the initial micellar structures and yellow emission. For some mechanochromic luminescent materials, the original photoluminescence properties are recovered upon exposure to vapors of organic solvents such as methylene chloride and tetrahydrofuran. In such cases, the effective vapors are those of good solvents of the mechanochromically luminescent compounds. In the case of compound **9**, water is a good solvent, and thus the hydrophilic properties of the dendrons attached to the luminescent core may enable compound **9** to respond to higher relative humidity.

Compound **9** was used as a mechano-sensing material for application in frictional wear testing for grease, owing to its immiscibility in non-polar solvents and its non-crystalline behavior. A water solution of compound **9** was applied to a stainless steel plate, and the water was allowed to evaporate, leaving a thin film of compound **9** on the plate. Subsequently, the surface was greased, leaving the layer of compound **9** on the plate under a thin film of grease (Figure 2.9C(a)). Most greases have no absorption in the visible region, so that the photoluminescence of compound **9** under the grease is easily detectable with the naked eye. Next, a stainless steel ball was placed in contact with the grease and rolled to-and-fro on the surface (Figure 2.9C(a)). Because compound **9** is immiscible with the grease, the thin film of compound **9** remained at the plate surface during this frictional wear test. As a result, the region where the thin film of grease was removed, *i.e.*, the contact area between the plate and the ball, where the film of compound **9** was exposed to mechanical stimulation, was detected in terms of green photoluminescence of compound **9** (Figure 2.9C(b–c)). In contrast, aliphatic compounds showing mechanochromic luminescence do not work in such frictional wear tests, as exemplified by pyrene-based liquid-crystalline compound **5a**.

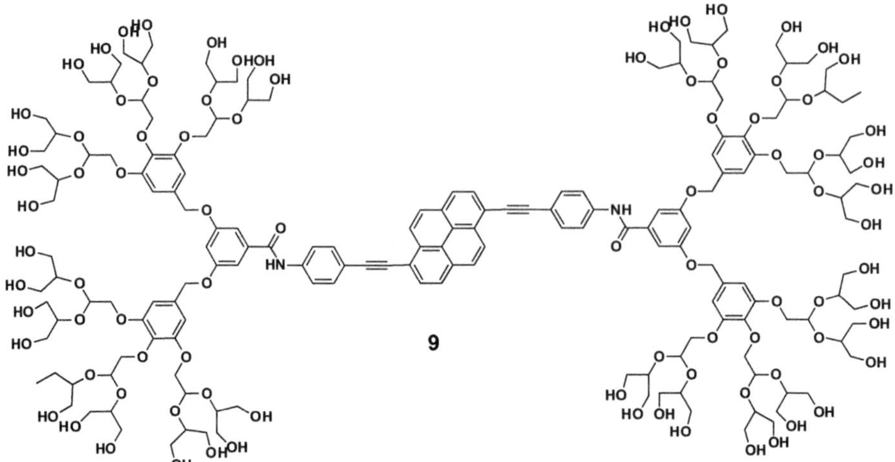

9

2.2.4 Heteropolycyclic D-π-A Compounds

A series of heteropolycyclic fluorescent dyes (dye **10** series) with donor-acceptor π-conjugation (D-π-A) have been reported by Harima *et al.* in 2010 and 2011.[38,39] The effects of the substituents and the chromophore skeletons on their mechanofluorochromic behaviors were systematically studied. The mechanofluorochromism ($\Delta\lambda_{em}$, change in emission wavelength before and after grinding) decreases with increasing the steric size of the substituent and with increasing the electron-accepting capability of the *p*-substituted phenyl group (Structure 2.10). The dyes containing the *tert*-butyl group, *i.e.*, **10i** and

10l, show almost no mechanofluorochromism. The authors proposed a possible mechanism that the mechanofluorochromism was attributed to a reversible switching between crystalline and amorphous states and with changes in dipole-dipole and intermolecular π-π interactions that occur by altering the densities of the solids before and after grinding (Figure 2.10). Moreover, they stressed the importance of the design of D-π-A fluorescent dye molecules with large dipole moments in developing mechanofluorochromic dyes.

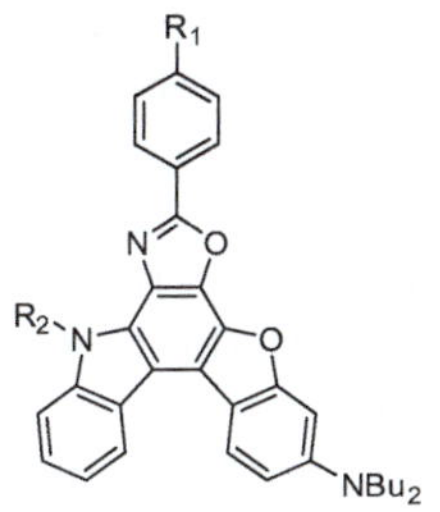

10

Code	R_1	R_2	Emission wavelength shift(nm)
10a	CN	H	53
10b	CN	*n*-butyl	53
10c	CN	benzyl	47
10d	CN	5-nonyl	25
10e	COOMe	H	41
10f	COOH	H	32
10g	COOH	*n*-butyl	40
10h	H	H	18
10i	t-Bu	H	3

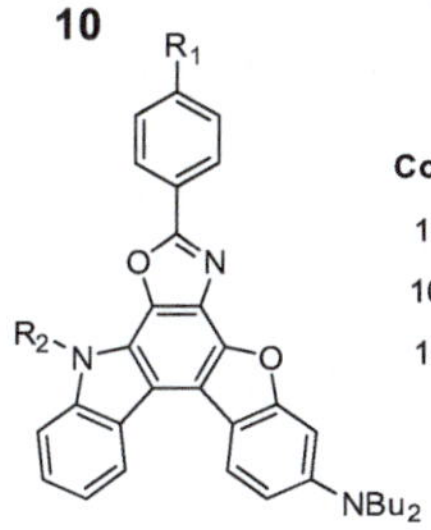

Code	R_1	R_2	Emission wavelength shift(nm)
10j	CN	*n*-butyl	39
10k	H	H	29
10l	t-Bu	H	4

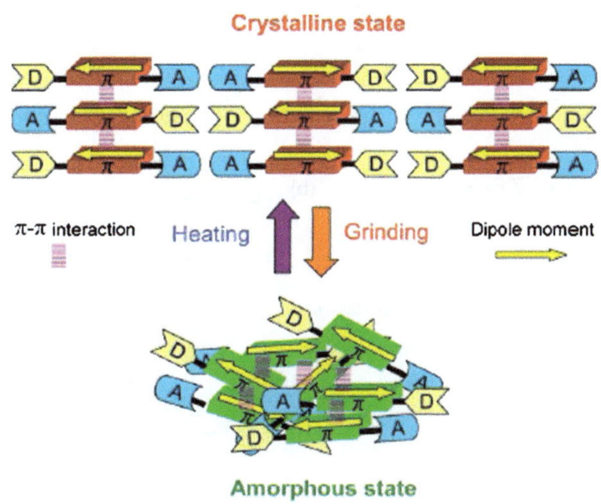

Figure 2.10 Proposed mechanism of mechanofluorochromism of the heteropolycyclic D-π-A fluorescent dyes.
Reproduced with permission from reference 39. Copyright 2011, The Royal Society of Chemistry.

2.2.5 Diphenylamine Carbazole Derivatives

In 2012, Ooyama *et al.*[40] designed and synthesized unsymmetrical carbazole-type D-π-A fluorescent dyes **11** and **12**. The dyes show a bathochromic shift-type mechanofluorochromism, and red-shifted 14 and 22 nm for dyes **11** and **12** after grinding, respectively. Compared with the dye **10** series, dyes **11** and **12** exhibit relatively small mechanofluorochromism. The difference may be caused by different dipole-dipole interactions. Dyes **11** and **12** have smaller dipole moments (μ_g approximately equal to 3.5 D) than those of the dye **10** series (μ_g approximately equal to 5 D). The XRD and DSC measurements demonstrate that the dyes inter-convert between microcrystalline and amorphous states by grinding and heating. The red-shifts in the emission spectra of the dyes after grinding are explained in terms of a continuous intermolecular hydrogen bonding for dye **11** and the intermolecular π–π interactions for dye **12** in the crystalline state. Moreover, the intermolecular hydrogen bonding for the ground samples for dye **11** become stronger than that in the as-recrystallized samples. Thus, the mechanofluorochromism for dyes **11** and **12**, which was observed by changing from the crystalline state to the amorphous state, is caused by the formation of stronger hydrogen bonding for dye **11** and intermolecular π–π interactions for dye **12** (Figure 2.11). Moreover, the fluorescence quantum yields of the dyes undergo very little change from the crystalline state to the amorphous state. A non-radiative decay route for the excited states is relatively discouraged by the hydrogen bonding and the π–π interactions in the amorphous states for this series of D-π-A fluorescent dyes.

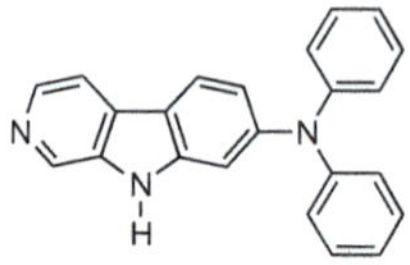

11

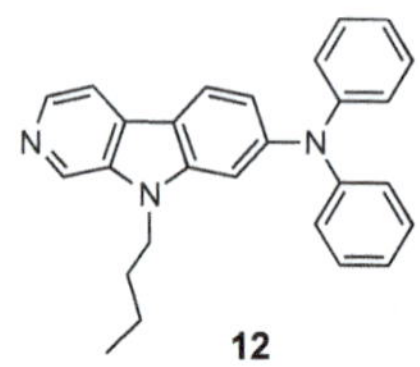

12

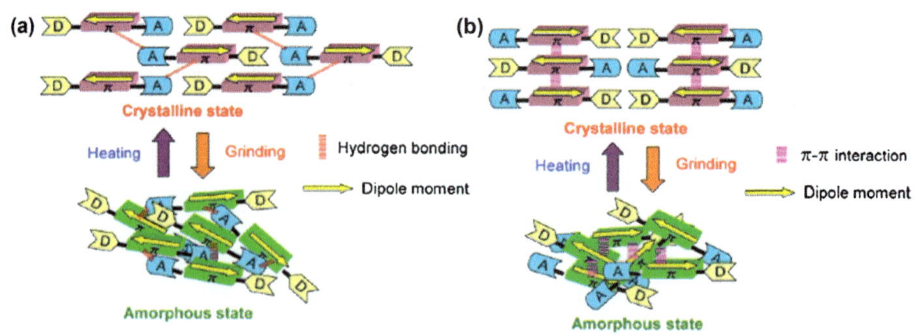

Figure 2.11 Proposed mechanisms of mechanofluorochromism observed with carbazole-type D-π-A fluorescent dyes **11** (a) and **12** (b).
Reproduced with permission from reference 40. Copyright 2012, Elsevier.

Ooyama *et al.* (2012)[41] also designed and synthesized carbazole-type donor-π-acceptor (D-π-A) fluorescent dyes **13** and **14**, each containing a diphenylamino system as an electron-donating group and a nitro moiety as an electron-accepting group, and investigated their photophysical properties in the solution and in the solid state. The absorption and fluorescence properties of dyes **13** and **14** in the solution are similar, and both dyes exhibit moderate fluorescence quantum yields. In the solid state, however, dye **14**, with a butyl substituent on the carbazole ring, exhibits red fluorescence at around 620 nm, whereas dye **13**, with no substituent on the carbazole ring, does not exhibit these solid-state fluorescence properties (Figure 2.12). A continuous intermolecular π-stacking between the fluorophores is observed in the crystal structure of dye **13**, in which π-overlaps are observed between the nitrophenyl group and the carbazole moiety in the adjacent molecules, but not in that of dye **14**. The crystalline structure of dye **14**, on the other hand, is made up of dimer units composed of pairs of fluorophores with the π-overlaps over the whole molecule from the nitrophenyl group to the carbazole moiety. The X-ray structural analysis therefore reveals that the formation either of a continuous intermolecular π-stacking or of intermolecular π–π interactions is responsible for large red-shifts in the absorption and fluorescence maxima and decreases in the solid-state fluorescence for dyes **13** and **14** on going from 1,4-dioxane to the solid state. However, the solid-state fluorescence quantum yield of dye **13** ($\Phi_F \approx 0$) is much smaller than that of dye **14** ($\Phi_F = 0.03$). It was concluded that the continuous intermolecular π-stacking between the fluorophores is a principal factor leading to the drastic solid-state fluorescence quenching of the crystal of dye **13**. These results also show that the introduction of a bulky substituent on a carbazole ring in a D-π-A fluorescent dye can effectively prevent the formation of continuous intermolecular π-stacking between fluorophores. On the other hand, for mechanofluorochromism of D-π-A fluorescent dyes, as mentioned above, it had demonstrated that newly developed unsymmetrical heteropolycyclic D-π-A fluorescent dyes with electron-withdrawing substituents such as cyano and carboxylate ester groups and thus relatively large dipole moments ($\mu_g \approx 5$–6 D) show pronounced mechanofluorochromism.[38–40,42–45] Dye **14** has a very large dipole character ($\mu_g \approx 8$ D), resulting in a weak mechanofluorochromism. The authors thought that the strong dipole-dipole interaction between the fluorophores might inhibit a switching between crystalline and amorphous states on grinding.

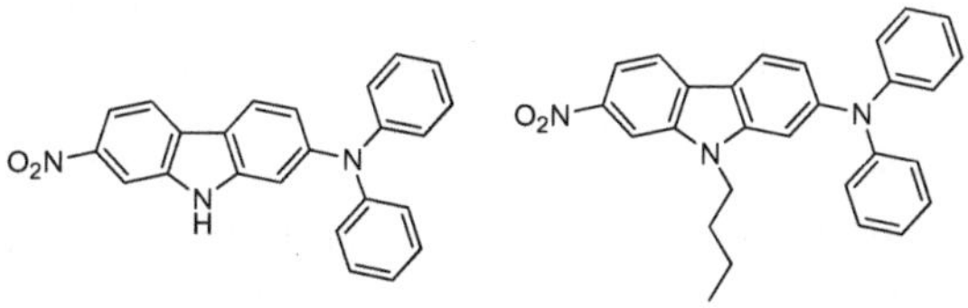

13 **14**

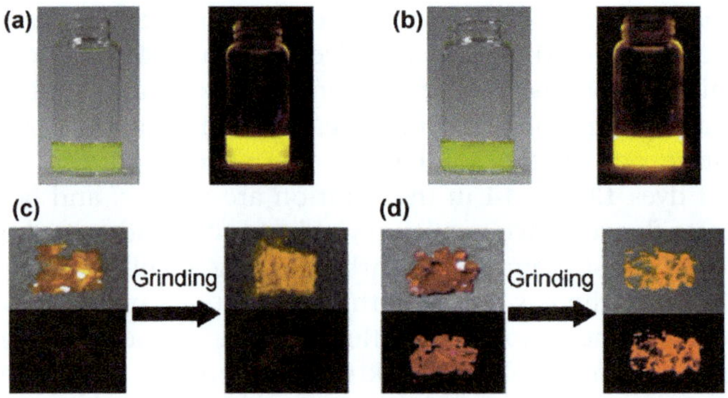

Figure 2.12 Color (left) and fluorescence (right) images of (a) **13** and (b) **14** in 1,4-dioxane. Photographs of the solids of (c) **13** and (d) **14** under room light (top) and under UV irradiation (bottom) before and after grinding of as-recrystallized dyes.

In 2013, Ooyama *et al.*[46] also reported that a pronounced bathochromic shift-type mechanofluorochromism was found for D-π-A pyridinium dyes **15a–c** having diphenylamino group as an electron donor and pyridinium ring as an electron acceptor linked by carbazole as a π-conjugated bridge, and bearing various counter anions ($X^- = Cl^-$, Br^-, or I^-). Both the absorption and fluorescence peaks of as-recrystallized dyes show bathochromic shift by grinding, and the degrees of the bathochromic shift depend on the type of counter anion of **15a–c**. By grinding the as-recrystallized dyes **15a–c** at a stress of *ca.* 50–100 N/cm^2 in a mortar with a pestle, the three dyes show a pronounced bathochromic shift-type mechanofluorochromism (Figure 2.13): the as-recrystallized dye **15b** changes its color to orange and exhibits yellowish-orange fluorescence by grinding. On the other hand, the as-recrystallized dyes **15a** and **15c** change their colors to reddish-orange by grinding, but the solids of **15a** and **15c** exhibit orange fluorescence and red fluorescence, respectively. After grinding, the λ_{max}^{em} are red-shifted by 69 nm for **15a**, 60 nm for **15b**, and 52 nm for **15c**, respectively. For the three dyes, the lowering of Φ_F value by grinding is observed. When the ground samples of **15b** and **15c** are heated to around 170 °C (beyond the recrystallization temperature), the colors and fluorescent colors of the dyes are restored almost perfectly to those of the original ones. Based on these results, the authors thought that the type of counter anion of D-π-A pyridinium dyes can certainly control the mechanofluorochromism, that is, the type of counter anion can conveniently tune the fluorescent color change by grinding. It is worth mentioning here that the D-π-A pyridinium dyes with a considerable degree of dipole moment (*ca.* 20 Debye) exhibit a pronounced mechanofluorochromism, despite the previous study suggesting that the dipole moment of over 8.0 Debye leads to reducing the

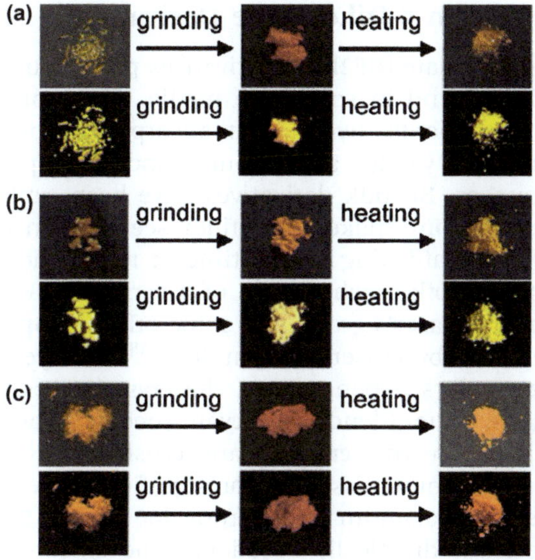

Figure 2.13 Photographs of (a) **15a**, (b) **15b** and (c) **15c** under available light (top) and under UV irradiation (bottom) before and after grinding, and after heating the ground solids.
Reproduced with permission from reference 46. Copyright 2013, Elsevier.

mechanofluorochromism properties of neutral D-π-A fluorescent dyes. The XRD measurements with the as-recrystallized dyes **15a–c** exhibit diffraction peaks ascribable to well-defined microcrystalline structures. The diffraction peaks for these three dyes almost disappear after grinding, showing that the crystal lattice is significantly disrupted. For the three dyes, the diffraction peaks of the ground solids after being heated are similar to those before grinding, which suggest recovery of the microcrystalline structure by heating. The DSC analysis for the dyes indicates that the ground solids undergo an endothermic glass transition, and then an exothermic recrystallization. The DSC traces of the ground powders are typical amorphous solids. The mechanism was proposed that the mechanofluorochromism of D-π-A pyridinium dyes bearing counter anions is attributed to a reversible switching between crystalline and amorphous states with changes of dipole-dipole interaction, intermolecular π-π interaction, and the interaction of pyridinium ring with counter anion before and after grinding.

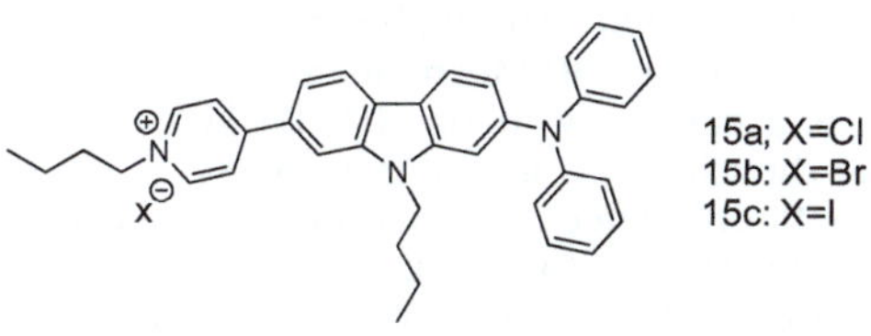

15

2.2.6 Difluoroboron β-Diketonate Derivatives

Difluoroboron β-diketonate (BF2bdk) derivatives possess impressive optical properties in solution and in solid-state media. These properties include large extinction coefficients, two-photon absorption cross-sections, high fluorescence quantum yields, and room-temperature phosphorescence. With these advantages, BF2bdk derivatives have been explored in various applications ranging from "naked-eye" fluorescent anion probes and imaging agents for cell trafficking to real-time *in vivo* tumor hypoxia nanosensors.[47] The mechanofluorochromism of a series of BF2bdk derivatives has been reported by Fraser's group and some other groups.

Dye **16** was reported by Fraser *et al.* in 2010.[48] The dye crystals possess different emission colors depending on the crystal types and processing conditions, that is, blue (dendritic), cyan (needle) and green (prism) (Figure 2.14a–c). When the crystals are crushed or when the solid samples are smeared, the emission color significantly red-shifts, the corresponding emission maximum shifts from 460 nm to 542 nm, and the fluorescence spectrum drastically broadens. The smeared materials can revert to the original green-blue emission either spontaneously at room temperature or by thermal annealing. The fluorescence "rewritability" of the dye **16** solid film on a piece of weighing paper is likewise observed (Figure 2.14A–D).

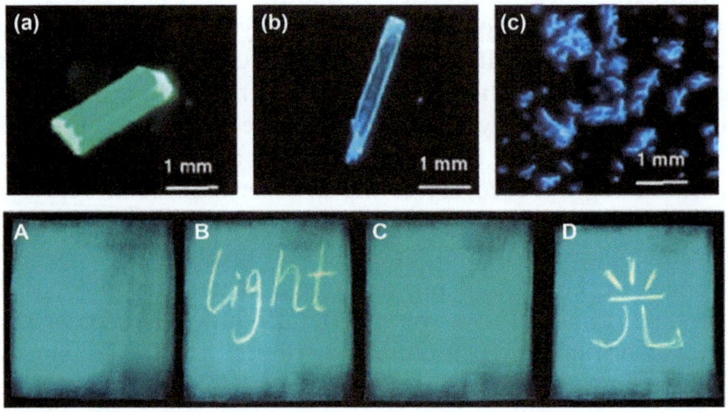

Figure 2.14 Photos showing (a) green and (b) cyan crystals and (c) the blue coral-like solid under UV excitation ($\lambda_{ex} = 365$ nm); (A) Fluorescence emission (λ_{ex} 365 nm) of a thermally annealed dye **16** solid film on a piece of weighing paper. (B) Mechanochromic fluorescence of "light" written with a cotton swab tip. (C) Background emission restored by heating the film with a heat gun for ~5 s. (D) Rewritable mechanochromic fluorescence demonstrated by the Chinese character "light" generated with a cotton swab tip.
Reproduced with permission from reference 48. Copyright 2010, American Chemical Society.

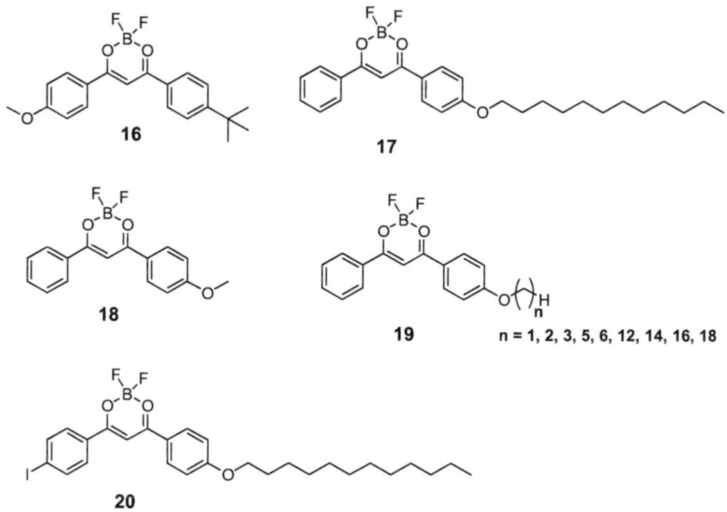

As mentioned above, dye **16** exists in at least three polymorphic forms, each with distinct emission colors under UV black light at 365 nm. Reddy *et al.*[49] obtained two single crystals with a size that is at least 2 mm^3 (a vital prerequisite for the studies) for green (form I) and cyan (form II) polymorphs of dye **16.** The X-ray crystalline structures of these polymorphs are orthorhombic, with space group P*nma* for green and monoclinic $P2_1/c$ for cyan emitting solids. It was found that form I is more brittle than form II. Therefore, the authors investigated the mechanical properties of the green form I and cyan form II dye **16** crystals by using qualitative deformation tests and nanoindentation (quantitative) techniques, and confirmed that the cyan form II is much softer and more compliant compared with the green form I. The X-ray crystalline structure analysis of the two forms suggests that the presence of slip planes in form II is responsible for its higher plasticity as compared to form I, which has a nearly isotropic structure, and hence is brittle. The nanoindentation experiments confirmed that the hardness of form II is significantly less than any face of form I. According to the single crystal X-ray analysis, the packing structure of form II is more anisotropic than that of form I, given stronger (multiple C–CH···F) and weaker interactions (*tert*-Bu, -OMe) in the former (Figure 2.15A and B).

The mechanofluorochromic behaviors in forms I (green) and II (cyan) were examined by crushing the single crystals separately in a clean and dry mortar with a pestle. When this was done in a warm mortar (taken out of an oven at 60 °C), gentle crushing results in a thin layer of powder particles (The authors thought that this process did not result in any change in the emission color for both the forms, probably because this did not result in a significant density of defects in the resulting particles; Figure 2.15C). However, upon scratching firmly with a pestle, the powders of the brittle green form I exhibit a slight color change, which is barely detectable. Careful inspection revealed that the material indeed showed a color change from green to yellow upon grinding, but faded quickly back to green along the path of pestle movement (resembling the fading tail of a satellite in the earth's atmosphere). At room temperature, the change in emission color is more prominent, but not dramatically different. As the mechanical properties of molecular crystals depend upon temperature, the authors repeated this at −98 °C by immersing the mortar into methanol, followed by cooling using liquid nitrogen. The material shows a clear color change from green to yellow and took some minutes to recover to the original green color. On the other hand, the cyan (bending) form II shows a prominent color change to yellow (Figure 2.15C) upon scratching firmly even above room temperature (≈55 °C), but it also heals back to cyan slowly at 55 °C, or more quickly by heating. The authors thought that the mechanofluorochromic experiments revealed three facts: (1) the perturbed yellow states, on recovery, emit the same color as their parent forms do, (2) the mechanofluorochromic behavior of form II is more prominent than that for form I at a same temperature and (3) the recovery of the green form I is faster than that of the cyan form II. The latter observation indicates a clear relationship between the reversible mechanochromism and mechanical properties of the crystalline forms, *i.e.*, the stiffer form recovers quickly from the metastable state to the original. For form I, all three faces show higher elastic recovery rate prior to the onset of elastic-plastic transition than that for form II. The fact that the perturbed states recover to emit the initial color suggests that these states must remain close enough to the crystalline structure of the initial form, but with defects that cause some changes in the molecular environment, so that they return to the original form over time or on heating. This means that the perturbed states do not convert to other stable solid state forms, but only revert to the parent form. To check if forms I and II can interconvert or transform to other forms, the authors carried out mechanical grinding (dry) experiments extended for about 30 minutes. A quick visual inspection revealed that the powders appear more yellowish just after grinding. They do not convert to any other forms, but with time or on heating recover to their parent forms: the stiffer form I returns quickly to its original state as compared to the softer form II, as confirmed by powder X-ray diffraction. Differential scanning calorimetry (DSC) of grown single crystals showed no phase transition endotherms before melting, confirming no detectable phase transition by thermal means. Another favorable factor for reversible mechanofluorochromic behavior may be that the original phase acts as a nucleating agent for the transformation of perturbed states back to the original state. These observations suggest that the solid-state packing in the perturbed state is likely to exist with some defects (*e.g.*, misalignment of layers, π-stacking interactions or tilting of molecules in the layers) that are large enough to change the emitting wavelength, but energetically small enough to recover. Hence, as supported by the nanoindentation results, the authors suggested that it is highly possible that the defects are formed mainly due to the breakage or rearrangement of weak interactions along the active slip

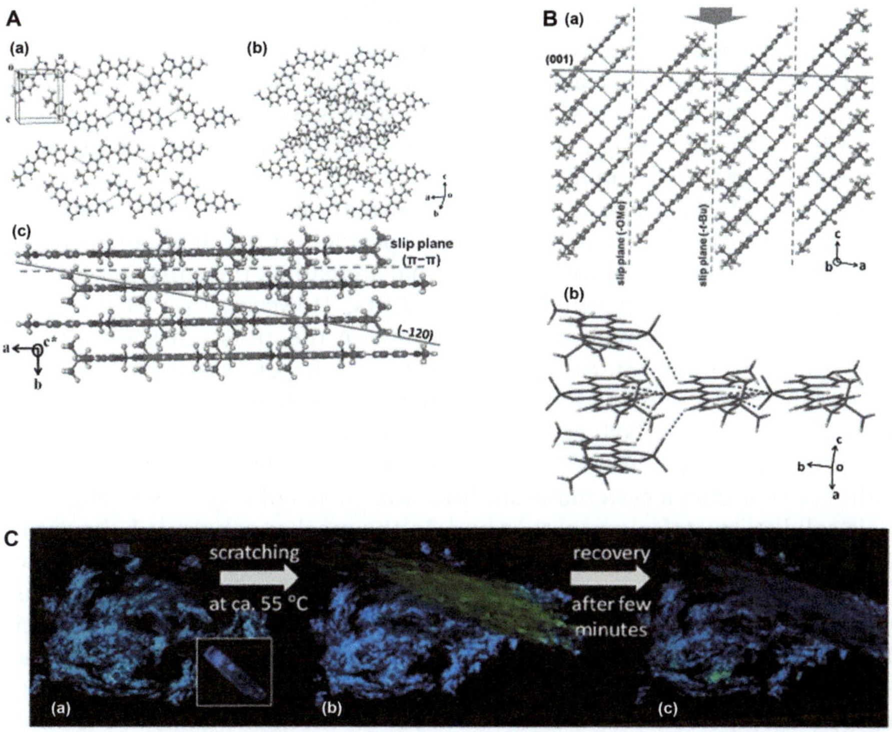

Figure 2.15 (A) Crystalline structure of the dye **16** green form I (brittle). (a) Top view of a 2D sheet to show 1D tapes formed *via* C–H···O interactions and the packing of similar groups of molecules on the same side. Crystal packing views perpendicular to (b) (011) and (c) (001) plane, *i.e.*, side view of 2D sheets. Notice the molecular packing with respect to the indentation directions perpendicular to (011) in (b), and (−120) and (001) planes in (c). (B) Crystal packing in cyan form II of the dye **16** (a) Thick molecular sheets (between two slip planes) formed *via* multiple C–H···F interactions. Arrow indicates the indentation direction, *i.e.*, perpendicular to (001) plane. (b) Antiparallel stacking and multiple C–H···F interactions between molecules that form the thick 2D sheet. (C) Dye **16** cyan form II. (a) Powder solid film obtained by crushing cyan crystals (inset) gently in a warm mortar at *ca.* 55 °C. (b) The film just after scratching firmly shows bright color change to yellow (top-right). (c) Film after heating with a hot-air gun for about a minute. Notice the prominent color change to yellow in (b) and recovery to the parent cyan in (c).
Reproduced with permission from reference 49. Copyright 2013, Wiley-VCH.

planes, for example, the π-stacking interactions in form I or the van der Waals interactions (*tert* -Bu, -OMe) in form II, rather than the breakage or disruption of stronger interactions. With external stimuli or with time, the material returns to the original form as moving molecules along these weak planes will not have high energy barriers. The study indicates that the experimentally established structure-mechanical property relationship is in excellent agreement with the mechanofluorochromic behavior of forms I and II. Thus, these observations with dye **16** polymorphs suggest that the introduction of active slip planes in solid-state structures of a fluorophore could be a valuable design strategy to enhance the mechanofluorochromic behavior for practical applications.

Dyes **17** and **18**, which are analogs of dye **16**, were also investigated by Fraser *et al.* (2011).[47] Smearing the dye **17** powder on a weighing paper red-shifts the emission color, and upon thermal annealing, the emission color of the smeared sample significantly blue-shifts. However, dye **18** does not possess mechanochromism, with no change in the emission spectra upon smearing or thermal annealing. In addition to the difference in the structures of dyes **16**, **17** and **18** in terms of their alkyl substituents, that is, a *tert*-butyl, a 12-carbon alkyl chain and a methyl group, respectively, their mechanochromisms are very different. These differences inspired the authors to conduct a systematic study on the effects of alkyl chain length on their solid-state optical properties and on the mechanochromic behavior of the dye **19** series.[50] All dyes, except for the dye with $n=1$, display mechanochromic fluorescence properties. After smearing, the mechanochromic samples exhibit red-shifts and broad structureless bands, whereas their fluorescence lifetimes become longer. Based on lifetime measurements, excimer emission is suggested to be activated upon smearing. At room temperature, the recovery time (smeared to thermally annealed sample emissions) of quartz substrates generally increases with increasing alkyl chain length. The recovery time ranges from minutes to days. Longer chain analogs ($n \geq 6$) do not completely return to the original annealed emissive state even after months of recovering quartz substrates.

Dye **20**, an iodo compound of dye **17**, exhibits significant red-shift fluorescence upon mechanical perturbation.[51] The emission of its solid film is quenched when the sample is smeared in the air conditions, but become orange when subjects to nitrogen. One possible mechanism for this experimental phenomenon was proposed. According to spectroscopic and lifetime studies, smearing causes the single excited state to be closer to the triple excited state, thereby increasing the coupling between the two states and inducing fluorescence quenching.

In addition, a series of aromatic substituents in BF2bdk dyes (the dye **21** series) with methyl, phenyl, naphthyl and anthracyl groups, were prepared by Fraser *et al.* to determine the effects of π-conjugation length and arene size on the mechanofluorochromic properties.[52] The materials display emissions at various wavelengths from blue to red, depending on the conjugation length. Dye **21a** is not mechanochromic, whereas dyes **21b** to **21d** exhibit mechanofluorochromic properties, the increased π-conjugation that is correlated with a more dramatic red-shift fluorescence, and more pronounced differences between as-spun and annealed states. Dye **21c**, which contains a naphthyl unit, showed unrecoverable emission after mechanical smearing, whereas the emissions of smeared dyes **21b** and **21d** move toward short wavelengths and resemble their initial annealed state. The amorphous film species recover faster than the powders, which is caused by the sample thickness effect. Therefore, thermal annealing is a very important step in observing the mechanochromism of samples. When unannealed samples are smeared, minimal change is observed in their spectra. The lifetimes of the smeared samples are close to those of the original amorphous film samples, which indicates that smearing likewise generates an amorphous state. The formation and disappearance of the aggregated species were considered to be the reasons for the fluorescence variation upon thermal annealing and smearing. Meanwhile, strong intermolecular interactions were considered to be the driving force for thermally induced aggregation and spontaneous recovery at room temperature after smearing.[52]

21

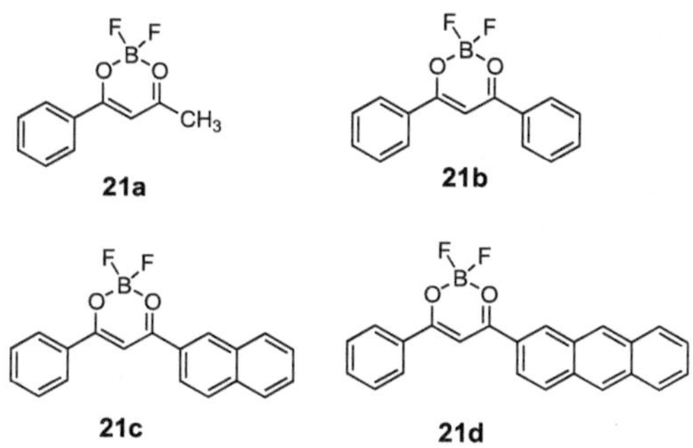

As stated above, the molecular solids of BF2bdk derivatives were found to exhibit reversible mechanofluorochromism at room temperature. However, the excited-state interactions among the solid-state BF2bdk molecules and their changes upon mechanical stimulation were not clear. Sun *et al.* first investigated the optical properties of monomeric, dimeric and polymeric BF2bdk derivatives (dyes **22a–e**) in 2012.[53] The experimental results showed that all compounds had mechanofluorochromic properties. By correlating the solution data with the solid-state characterization results, the authors drew the following conclusions:[53] (1) BF2bdk fluorophores have a propensity to form H-dimers or H-aggregates in solution and in amorphous solids, such as polymeric nanoparticles or smeared BF2bdks crystals; (2) mechanical stimuli can cause the formation of H-dimers or H-aggregates of BF2bdk that exhibit a low-lying emissive state around 550 nm; and (3) an energy transfer may happen from other higher excited states to the lowest H-aggregate excited states, which eventually makes the entire solid emit at approximately 550 nm. In short, two coupled processes, namely, force-induced emissive H-aggregate formation and energy transfer to the emissive H-aggregates, are responsible for the observed BF2bdks mechanofluorochromism in the solid state. Figure 2.16 illustrates the mechanofluorochromic mechanism of this series of compounds. The authors considered that the mechanism is also of great usefulness in understanding the mechanofluorochromism of all BF2bdk compounds that were previously mentioned.

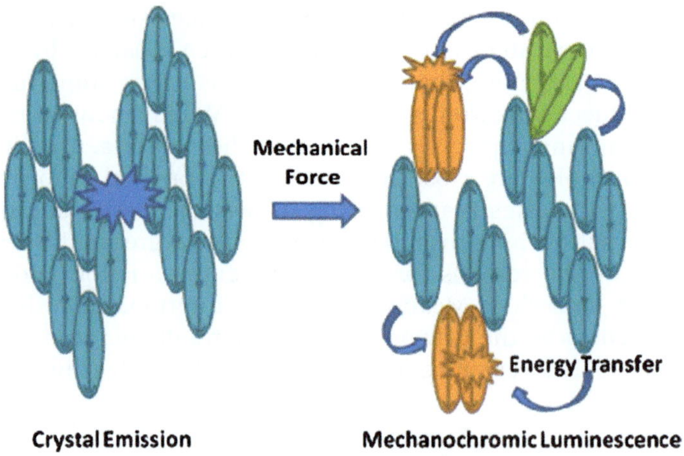

Figure 2.16 Pictorial representation of BF2bdks mechanofluorochromism process: the crystals are ordered and emit green-blue fluorescence (blue molecules); when the order is disrupted by mechanical force, limited numbers of H-aggregate forming sites (orange molecules) can give the entire solid a red-shifted emission mediated by energy transfer (exciton migration). Green molecules in the figure represent any other type of intermolecular interaction with an excited-state energy level above that of the H-aggregates but below the crystals.
Reproduced with permission from reference 53. Copyright 2012, The Royal Society of Chemistry.

22

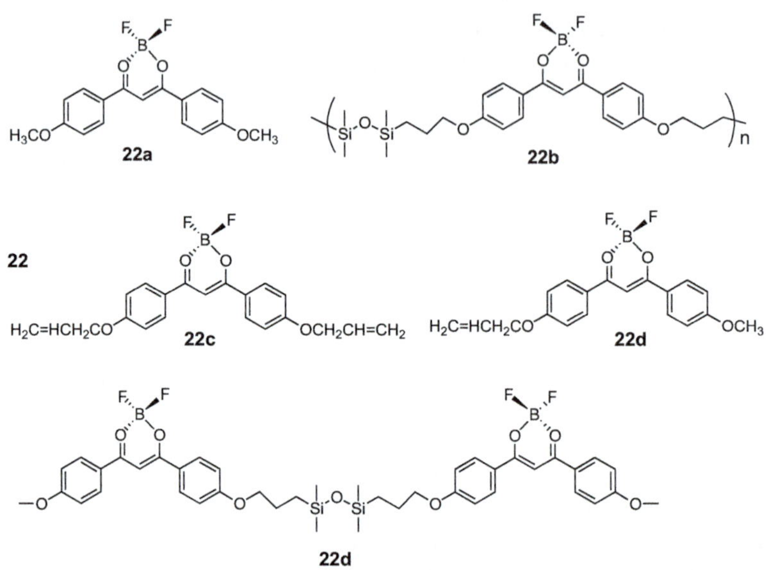

Organic molecular solids or molecular assemblies are typically based on van der Waals interactions, such as π–π interactions or hydrogen bonding interactions. These interactions promote the development of novel functions and properties, which differ from those in the monomer state. In the development of luminescent materials based on organic molecules, it is important to understand the properties and functions, not only for the monomer species, but also for the assembled species. The BF2bdk complex exhibits a fluorescence spectral change by mechanical perturbation. Fraser *et al.* reported a reversible mechanochromic fluorescent organic solid based on the 4-*tert*-butyl-4′-methoxydibenzoylmethane boron difluoride complex, which shows different emissions depending on the crystal type.[48,49] The emission color significantly red shifts upon rubbing/smearing the samples. The samples then revert back to the original emission species over time. Ito and Sagawa (2013)[54] quantitatively evaluated, for the first time, the thermodynamic parameters for thermal-back type fluorescent mechanochromism phenomenon of dyes **16**, **22a** and **23** after mechanically induced fluorescence change. Figure 2.17A shows this phenomenon: the drop-casted dye **16** powder exhibited a blue emission that peaked at 460 nm. After rubbing with a spatula, a new fluorescence band appeared around 500 nm, with a shoulder at 550 nm, originating from the amorphous state of dye **16**. The fluorescence intensity up to 530 nm decreased as time elapsed. After 1030 min, the fluorescence spectrum showed peaks around 460 and 500 nm.

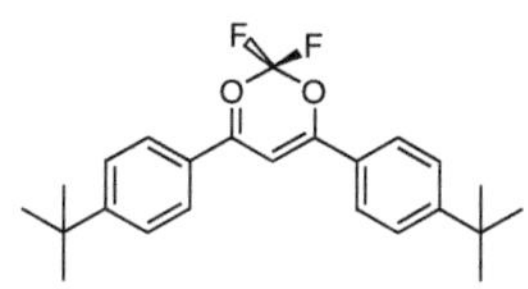

23

In order to conduct a quantitative kinetic analysis for the thermal back-reaction after mechanical perturbation, changes in fluorescence intensity at 550 nm excited at 370 nm as a function of time were obtained. The temperature was maintained during the measurement at (a) 296, (b) 303 and (c) 313 K (Figure 2.17B). The fluorescence intensity at 550 nm decreased with elapsed time, which can be reproduced with a double-exponential decay function obeying first-order kinetics. The rate constants of the faster (k_F) and slower (k_S) components obtained by least-squares fitting are listed in Table 2.1. The two rate constants increase with increasing temperature. The changes in fluorescence intensity quantitatively exhibit exponential decay from the manual smearing. The temperature dependence of the rate constants provides a convenience to determine the activation parameters of the amorphous–crystal phase transition of the BF2bdk complexes. The Arrhenius kinetics equation $k = A\exp(-E_a/RT)$ can be evaluated and used to provide the activation parameters, where A, Ea, R and T are the pre-exponential factor, activation energy, gas constant and temperature, respectively. By least-squares fitting, the A and Ea values were estimated to be 1.05×10^6 s^{-1} and 45.8 kJ mol^{-1} for k_F, and 2.97×101 s^{-1} and 27.2 kJ mol^{-1} for k_S, respectively. The enthalpy and entropy of activation energies ($\Delta H^{\ddagger}$ and $\Delta S^{\ddagger}$) for the thermal back-reaction of dye **16** were estimated using the

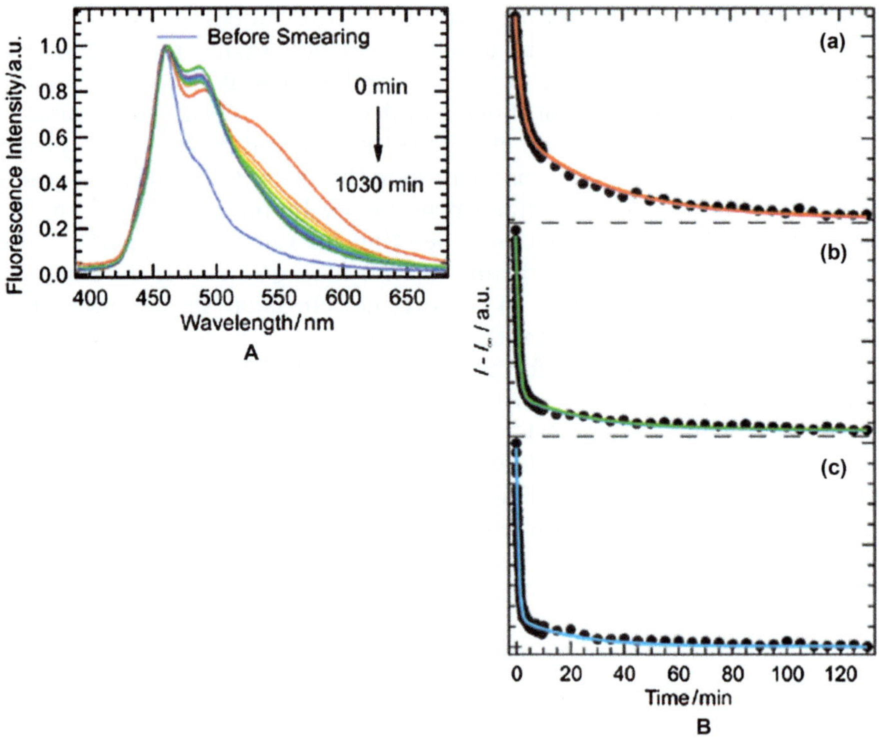

Figure 2.17 (A) Fluorescence spectra of dye **16** at 303 K. The excitation wavelength is 370 nm. (B) Changes in fluorescence intensity of dye **16** monitored at 550 nm as a function of time after mechanical perturbation. The temperature is (a) 296, (b) 303 and (c) 313 K. The solid lines indicate the best-fitting curves based on a double- exponential decay function. Reproduced with permission from reference 54. Copyright 2013, The Royal Society of Chemistry.

Table 2.1 Rate constant of thermal back-reaction for dye **16** from fitting of fluorescence decay.[a]

	Rate constant/s^{-1}	
Temperature/K	k_F	k_S
296	8.06×10^{-3}	5.03×10^{-4}
303	1.70×10^{-2}	6.43×10^{-4}
313	2.25×10^{-2}	7.83×10^{-4}
323	3.05×10^{-2}	9.72×10^{-4}
333	8.52×10^{-2}	1.95×10^{-3}

[a]The fluorescence intensity at 550 nm excited at 370 nm. The k_F and k_S correspond to faster and slower rate constants. Reproduced with permission from reference 54. Copyright 2013, The Royal Society of Chemistry.

Table 2.2 Kinetic and activation parameters of the thermal back-reaction.

Compound	*Component*	*Ea* (*kJ mol*$^{-1}$)	*A* (s^{-1})	$\Delta H^{\ddagger}$ (*kJ mol*$^{-1}$)	$\Delta S^{\ddagger}$ (*J K*$^{-1}$)
Dye **16**	k_F	45.8	1.05×10^{6}	43.2	−104
	k_S	47.2	2.97×10^{1}	24.6	−191
Dye 22a	k_F	44.1	2.36×10^{5}	41.6	−116
	k_S	33.6	2.82×10^{2}	31.1	−172
Dye 23	k_F	23.1	2.65×10^{1}	20.6	−192
	k_S	17.1	2.70×10^{-1}	14.5	−230

equation $k=(k_BT/h)\exp(\Delta S^{\ddagger}/R)\exp(-\Delta H^{\ddagger}/RT)$, where k_B is the Boltzmann constant. The resulting $\Delta H^{\ddagger}$s are 43.2 kJ mol^{-1} for the faster component and 24.6 kJ mol^{-1} for the slower component, and the $\Delta S^{\ddagger}$s are −104 J K^{-1} for the faster component and −191 J K^{-1} for the slower component. The data were summarized in Table 2.2. The estimated $\Delta S^{\ddagger}$ values are negative, suggesting the order parameter of the activated complex is higher than that of the initial amorphous phase. The thermodynamic findings confirmed that the thermal back-reaction of dye **16** originates from the amorphous–crystal transformation. The $\Delta S^{\ddagger}$ value for the faster component is lower than that for the slower component. The free energy barriers ($\Delta G^{\ddagger}$) of dye **16** at 303 K for faster and slower components were 74.7 kJ mol^{-1} and 82.5 kJ mol^{-1}, respectively. These findings indicate that the reaction *via* the lower-order activation complexes (transition state) is the dominant pathway for thermal back-reaction after mechanical perturbation.

2.2.7 Dipeptide-based Dyes

In 2011, Jia *et al.*[55] designed and prepared a novel dendritic molecule, which is a polypeptide-based dendron from glycine and aspartic acid with a pyrenyl group as the focal fluorophore (dye **24**). This dye shows a reversible mechanochromic luminescence property. Their results first demonstrated that the polypeptide, which can provide multiple noncovalent molecular interactions, is an effective framework for preparing mechanochromic materials. Dye **24** exhibited both reversible mechanofluorochromic and thermal history-dependent fluorescent properties (Figure 2.18). Small-angle X-ray scattering and spectroscopy measurements demonstrated that the reversible mechanofluorochromic property of dye **24** originated from the reversible self-assembly between the hexagonal packing and lamellar arrangement, thereby generating different emissions. The new mechanism was proposed based on the excimer-to-excimer transition that facilitated the conversion of emission colors. Unlike the previously reported monomer-to-excimer transition mechanism,[13,17,48,56] dye **24** is an example of rarely reported mechanochromic compounds and is based on the mechanism of excimer-to-excimer transition. This mechanism offers a promising strategy for tuning

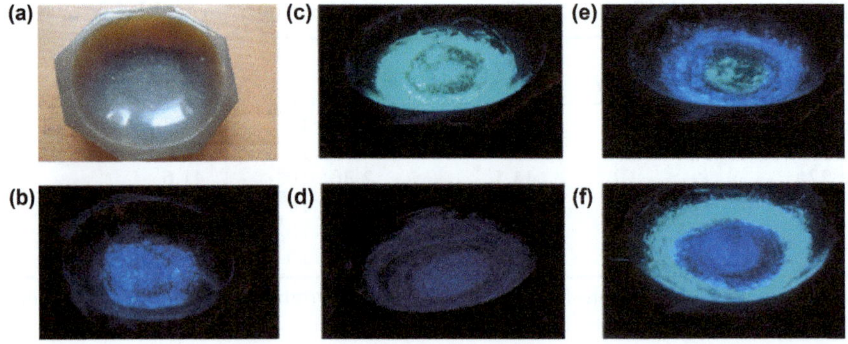

Figure 2.18 (a) Drop cast aggregates of dye **24** under ambient light; (b) the same sample under illumination of 365 nm; (c) ground sample; (d) reversion to blue emission after simple CH_2Cl_2 treatment by adding solvent to the sample; (e) further scratching the solvent-treated sample at the center; (f) scratching the solvent-treated sample on the periphery.
Reproduced with permission from reference 55. Copyright 2011, The Royal Society of Chemistry.

the luminescent property. This newly discovered amino acid-based smart material might enrich the limited family of mechanochromic materials and might further establish a deep understanding of the relationship between the molecular self-assembled structures and luminescent properties. The authors suggested that the design strategy of combining the polypeptide structure and the aromatic groups in the same molecule can be applied to other systems because of the wide applicability of polypeptides.

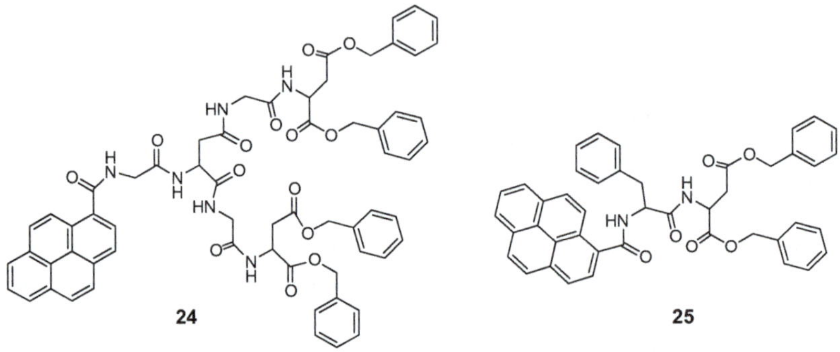

As suggested above, an analog of dye **24** was reported in 2012. This compound (dye **25**) had dual-switching modes, whose emission color and emission intensity could be separately modulated by external force (Figure 2.19A).[57] The force-induced emission enhancement, which is a "turn-on" mechanism, presents a useful strategy to convert the macroscopic mechanical energy to photon luminescence. The authors boldly presented a unique hypothesis for fabricating force sensors designed for special populations, such as color-blind people who are unable to perceive color changes but can distinguish bright from dark luminescence, given the unique

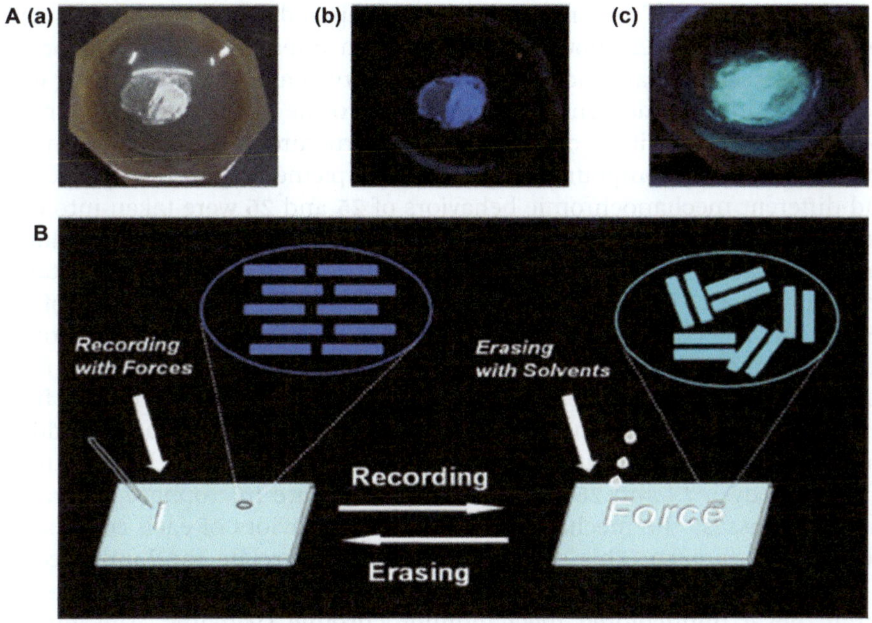

Figure 2.19 (A) The images of dye **25** (a) as-prepared powder under ambient light; (b) as-prepared powder under UV light; (c) shearing powder under UV light. (B) Illustrations of luminescent color variation from blue to bright green.
Reproduced with permission from reference 57. Copyright 2012, Wiley-VCH.

advantages of dye **25**. According to theoretical simulation and experimental data, a plausible mechanism for the dual-switching feature of dye **25** was proposed (Figure 2.19B):[57] (1) freedom in molecular motion is crucial in luminescent color switching. In an as-prepared powder state, the complex interactions between branched groups, especially hydrogen bonding interactions, can inhibit molecular mobility and confine the orientation of pyrene groups to a partially overlapped architecture in the excited state. After shearing the sample, the force perturbation weakens the intermolecular interactions, which leads to the disruption of ordered structure. Such a disordered state provides more freedom by lowering the energy barrier of molecular motion, which eliminates the molecular pre-organization and increases the possibility of sandwich-type excimer formation. (2) The dynamic fluorescent behavior may be derived from the rich noncovalent interactions of dye **25** and its strong tendency towards self-crystallization. The variability of the H-bonding interaction allows the simultaneous reorganization of the molecular assemblies and the dislocation of the pyrene groups, which account for the quenching and re-enhancing emission (*i.e.*, dark and bright emission behavior). In other words, the dye displays reversible mechanochromic behavior between blue and green because of the formation of different pyrene excimers in the crystalline and amorphous states.

In 2012, Jia *et al.*[58] reported that the as-prepared powder **26** with lactam rhodamine B moieties showed a low emission intensity. However, a bright red luminescence at around 580 nm was observed after heavy shearing with a spatula. This luminescence was attributed to the emission of rhodamine B-opening isomers with a more conjugated structure. Dye **26** is a mechanochemical reaction compound. When the complementary optical properties and different mechanochromic behaviors of **25** and **26** were taken into consideration, a new mechanofluorochromic material (MC-25/26) was produced by incorporating dye **26** into dye **25** *via* self-assembly. In a representative procedure, dye **25** is dissolved in chloroform, followed by the addition of dye **26** powder. The co-assemblies are obtained by diluting the stock solution in excess cyclohexane. The aggregates initially display a blue color. After force perturbation, such as slight shearing with a spatula, the color transition from blue to a distinct green occurs. The green emission switches to a new reddish color with further mechanical action (Figure 2.20). Further investigations reveal that dyes **25** and **26** in the co-assemblies are microphase separated. The dyes possess the mechanofluorochromic behaviors of each component without interference. This study is the first to correlate the luminescence property with molecular assembly and the mechanochemical reaction to obtain novel multicolored mechanofluorochromic switching. The authors suggested that such a binary system displays unique advantages not only in the multiple outputs, but also in the facile preparation method, by exploiting already available dyes. This system does not involve tedious molecular synthesis, and is therefore of great importance in practical applications.

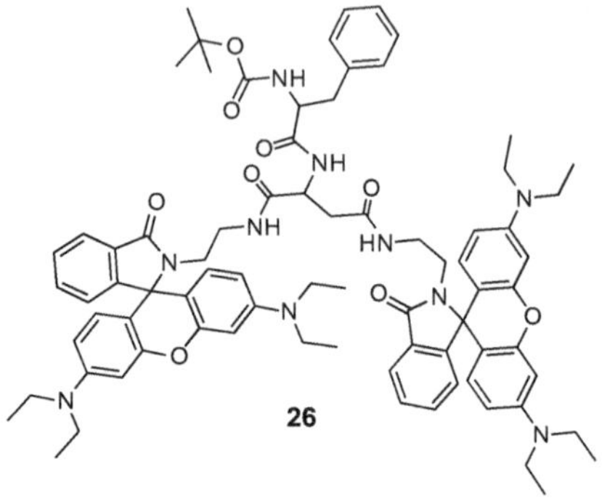

26

Jia *et al.* (2013)[59] reported an organic molecule **27** with tetraphenylalanine as a spacer to link the chromophores of pyrene and rhodamine B. Compound **27** can change color from blue (440 nm) to bluish-green (480 nm) and further to a reddish (583 nm) color (Scheme 2.1A). This color transition can be reversed by heating and treatment of the sample with solvents. On the basis of the results and analysis of the differential scanning calorimetry (DSC), small-angle X-ray scattering (SAXS), circular dichroism (CD), Fourier transform infrared (FTIR) and UV/Vis and fluorescence spectroscopy, the mechanism for mechanically induced multicolored switching of **27** was proposed as follows (Scheme 2.1B).

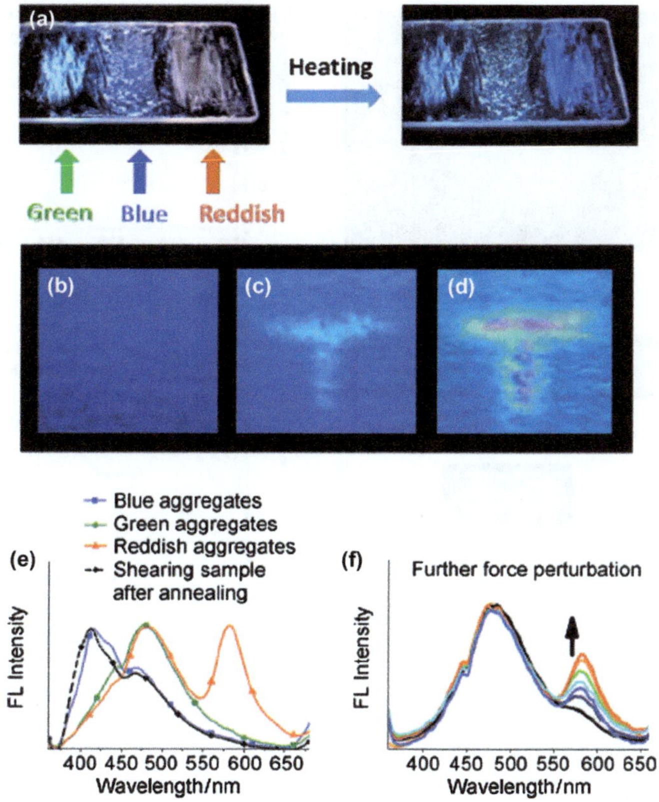

Figure 2.20 (a) MC-25/26 aggregates showing different emission colors under force perturbation (irradiated by 365 nm UV light) and their color restoration by heating. Blue: as-prepared aggregates without force perturbation; green: slight shearing of the aggregates with a spatula; reddish: further shearing of the green aggregates. (b–d) Fluorescence images of (b) MC-25/26 on a filter paper with blue luminescence; (c) force-induced green luminescence pattern; (d) force induced reddish luminescence pattern and three different emission colors in one paper. (e, f) Emission spectra of (e) the samples with blue, green and reddish emission colors and the sheared sample after annealing, and (f) the tunable fluorescence property under force perturbation. Reproduced with permission from reference 58. Copyright 2012, Wiley-VCH.

The initial powder of **27** emits blue luminescence (440 nm) because of the partially overlapped packing of excimer in a helical-like hexagonal column. Upon slight grinding, the assembled structures are destroyed, which leads to a sandwich stacking of the pyrene moieties and results in a change in fluorescence from blue to bluish-green. Extreme force perturbation induces the ring-opening reaction of rhodamine B from a spirolactam to a ring-opened amide with a reddish color. The emission switches from blue to bluish-green to reddish in sequence as the external force increases.

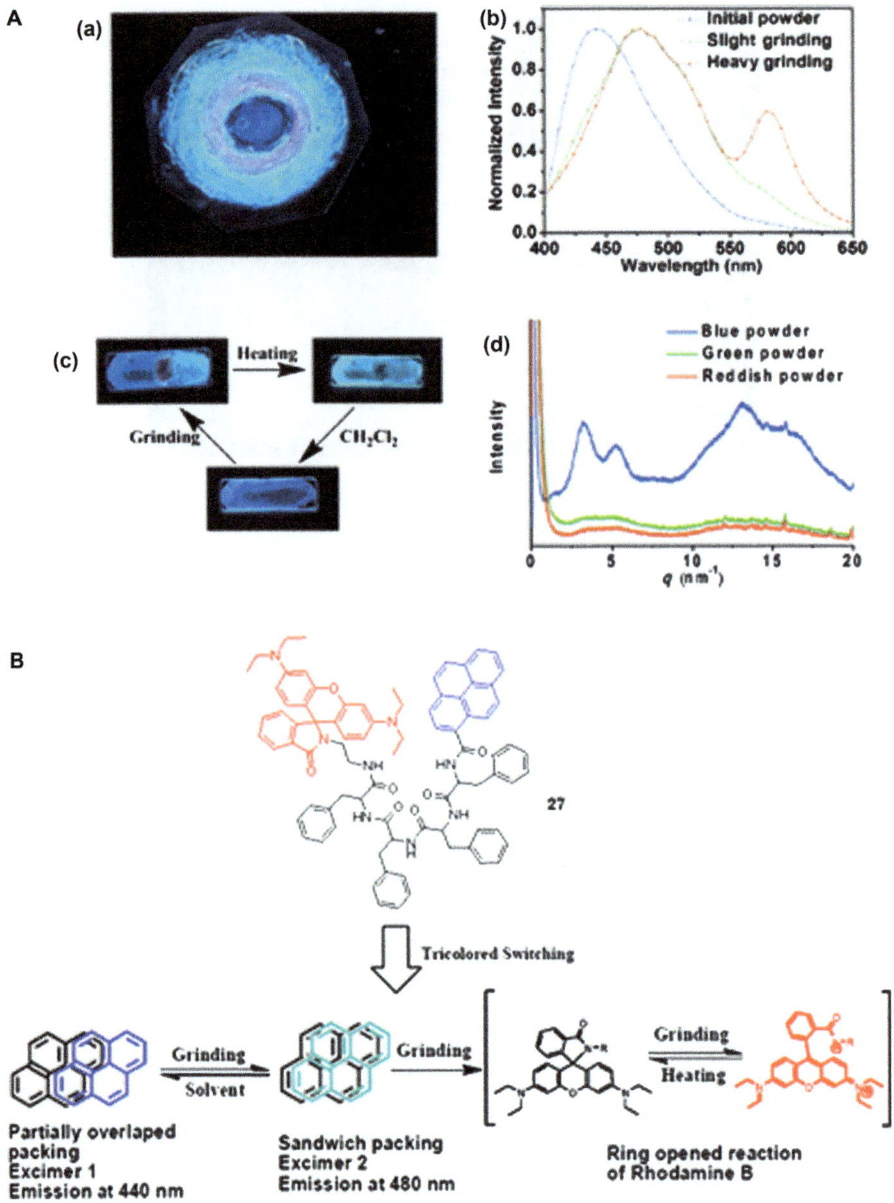

Scheme 2.1 (A) (a) Different emission colors (upon irradiation with 365 nm UV light) of **27** on an agate mortar upon grinding: original blue (center), reddish (middle ring) and bluish-green (external ring). (b) Fluorescence spectra of the initial powder and the same sample upon grinding ($\lambda_{ex} = 365$ nm). (c) Reversible change in the emission colors *in situ*. (d) SAXS patterns of the blue powder (blue line), bluish-green powder (green line), and reddish powder (red line). B) The tricolored switching mechanisms of **27**.

The authors emphasized that compound **27** is a rare example of a mechanofluorochromic material with multicolored emission composed of a single organic molecule and its multicolored emission can be tuned simply by different packing of the chromophores and a force-induced ring-opening reaction.

2.2.8 Triphenylamine Formaldehyde Derivatives

Nakano *et al.* (2011)[60] reported a new example of mechanochromic luminescent materials, dye **28**. The glass-transition temperature of the dye is very low at only 8 °C. Thus, obtaining different morphologies (crystal and supercooled liquid) at room temperature is easy. The emitted colors are light blue and greenish yellow for the crystalline and supercooled liquid states, respectively. The spectrum is red-shifted when the morphology changes from crystalline to a supercooled liquid state. The emission color of the crystalline sample changes from light blue to greenish yellow upon grinding with a spatula, and then gradually returns to its original light blue (Figure 2.21). The change in the emission color was attributed to the formation of the supercooled liquid state by grinding and subsequent recrystallization.

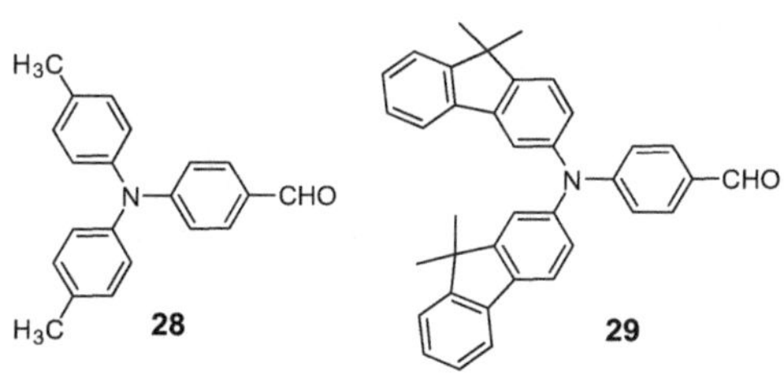

28 **29**

As stated above, the glass-transition temperature (*Tg*) of dye **28** is very low, only 8 °C. Thus, at room temperature, the amorphous state of dye **28** obtained by grinding the crystal sample is not so stable due to rapid re-crystallization and hence the fluorescent color returns to the original soon after grinding. It can be expected that the increase in *Tg* of the material makes the amorphous state more stable to prevent crystallization at room temperature. Therefore, Nakano *et al.* (2013)[61] designed and synthesized 4-[bis(9,9-dimethylfluoren-2-yl)amino]benzaldehyde (dye **29**) whose *Tg* was expected to be higher than that of the parent compound, dye **28**. As intended, DSC results show that the Tg of dye **29** is 86 °C, distinctly higher than that of dye **28** (8 °C). The crystalline powder sample of dye **28** emits a light blue color upon UV irradiation. By grinding the crystalline sample, the emission color is changed to greenish yellow, and therefore letters can be recognized upon UV irradiation, as shown in Figure 2.22A. Whereas the amorphous sample is obtained by grinding the crystals of dye **29**

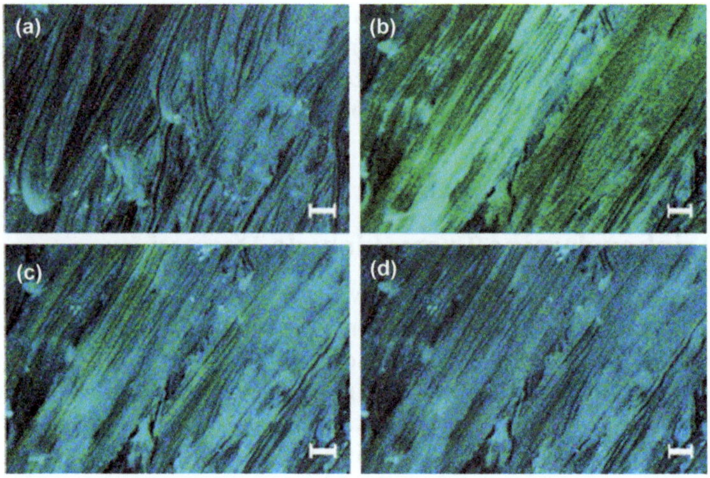

Figure 2.21 Photographs of dye **28** powder on a glass substrate upon irradiation with 365 nm UV light. (a) Before grinding. (b) Just after grinding with a spatula. (c) About 20 s after grinding. (d) About 40 s after grinding. Scale bar: 0.2 mm.
Reproduced with permission from reference 60. Copyright 2011, Elsevier.

re-crystallized soon, the state obtained by grinding the dye **29** crystals was found to be quite stable at room temperature, as intended. The written letters are recognizable under UV irradiation for more than three weeks at room temperature. It was suggested that the change in emission color by grinding the crystalline sample is due to the morphological change from crystalline state to amorphous state by grinding and the amorphous state is stabilized by increasing the *Tg* of the material. When the sample is heated above *Tg*, the letters disappear due to re-crystallization of the sample. The absorption spectrum for amorphous film obtained by spin-coating method can be measured, being almost similar to that for solution. The result suggested that the effect of intermolecular interaction of dye **29** molecules in the ground state is negligible in solid states. It was found that the emission spectra for the crystalline powder sample and amorphous glass sample obtained by cooling the melt were different from each other. The emission spectrum is red shifted when the morphology is changed from the crystalline state to the amorphous one. In addition, the spectrum in the amorphous state is confirmed to be recovered to that in crystalline state when the re-crystallization is induced by poking the sample with a spatula on heating. Thus, the fluorescence spectra of dye **29** in condensed state depend upon the morphology (Figure 2.22B). The fluorescence spectrum of the sample obtained by grinding the crystalline sample was also found to be red shifted from that for the crystalline sample. The result suggested that the sample obtained by grinding the crystalline sample is in amorphous state. However, the spectrum is not completely coincident with that for the amorphous

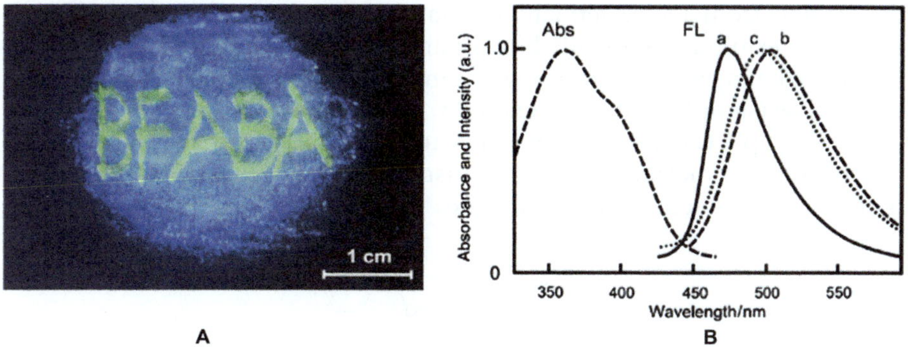

Figure 2.22 (A) Photographs of dye **29** powder on a parchment paper upon irradiation with 365 nm-light. After writing the letters "BFABA" by grinding with a spatula. (B) Electronic absorption spectrum of spin-coated amorphous film of dye **29** (Abs) and fluorescence spectra of dye **29** (FL) in solid states. (a) Crystalline sample. (b) Amorphous sample obtained by cooling the melt. (c) Powder sample obtained by grinding the crystal. Excited wavelength: 365 nm.
Reproduced with permission from reference 61. Copyright 2013, Elsevier.

sample obtained by cooling the melt. It was suggested that the crystalline sample is not completely altered to amorphous state by grinding, which was also suggested by the DSC and optical microscopy of the sample obtained by grinding the crystals. When the sample was heated, clear glass-transition phenomenon was not detected, however, exothermic peaks due to crystallization were observed just above the *Tg* of the glass sample. The result suggested that the seeds of crystal of dye **29** remained after grinding, which induced the crystallization of the sample by heating above the *Tg*. When the amorphous sample obtained by cooling the melt is heated, no crystallization is observed, maybe because of the absence of the seed of the crystal in the sample. The difference in emission color between the crystalline state and the amorphous one was suggested to relate the TICT structure of the molecules in the excited state, that is, the conformational change of molecules to form the TICT structure were restricted in the crystal, whereas the existence of free volume may permit the conformational change to form the TICT structure in the amorphous state. It was suggested that fluorescent molecular materials with both abilities of glass formation and of taking TICT structure in the excited state are candidates for the new mechanofluorochromic materials.

2.2.9 Benzodifuran Derivative

Dye **30** was reported by Tsuji *et al.*[62] in 2011. Its sublimed sample (sample S) showed blue light emission with a maximum wavelength of 467 nm, which is similar to that of the diluted solution. The recrystallization of the sublimed dye crystals from 1,1,2,2-tetrachloroethane produced a yellow solid (sample R), which exhibits a red-shift broad band with a maximum wavelength of 565 nm. The mechanical grinding of sample R causes the color to change to pale yellow (sample G) in ambient light. Under UV irradiation, the emission color of sample G changes from yellow to bluish (approximately 485 nm). In conclusion, the mechanofluorochromism is probably caused by alternating molecular alignment in the solid state.

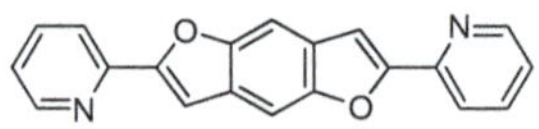

30

2.2.10 Anthrylpyrazole Derivatives

Anthrylpyrazole derivatives could produce polymorphs with various emission colors from deep blue to green because of different intermolecular π-stacking degrees between the anthracene rings.[63,64] Wang *et al.* (2011)[65] reported on the mechanochromism of dyes **31** and **32** in 2011. The crystals of the two dyes, namely **31a** and **32a**, were prepared by vacuum sublimation and solution growth methods, respectively. Crystals **31a** and **32a** emit deep blue and green fluorescence in UV light, respectively. Both crystals exhibit significant mechanofluorochromism when they are ground or pressed. However, the trends of the emission color changes are completely different, that is, **31a** changes from deep blue (430 nm) to green (500 nm), whereas **32a** changes from green (515 nm) to blue (460 nm) (Figure 2.23A). When the ground samples of **31a** and **32a** (named **31b** and **32b**, respectively) are heated to 90 °C and 110 °C, respectively, for 5 min and subsequently cooled at room temperature, they exhibit the emission colors of the corresponding

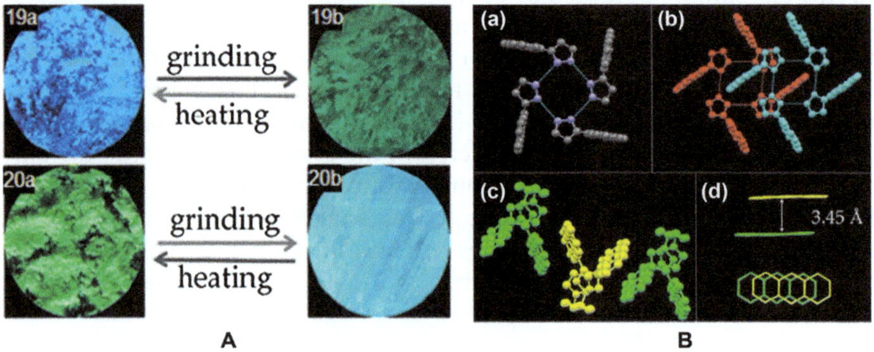

Figure 2.23 (A) The fluorescence variation of **31a** and **32a** upon grinding and subsequent heating. (B) (a) Tetramer structure in **31a**; (b) stacked tetramers in **31a**; (c) π-stacking structure in **32a**; (d) top and side views of the π-stacked anthracene rings in **32a**.
Reproduced with permission from reference 65. Copyright 2011, The Royal Society of Chemistry.

un-ground crystalline samples, which show good reversibility. The XRD patterns reveal that **31a** and **32a** are crystalline and **31b** and **32b** are amorphous, which indicates that grinding converts the aggregative phases from crystalline to amorphous. The DSC curves for the heating of **31a** and **32a** reveal melting endothermic peaks at 195 °C and 223 °C, respectively. By contrast, amorphous **31b** and **32b** show exothermic recrystallization peaks (approximately 81 °C for **31b** and 102 °C for **32b**) and endothermic melting peaks at the melting points of the corresponding un-ground crystals. The peaks indicate that grinding can convert thermodynamically stable crystals **31a** and **32a** to metastable states, such as of **31b** and **32b**, which can be restored to their original crystalline states *via* cold crystallization. Based on the single crystal analysis, a possible mechanism was proposed.[65] In **31a**, the molecules form propeller-type tetramers *via* strong intermolecular N–H···N hydrogen-bonding interactions. Moreover, such tetramer structures render unfavorable π-stacking between the anthracene rings (Figure 2.23B), which results in a deep-blue emission. The grinding process increases the π-stacking degree for several anthracene rings in **31b**, which results in green emission. In **32**, the elimination of intermolecular N–H···N hydrogen bonds is attributed to the substitution of protons on nitrogen atoms in the methylene groups. Furthermore, the contribution of the methylene groups to the molecular flexibility enhances π–π interactions between the anthracene rings, which results in a strong face-to-face π-stacking of the anthracene rings in **31a** (Figure 2.23B) and produces a green emission. The grinding process possibly results in the destruction of the strong π-stacking mode of anthracene rings, which weakens the π–π interactions and results in the blue-shifted emission. In addition, dyes **31** and **32** possess photochromism, which is achieved *via* the photocycloaddition of the face-to-face π-packed anthracene rings in UV irradiation. These results show that the dyes have multi-stimuli-responsive function.

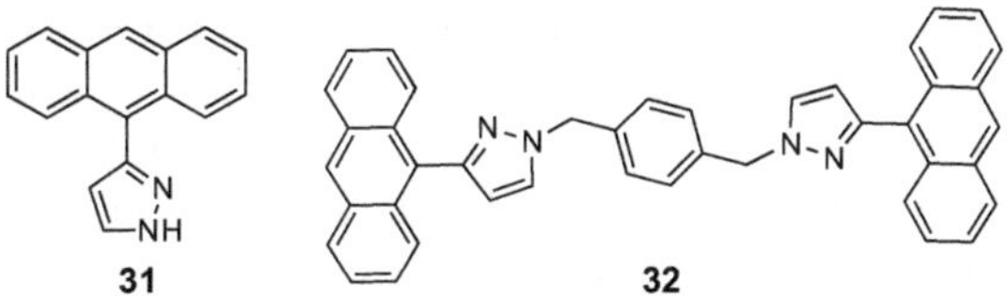

31 **32**

2.2.11 Benzothialdiazole-cored Phenylene Vinylene Derivative

In 2011, Wang *et al.*[66] synthesized and investigated another multi-stimuli-responsive compound, dye **33**. The recrystallization of dye **33** from a mixture of chloroform and petroleum ether (1 : 1, v/v) or *via* vacuum sublimation results in an orange powder with bright orange emission. After grinding using a spatula or pestle, the sample changes into a yellow solid with green emission (Figure 2.24a and b). This solid exhibits significant mechanofluorochromic behavior. The ground sample returns to its original orange

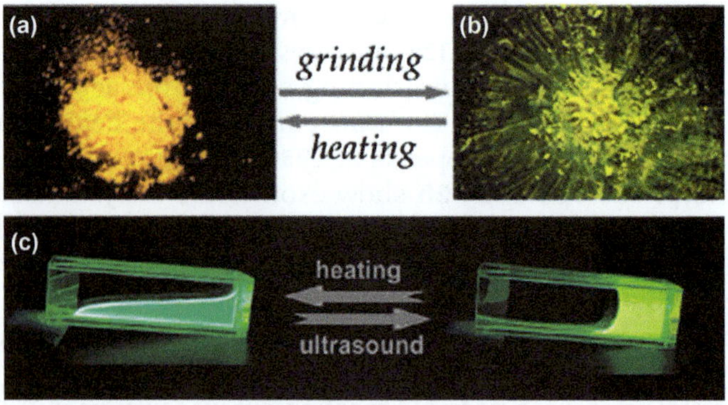

Figure 2.24 Stimuli-responsive behaviors of dye **33** upon (a) heating and (b) grinding treatments (under UV light, $\lambda_{ex} = 365$ nm); (c) stimuli-responsive gelation properties of dye **33** in DMF solution upon heating and ultrasound treatments.
Reproduced with permission from reference 66. Copyright 2011, American Chemical Society.

color with a bright orange fluorescence after heating at approximately 100 °C for 1 min. The XRD results reveal that the unground solid possesses a microcrystalline-like structure, whereas the ground solid has an amorphous feature. The conversions between the orange and green emission samples are achieved by changing the molecular packing modes between the microcrystalline-like and amorphous phases. The DSC results of dye **33** are similar to those of dyes **31** and **32**. The amorphous ground sample undergoes exothermic recrystallization at approximately 95 °C. In addition, these findings suggested that grinding could convert the thermodynamically stable crystalline sample into a metastable state, which could be restored to the original crystalline state *via* an exothermic process.[66] The dye possesses supramolecular assembly characteristics in a solution. When the N,N-dimethylformamide (DMF) solution of the dye at 80 °C was kept at room temperature for over 0.5 h, some orange fluorescent fibers are gradually formed in the solution. However, if the hot DMF solution is quickly cooled to room temperature in a water bath at 25 °C and sonicated for 30 s, the solution turns into a stable organogel with a strong yellow emission (Figure 2.24c). After the ground sample is exposed to saturated ether vapor for approximately 1 min or longer, the fluorescence changes from green to orange, with good reversibility, which indicates that the sample has pronounced vapofluorochromism.

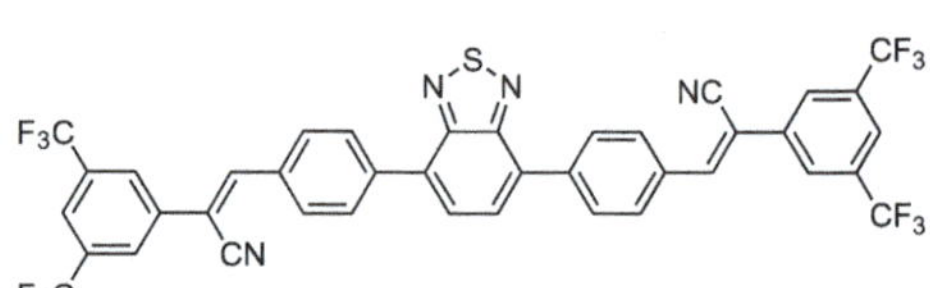

33

2.2.12 Salicylaldehyde Azine Derivatives

Tong *et al.* (2011)[67] reported a series of salicylaldehyde azine derivative (dyes **34** to **37**) in 2011. When the crystals of dye **34**, with an emission of 553 nm and obtained *via* crystallization from tetrahydrofuran solution is annealed at 115 °C or ground, the emission wavelength decreases to 529 nm (Figure 2.25). On the contrary, no such stimuli-responsive property is observed in other dyes. The planar fluorescent core and the attached donor-acceptor (D-A) groups within the same molecular structural core are significant in the construction of salicylaldehyde azine derivatives with fluorescence-switching properties in the solid state.

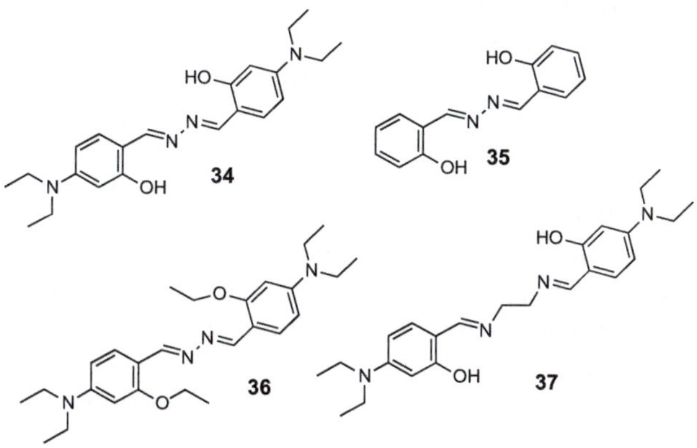

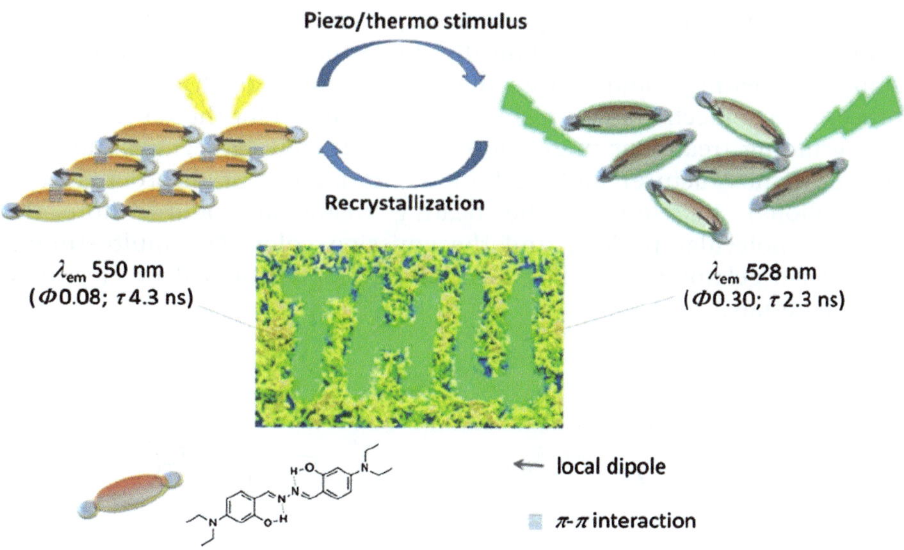

Figure 2.25 Schematic illustration of mechanofluorochromic response of dye **34**. Reproduced with permission from reference 67. Copyright 2011, American Chemical Society.

2.2.13 Cholesterol-appended Quinacridone Derivatives

Wang *et al.* (2011)[68] reported a series of cholesterol-appended quinacridone derivatives (dye **38**). Dyes **38b** and **38c** form a stable organogel in a wide range of organic solvents upon ultrasound irradiation. The formed xerogel film of dyes **38b** and **38c** exhibits a mechanochromic property with reversible fluorescence changes between yellowish green and orange when ground or when subjected to heating treatments. Upon grinding, the orange luminescent xerogel of **38b** could convert into yellow fluorescent film (Figure 2.26a). More interestingly, the ground sample recovers its original orange color by heating at about 120 °C for 5 min. The photoluminescent spectra show that the emission maxima of the ground and heated samples appear at 553 and 586 nm, respectively. The UV absorption spectra of the conversion between different states are recorded in Figure 2.26b. The xerogel shows the main absorption peak at 539 nm, with two shoulder peaks at 498 and 465 nm, which is very similar to the gelation system, suggesting the aggregation packing model in the xerogel. The ground sample displays the blue-shift absorption spectrum with the main peak at 518 nm and a shoulder at 485 nm, which is similar to that of **38b** cool solution. Therefore, grinding treatment destroys the ordered aggregated state and leads to the formation of disordered state. Heating treatment of the ground sample at 120 °C could restore the ordered molecular packing. The differential scanning calorimetry (DSC) experiments (Figure 2.26c) show that the xerogel, ground and heated samples all had similar melting points around 175 °C. When the ground solid is heated, it experiences an exothermic process at a relatively low temperature around 105 °C before melting, which suggested that the grinding process could convert the thermodynamically stable xerogel film to a metastable state and heating could restore the stable state through an exothermic process. Such mechanochromism was also observed for the xerogel of **38c**. The grinding treatment might destroy the ordered aggregated state (1D fibrous nanostructure) and lead to the formation of disordered state (amorphous). By contrast, the heating treatment could restore the ordered molecular packing and the emission color. The multi-stimuli-responsive gelator system is crucial in developing optical or electronic nanomaterials.

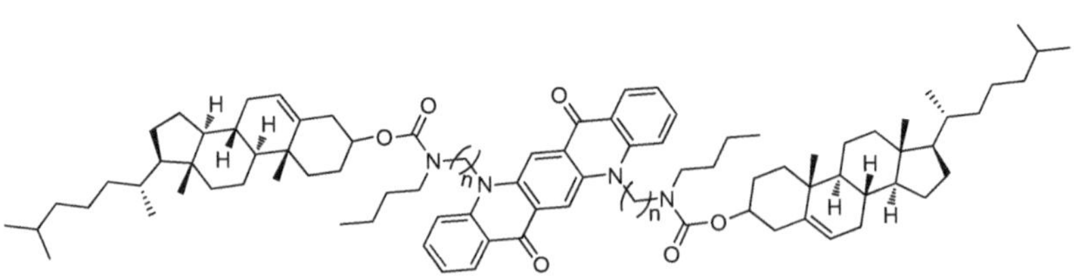

38 **38a**: $n = 4$ **38c**: $n = 8$
38b: $n = 6$ **38d**: $n = 10$

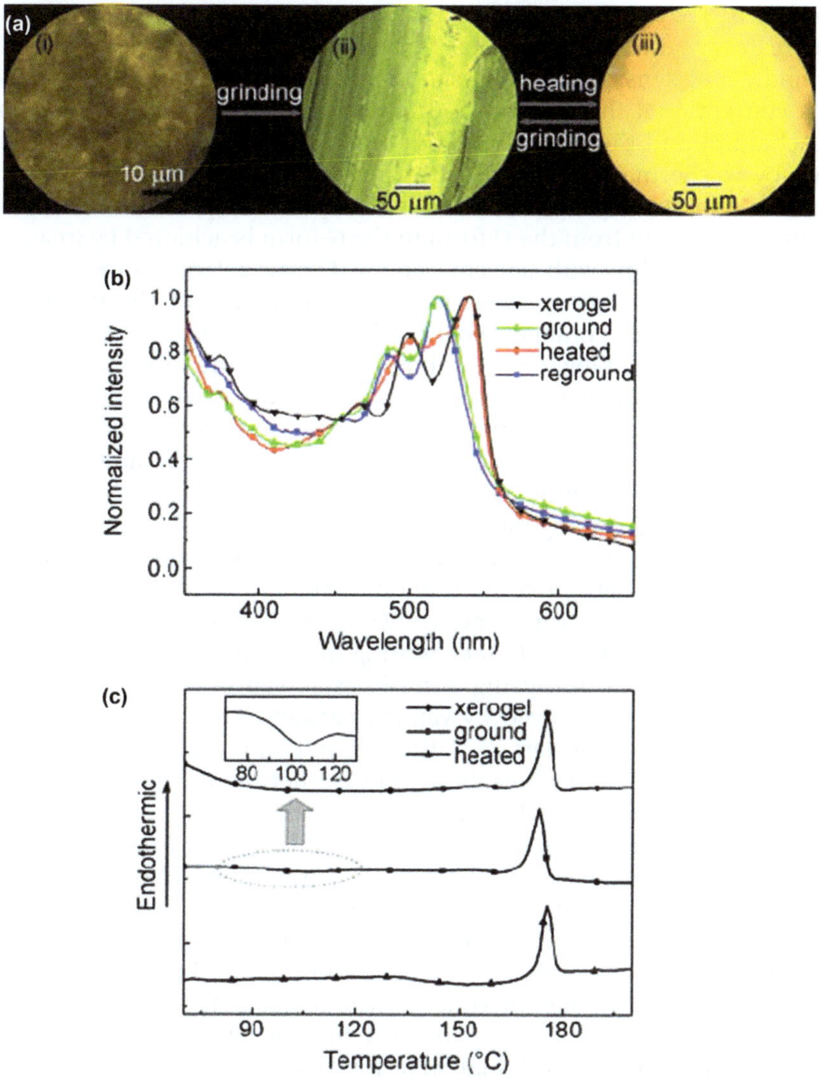

Figure 2.26 (a) Mechanochromism of **38b** upon grinding and heating stimuli; (b) UV/Vis absorption and (c) DSC profiles of **38b** at different states. Reproduced with permission from reference 68. Copyright 2011, Springer.

2.2.14 Siloxane-based Perylene Bisimide Derivative

Mizoshita *et al.* (2012)[69] designed a perylene bisimide (PBI) (dye **39**), in which the fluorescent PBI core is surrounded with siloxane-based bulky and flexible substituents at the 2, 5, 8, 11-positions to diminish intermolecular interactions. They expected that the PBI dye, which would be packed through easily rearrangeable intermolecular forces, would respond to external stimuli rapidly and sensitively in the solid state. As predicted, the PBI dye exhibits conventional fluorescence mechanochromic behavior that responds to mechanical and thermal stimuli (Figure 2.27A). The fluorescent patterns drawn by isothermal-mechanical shearing of dye **39** could be erased isothermally and quickly by treatment with solvent vapors. An isothermal structural transition from the *O*-form to the *G*-form is achieved by treatment of the solid dye film with solvent vapors. Figure 2.27A(a) shows the isothermal erasure of a fluorescent pattern by exposure of the dye film to chloroform vapor. The orange-colored *O*-form regions turns to the *G*-form in a moment (<1 s) by blowing the vapor. Similar erasure of fluorescent patterns could be performed using other good solvents, such as hexane, ethyl acetate and THF. However, exposure of the film to vapors of poor solvents (water, methanol, ethanol and acetonitrile) induces no structural transition. These results suggest that the permeation of vapors of good solvents into the dye film leads to increasing mobility of the dye molecules, in particular, the siloxane-based substituents, which promotes rearrangement of the dye molecules from a metastable amorphous structure to a more stable crystalline *G*-form state. The vapor-induced *O*-form – *G*-form transition can also be applied to isothermal and quick writing of fluorescent patterns onto the surface of the *O*-form film using a mask, as shown in Figure 2.27B. While structural transitions induced by vapor treatment have been reported for several organic dye powders,[70–73] the present approach enables the induction of both long-term bi-stability and isothermal rewritability for film-shaped mechanoresponsive fluorescent materials.

The self-erasing properties are also successfully induced on the dye films by adding a small amount of silicone oil. It was found that the dye films containing a small amount of silicone oil could form self-erasing patterns. The orange-fluorescent square pattern formed on the solid film of **39** alone shows no change for at least one week. The addition of silicone oil to the film leads to the tuning of the "lifetime" of the metastable *O*-form from a few minutes to several hours. The patterned film of **39** containing 2 wt% of silicone oil maintains its original state for 1 h, but the orange-fluorescent *O*-form region gradually turns to the green-fluorescent *G*-form in several hours and the orange square pattern completely disappears in 24 h. The increasing amount of the added silicone oil accelerates the spontaneous *O*-form – *G*-form structural transition. The fluorescent pattern of the film of **39** containing 10 wt% of silicone oil disappears in a few minutes. The added silicone oil probably enhances the mobility of the siloxane-based shell part of **39** and destabilizes the metastable amorphous state, which promotes rearrangement of the dye molecules from the amorphous *O*-form to the crystalline *G*-form. The gradual development of crystalline regions at room temperature was confirmed by polarizing microscope observation of the *O*-form films containing the silicone oil. Although most of the previously reported self-erasing media have relied on the use of photochromic

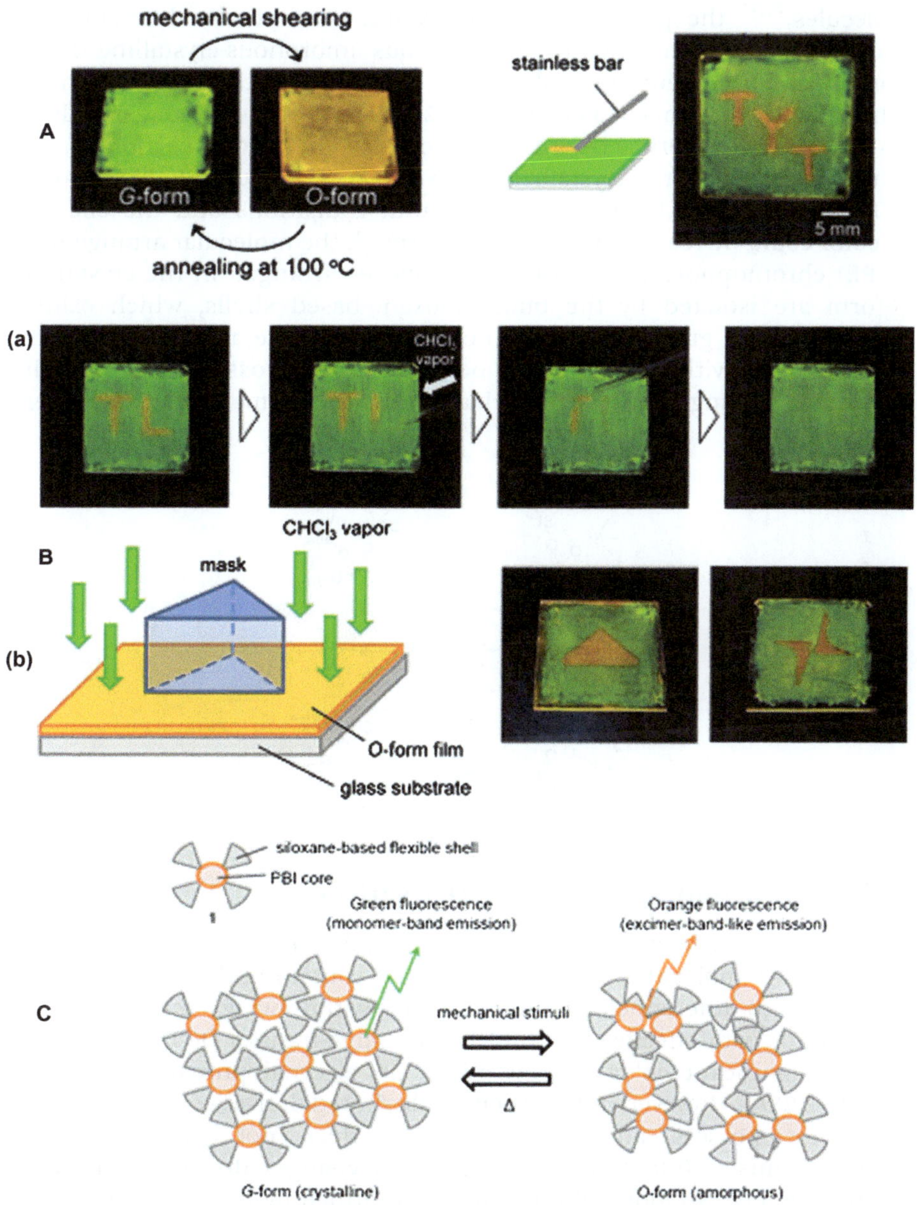

Figure 2.27 (A) (a) A photograph of an annealed film (left, *G*-form) and a mechanically sheared film (right, *O*-form) of **39** under UV irradiation. (b) A photograph of a patterned film formed by rubbing the *G*-form with a stainless steel bar. (B) (a) Isothermal erasure of the fluorescent patterns by exposure to $CHCl_3$ vapor. (b) Pattern formation on the *O*-form film by vapor treatment through a mask. (C) Schematic illustration of the structural transition of **39** in the solid state.
Reproduced with permission from reference 69. Copyright 2012, Wiley-VCH.

molecules,[74,75] the present dye films exhibit conceptually different self-erasing properties in which the spontaneous amorphous-crystalline transition of the fluorescent dye molecules is controlled by the additive. On the other hand, solid films of **39** containing 20–30 wt% of the silicone oil exhibit no spontaneous *O*-form – *G*-form transition for more than three months, due to the inhibition of the crystallization by the large amount of silicone oil.

The relationship between the structural transitions and the changes in color could be reasonably explained through the molecular arrangement of PBI chromophores. The PBI chromophores arranged in the crystalline *G*-form are isolated by the bulky siloxane-based shells, which exhibit monomer-band green emission. The PBI cores in the amorphous *O*-form film interact with one another because of the loss in the periodic molecular arrangement, which leads to the formation of excimers (Figure 2.27C).

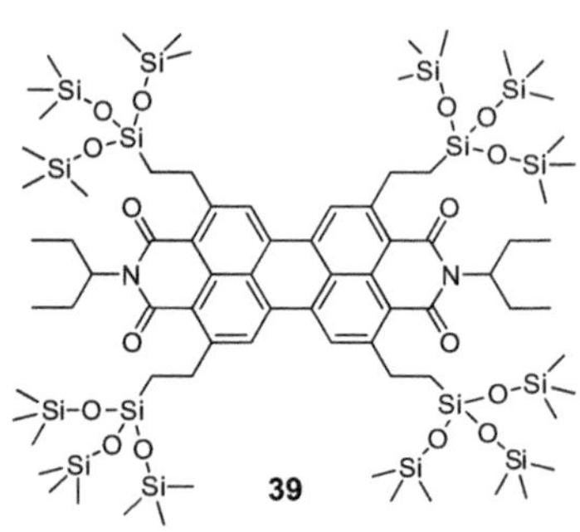

39

2.2.15 Anthraquinone Imide Derivatives

With the assistance of molecular simulations, Wan *et al.* (2012)[76] designed a series of D-A molecules (dyes **40** to **43**) that contain anthraquinone imide moieties with different electron-donating or -withdrawing substituents. The compounds with effective intramolecular charge-transfer (ICT) interactions exhibit significant color changes when an external pressure is applied (Figure 2.28A). However, 4-nitrophenyl-substituted derivative (dye **42**), which does not possess an ICT interaction, does not show mechanofluorochromism. Intermolecular interaction changes do not significantly contribute mechanofluorochromism, as confirmed by UV spectra and X-ray measurements. Thus, they proposed a new mechanism, in which the mechanofluorochromism of the anthraquinone imide derivatives results from the decrease in the torsion angle between the electron-donating and -withdrawing segments to form a more coplanar molecular conformation. The pressure-induced coplanarization might lead to a better conjugation between the D and A moieties, thereby red shifting the emission spectrum (Figure 2.28B). The proposed mechanism is different from the previously discussed mechanism.[40]

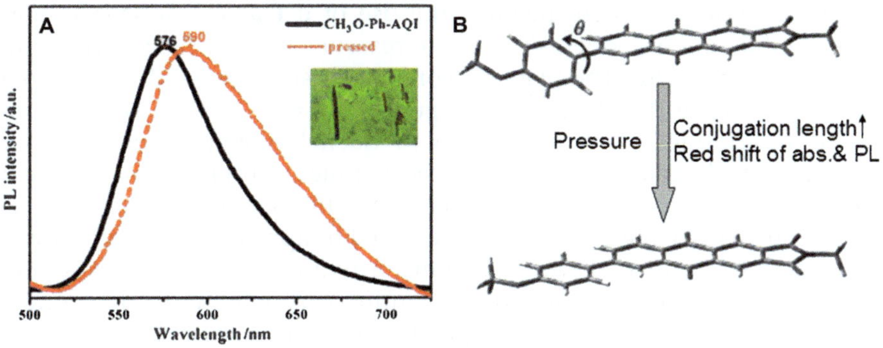

Figure 2.28 (A) Emission spectra of **40** excited at 310 nm before and after the employment of pressure (Inset: photograph of the partially pressed film under UV radiation). (B) An illustration of the proposed mechanochromic mechanism.
Reproduced with permission from reference 76. Copyright 2012, Wiley-VCH.

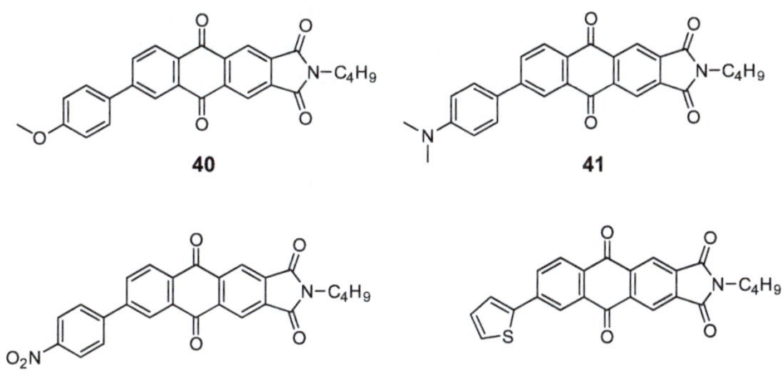

40 **41** **42** **43**

2.2.16 Distryrylfuran-substituted Maleimide Complexes

High-contrast luminescence recording requires a severe change in the luminescence intensity that can not be easily obtained by a single component luminescence switching system. Few examples of mechanochromic on/off luminescence switching have been reported.[77–80] Song *et al.* (2011)[80] developed a unique luminescence switching system with a high contrast ratio (approximately 10^2) based on co-crystallized D-acceptor and A-binary complexes (**44**). In this system, fluorescent D (Donor) and nonemissive A (acceptor) were mixed to form a binary complex. A binary complex helps to obtain mechanochromic luminescence that switches from quenching (black) to emitting (color) state once a mechanical force is applied. This D-A complex strategy provides a new way to obtain mechanochromic luminescence, which does not require the use of special polymorphic materials.

The binary complex system involves the use of **d-1** as an electron donor, which exhibits intense greenish blue fluorescence both in the solution and in the crystalline state. The planar π-conjugated structure of **d-1** leads to a strong tendency of the electrons to aggregate by π-π stacking. N-alkyl substituted maleimides **a-1-3** are chosen to quench donor-fluorescence *via* photoinduced electron transfer (PET). The complex (1:1) of **d-1** and **a-1** or **a-2** is deposited on the substrate, in which the D-fluorescence intensity is severely quenched. The fluorescence is recovered by applying mechanical force (Figure 2.29A). Compared with the transformation between two emission colors in traditional mechanochromic luminescent materials, the contrast ratio of this fluorescence is very high. However, the lack of flexible chains (**a-3**) leads to ineffective quenching fluorescence of **a-1**, and an obvious phase separation in pristine complex film is observed when no external force is applied. The strong packing of **a-3** breaks the balance between the individual self-assembly and the co-assembly of D-A. This balance must always be ensured. Mechanofluorochromism could not be obtained in such a binary complex system. Moderate co-assembly interaction causes the D-A complex to remain in a metastable quenching state. However, the stronger self-assembly effect (self-crystallization) of D promotes phase separation and enables fluorescence recovery by applying mechanical forces. The flexible alkyl chains in the nitrogen atoms of the maleimides are indispensable for supplying such a balance in the system.

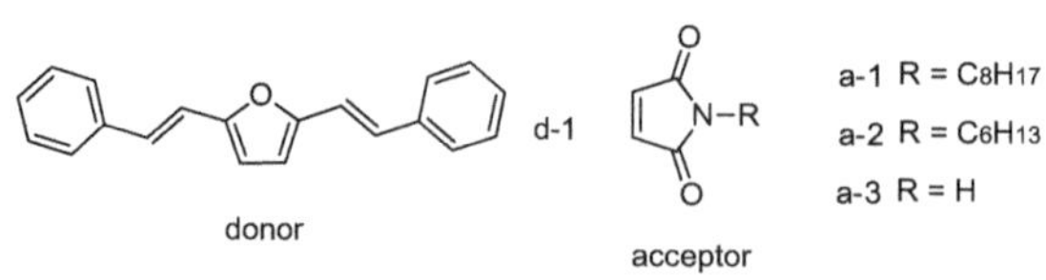

44

In addition, the erasing-writing cycles turn out to be stable by monitoring their fluorescence intensity (Figure 2.29B(a)). The metastable co-crystal gradually breaks down without applied force because of donor crystallizing and the relationship between the stable time of quenching state and ratio of acceptor **a-1** has been investigated. The stability time is first defined as the time elapsed when the ratio of fluorescence intensity between the self-recovered sample and the pristine sample surpasses one order of magnitude. For 1 : 1 complex, only several to tens of minutes need to observe the phase separation without applied forces. However, the lifetime can be elongated significantly by addition of acceptor molecules without loss of their mechanochromic luminescent properties. As illustrated in Figure 2.29B(b), once the ratio of **a-1** surpasses 80% (1 : 4 for 1: **a-1**), the self-decomposition rate becomes very slow under no external pressure and the lifetime is close to 5 h. A possible reason is that one molecule **d-1** may be completely surrounded by five maleimide **a-1** molecules and it is difficult to approach another **d-1** molecule to form stable assemblies. A stable pattern benefits from this simple strategy. Notably, a high contrast ratio of this

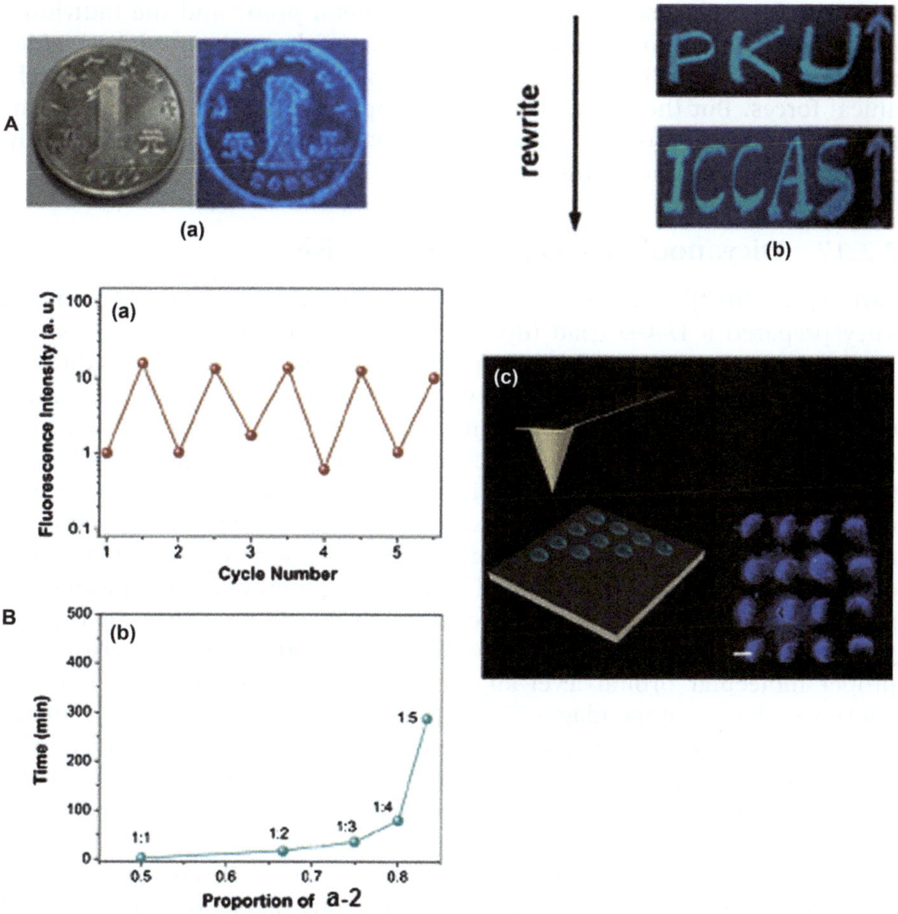

Figure 2.29 (A) Mechanochromic luminescence from the binary complex. (a) Transferring the protrudent pattern of a coin by stamping. The image mirrors the pattern of the coin. The substrate is paper adsorbed with 1 : 1 complex of d1 and a-1. (b) Rewritable mechanochromic fluorescence ($\lambda_{ex} = 365$ nm) demonstrated by writing "PKU" and "ICCAS" at the same position (bottom image). The "PKU" was erased by dichloromethane vapor and the arrow marked the same paper used. (B) (a) Quenching-recovering cycles by dissolving-pressing procedure on the quartz plate with complex of **d-1** and **a-1**. (b) Relationship between the stable time of quenching state and the proportion of acceptor **a-1**. (c) Fluorescence microscopy of a 4×4 mechanofluorochromic array on paper generated by manipulated metal probe and the scale bar is 400 mm. The mechanofluorochromic material is the 1 : 4 complex of **d-1** and **a-1**. Reproduced with permission from reference 80. Copyright 2011, Wiley-VCH.

mechanofluorochromic strategy ensures easy observation under stimuli at very small areas and high-resolution pattern can be obtained using non-interfering small point arrays. Figure 2.29B(c) demonstrates a 4×4 mechanofluorochromic array generated by metal probe and the individual point has similar diameter (*ca.* 400 mm) to that of the probe used. It provides a promising direct display/storage technology using easily available mechanical forces. But the molecular mixture approach is practically limited by the ill-defined composition, molecular packing and precise control mechanisms of fluorescence turn-on.

2.2.17 Dicyanodistyrylbenzene Derivative

Park *et al.* (2012)[81] suggested a novel material based on the D-A strategy. They prepared a D-A-D triad (dye **45**) that incorporates a dicyanodistyrylbenzene (DCS) unit as a key component. Compared with the traditional luminescence color switching, this unique covalently linked and structurally well-defined D-A-D system showed intriguing luminescence on/off switching behavior in response to a mechanical stimulus in the solid state (Figure 2.30). Park *et al.* utilized this phenomenon to demonstrate a rewritable mechanochromic fluorescence recording media with a high contrast ratio (around 10^3) and a high stability (stable over several weeks in ambient conditions) by using a conventional spin coating method. The experimental results indicate that the unique properties of dye **45** are driven by the targeted design of the D-A-D triad, which possess three characteristics: (a) proper molecular orbital level alignment between the DCS core and the chosen carbazole appendages that provide electronic basis for resonant energy transfer and electron transfer (eT); (b) the ability of carbazole to form π-stacks with DCS; and (c) the flexibility of the linker that allows switching between the conformations in the solid state. In the crystalline phase, dye **45** forms a characteristic architecture, which leads to the instantaneous formation of a non-luminescent charge-transfer state. By contrast, the featured architecture is destroyed in the amorphous phase, which is induced by external shearing forces. Thus, eT rate significantly slows down, and the fluorescence is recovered. This study was considered to pave the way for the practical applications of covalently linked D-A-D molecules in sensitive pressure sensor, high-contrast optical memory, and data storage devices.

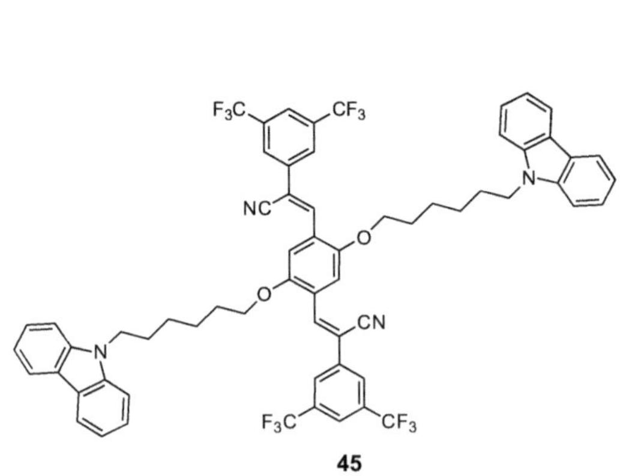

45

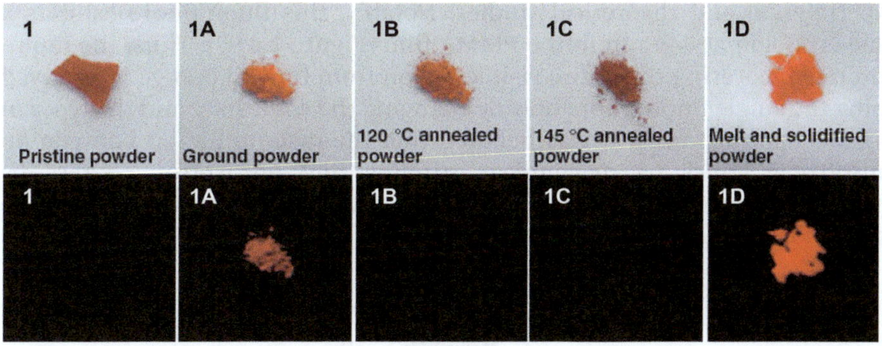

Figure 2.30 Color (top) and fluorescence (bottom) changes of various dye **45** powders.
Reproduced with permission from reference 81. Copyright 2012, Wiley-VCH.

2.2.18 Tetrathiazolylthiophene

From the discussion above, the molecular orientation and intermolecular interactions are perturbed by mechanical forces, namely, shearing, grinding, tension or hydrostatic pressure, to undergo a drastic emission color change. However, there are only a few reports that describe the responses to both the mechanical grinding and hydrostatic pressure,[82,83] despite the importance of these factors when considering the applications of these materials. Since the molecular systems that show distinct spectral responses to these stresses are rather scarce, the difference between the anisotropic grinding and the isotropic compression has not been clearly discriminated. Saito *et al.* (2013)[84] reported the distinct luminescent responses to these stresses using a new fluorophore, tetrathiazolylthiophene **46** and the origin of this difference has been experimentally and theoretically elucidated in detail (Figure 2.31A). The excimer formation in the 3D hydrogen-bond network is responsible for the yellow emission at 556 nm in the crystalline state of the tetrathiazolylthiophene fluorophore (Figure 2.31B). Grinding the crystals leads to an increase in the disordered phase in which the excimer formation is suppressed and thereby a blue-shifted green-color emission emerges with the λ_{em} at 490 nm in the powder state. In stark contrast, the high-pressure experiments using a diamond anvil cell (DAC) demonstrated that the yellow excimer emission of the crystals is changed to an orange fluorescence with the λ_{em} at 609 nm at 3.2 GPa. A high-pressure single-crystal XRD analysis unambiguously determined the closer arrangement of the face-to-face dimer of the fluorophore, which is the origin of the more red-shifted excimer emission, because under high pressure the molecular packing structure takes closer distance and more straight arrangement of the C–H···N geometry. The reversible structural deformation of the hydrogen-bond network upon compression/decompression is confirmed from both experimental and theoretical studies. Notably, this fluorophore exhibits a blue emission at 449 nm in the PMMA film (Figure 2.31C). Thus, the multichromism covering the entire visible region from blue to orange is achieved using the single fluorophore only by changing the solid state and the types of the mechanical forces. It was believed that the discrimination between an anisotropic stress by mechanical grinding and an isotropic stress by hydrostatic pressure would accelerate, and thus enrich the chemistry of the mechanochromism.

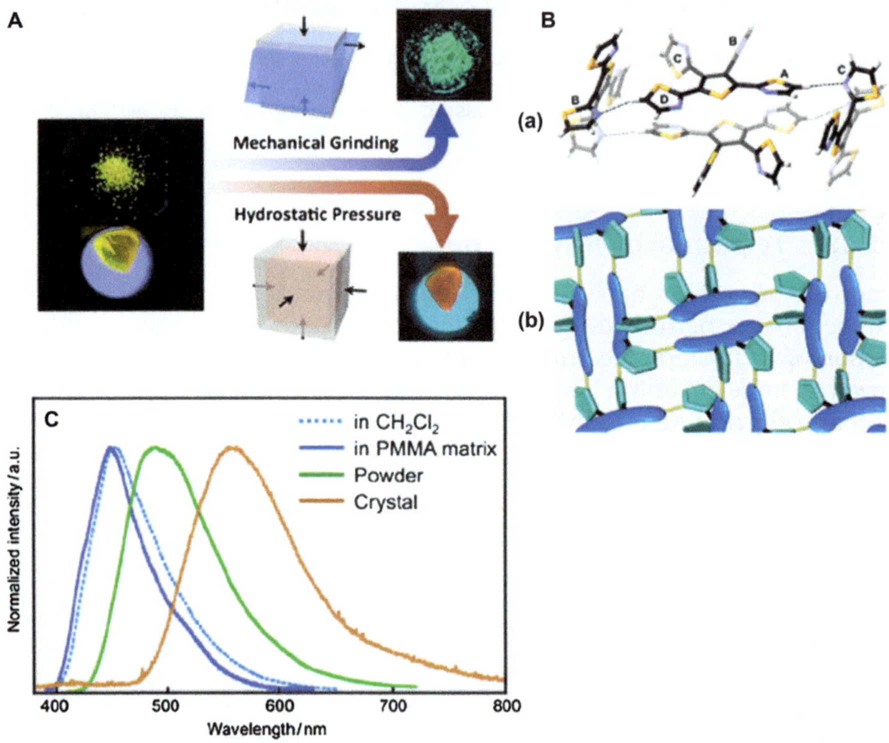

Figure 2.31 (A) Tetrathiazolylthiophene **46** and its yellow-emissive crystals (left) and distinct luminescent responses to mechanical grinding, producing a green-emissive powder, and to hydrostatic pressure, affording an orange-emissive crystal (right). (B) Crystal packing structure of **46**: (a) face-to-face dimer fixed with weak C–H···N hydrogen bonds (dotted line) and (b) representation of the crystal packing. Blue boards, green pentagons and yellow cylinders demonstrate teraryl fluorophores, twisted thiazole rings B and C, and C–H···N hydrogen bonds between the thiazole rings, respectively. (C) Fluorescence spectra of **46** in a CH_2Cl_2 solution (dotted line), in a PMMA film with 10 wt% of the sample (blue), in a ground powder (green) and in the crystalline state (dark yellow).
Reproduced with permission from reference 84 Copyright 2013, American Chemical Society.

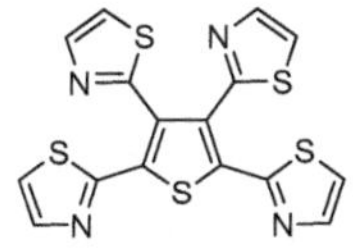

46

2.2.19 Multi-carbazole Substituted Quinacridone Derivatives

For mechanofluorochromic molecule systems, their aggregation states can be easily alternated by mechanical grinding. Therefore, the organic dendrimers with variable conformation structures should be desired candidates for the study of mechanofluorochromic phenomenon. Thus, a series of luminescent dendrimers **47a–f** composed of quinacridone core and carbazole dendrons have been designed and synthesized by Wang's group (2012).[85] The relationships between the molecular structures and properties of dendrimers have been established. Table 2.3 presents the powder X-ray diffraction (XRD) patterns of the as-synthesized samples and their grinding induced emission variation properties. The typical crystalline (**47a** and **47b**) and amorphous (**47f**) samples do not show mechanochromic luminescence feature. A possible explanation for these results is that these samples adopt a thermally stable phase and grinding operation cannot alter their original phases. Upon grinding and fuming with CH_2Cl_2, the emission spectrum of **47d** displays slight variation and poor reversibility, suggesting that **47d** is not a material with typical mechanochromic luminescence property. The as-synthesized **47c** and **47e** exhibit moderate crystallinity and mechanoflurochromic behavior. Upon grinding with a pestle, the yellow emissive as-synthesized **47c** changes into an orange-red emissive amorphous solid. The as-synthesized crystalline and ground amorphous **47c** samples show emissive maxima at 588 and 606 nm, respectively, and upon fuming the ground sample with CH_2Cl_2 vapor for 2 min, the emission maximum displays a blue-shift from 606 to 582 nm (Figure 2.32A). The reversible emission switch between 582 and 606 nm can be easily repeated by grinding and fuming. The ground sample exhibits an orange-red emission and the fumed sample shows a yellow emission. The powder X-ray diffraction (XRD) patterns (Figure 2.32B) demonstrate that the ground sample is an amorphous phase, while the as-synthesized and fumed ones have remarkable crystalline feature. Although the as-synthesized and fumed samples display similar X-ray diffraction pattern, they do not belong to the same crystalline phase. Therefore, the two crystalline phases show different profiles of emission spectra as shown in Figure 2.32A. The DSC curves of **47c** in different solid states confirm that the three samples possess different metastable aggregation structures. The dendrimer **47e** exhibits similar mechanochromic luminescence property to **47c**.

Table 2.3 Mechanochromic luminescence properties and phase states of the dendrimers.

Dendrimer	***47a***	***47b***	***47c***	***47d***[a]	***47e***	***47f***
Mechanochromic (yes/no)	no	no	yes	partial	yes	no
Phase states	Typical crystalline phase	Typical crystalline phase	Partial crystalline order	A little crystalline order	Partial crystalline order	Typical amorphous phase

[a]The emission of dendrimer **47d** changes slightly after grinding and the fuming treatment with CH_2Cl_2 vapor cannot lead complete recovery of emission, so **47d** is not a compound with typical mechanochromic luminescence property. Reproduced with permission from reference 85. Copyright 2012, American Chemical Society.

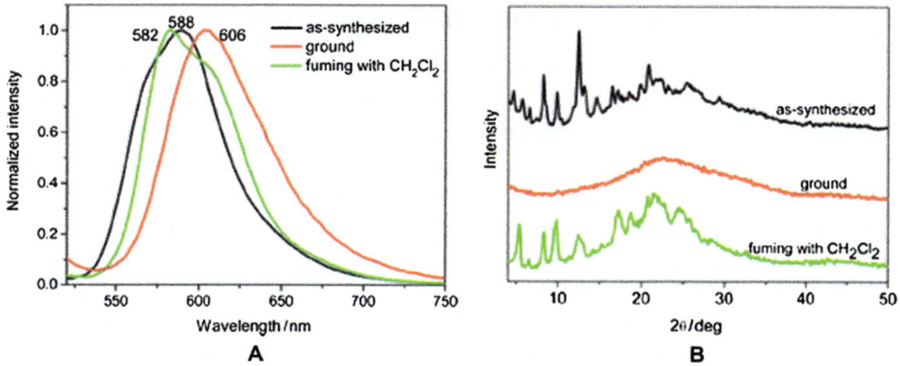

Figure 2.32 (A) Fluorescence spectra of **47c** in different solid states. (B) XRD patterns of **47c** in different solid states.
Reproduced with permission from reference 85. Copyright 2012, American Chemical Society.

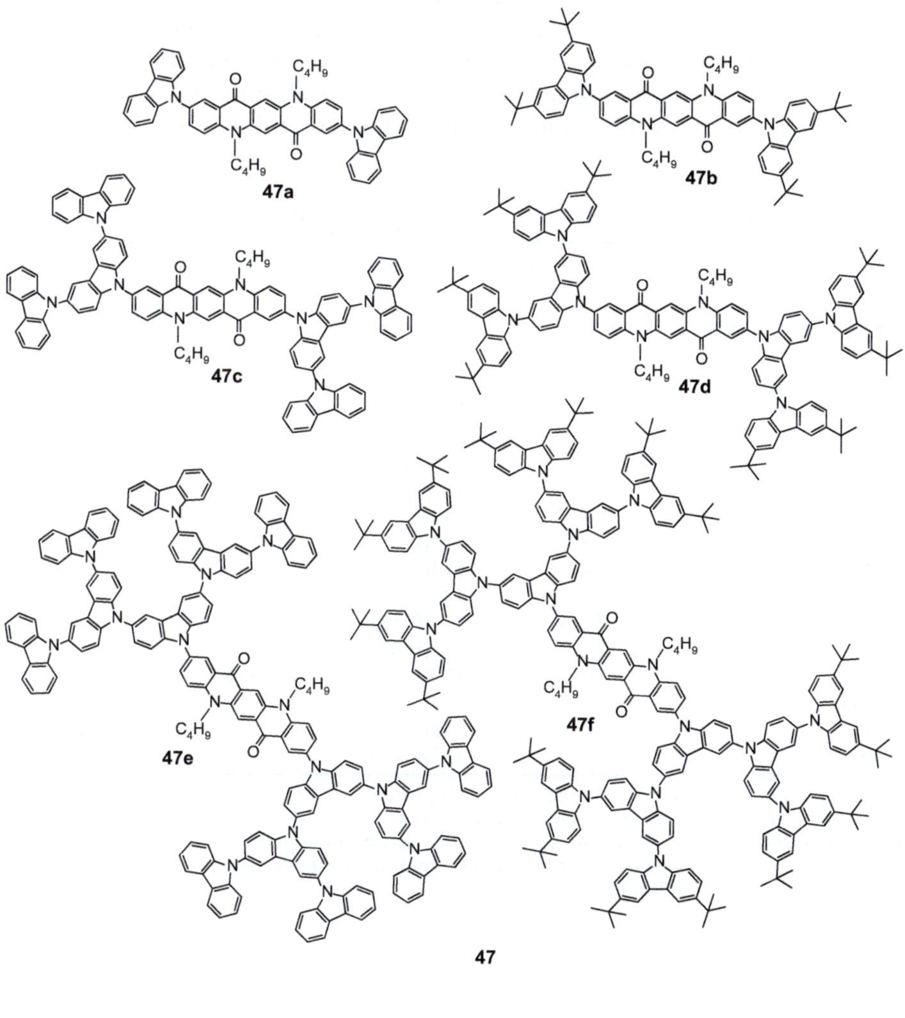

47

2.2.20 N-heteroaromatic Onium Compounds

You *et al.* (2013)[86] reported a strategy of molecular design for mechanochromic luminescence, based on the cation–anion interaction-directed switching of molecular stacking (**48**). On the basis of this strategy, a series of common N-heteroaromatic onium fluorophores such as imidazolium, 1,2,4-triazolium, triazolopyridinium, benzoimidazolium, γ -carbolinium and pyridinium salts have been designed and proved to have striking reversible mechanofluorochromic behaviors (Table 2.4; Figure 2.33 for **1** · $\mathbf{NTf_2}$). In addition, the simple attachment of a non-fluorescent imidazolium unit to the pyrene scaffold through a flexible carbon chain can even trigger the mechanofluorochromic phenomenon, which gives a consummate interpretation that the cation–anion interaction can be considered as an important general tool to design organic mechanochromic luminescent materials.

After surveying a series of common counter anions such as NTf_2^- (bis-((trifluoromethyl)sulfonyl)amide), PF_6^- , OTf^- , BF_4^- and $N(CN)_2^-$, the NTf_2^- anion was proven to be an ideal counter anion to enable excellent luminescent properties. Moreover, the emission wavelengths varied with different counter anions, which implied that different molecular packing modes could be gained by tuning anions. Owing to the low emission quantum yield

Table 2.4 Physical and photophysical properties of onium salts.

Compd.	*M.P. [°C]*[a]	λ_{abs} *[nm]*[b] *(in $CHCl_3$)*	λ_{em} *[nm]*[c] *(Φ_F)*[d] *(pristine)*	λ_{em} *[nm]*[c] *(ground)*	$\Delta\lambda_{em}$ *[nm]*
1 · PF_6	134	299	371(0.12)	461	90
1 · OTf	59	300	423(0.59)	501	78
1 · BF_4	157	302	433(0.45)	469	36
1 · $N(CN)_2$	109	298	418 (0.53)	425	7
1.NTf_2	86	299	407(0.76)	518	111
2.NTf_2	–[e]	293	479(0.05)	–	–
3.NTf_2	–[e]	302	481(0.03)	–	–
4.NTf_2	–[e]	307	484(0.10)	–	–
5.NTf_2	70	290	430(0.45)	472	42
6.NTf_2	86	289	414(0.45)	504	90
7.NTf_2	80	299	425(0.38)	492	67
8.NTf_2	122	253	453(0.13)	532	79
P.NTf_2	60	375	415(0.07)	486	71
Bz.NTf_2	67	316	475(0.55)	511	36
Pz.NTf_2	93	345	434(0.06)	472	38
Py.NTf_2	147	378	648(0.09)	712	64
C.NTf_2	164	338	456(0.23)	485	29

[a]Melting points were obtained from the first heating run using DSC under N_2 , rate = 5 °C min $^{-1}$.
[b]Only the maximum absorption wavelengths are shown.
[c]Emission of samples excited at the corresponding maximum excitation wavelength.
[d]Absolute emission quantum yields estimated by calibrated integrating sphere system.
[e]Room temperature ionic liquid.

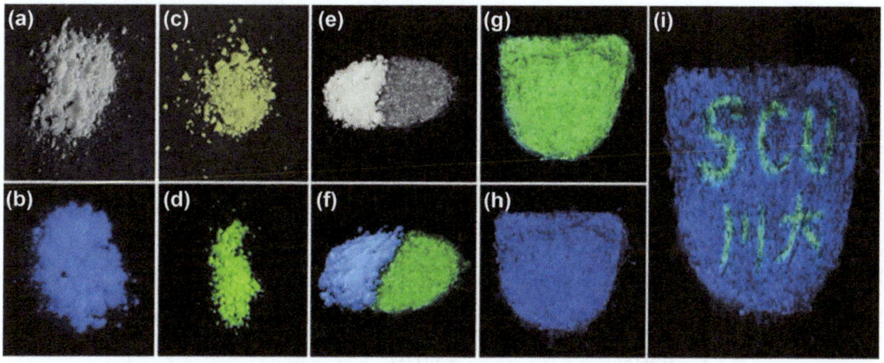

Figure 2.33 Photographs of **1 · NTf** 2 under ambient light and UV irradiation at 365 nm: (a) the pristine crystalline powder under ambient light, and (b) under UV light; (c) the solid under fast cooling from its melting state under ambient light, and (d) under UV light; (e) the pristine solid after grinding the right half with a spatula under ambient light, and (f) under UV light; (g) the entirely ground solid, and (h) the sample after thermal annealing under UV light; and (i) the regeneration of the yellowish-green emission demonstrated by the letters "SCU" and the Chinese characters "Sichuan University" written with a spatula.
Reproduced with permission from reference 86. Copyright 2013, Wiley-VCH.

of **1 · I** in the solid state, five counter anions (*i.e.*, NTf_2^-, PF_6^-, OTf^-, BF_4^- and $N(CN)_2^-$) were chosen to illustrate the correlation between the cation–anion interaction and the mechanochromic performance of **1 · X** (X = counter anion). After three minutes of pressing under 6 MPa using an oil press, **1 · NTf_2**, **1 · PF_6** , **1 · OTf**, **1 · BF_4** and **1 · $N(CN)_2$** undergo redshifts of the peak emission wavelengths (approximately at 111 nm, 90 nm, 78 nm, 36 nm and 7 nm, respectively) (Table 2.4), which is in good agreement with the known order of the strength of the cation–anion interaction as following: $NTf_2^- < PF_6^- < OTf^- < BF_4^- < N(CN)_2^-$. It hints that the strength of the cation–anion interaction could determine the mechanochromic luminescence property to a certain extent. The NTf_2^- anion is well demonstrated as an ideal counter anion to induce mechanochromic luminescence. It was considered that the weak interactions between the imidazolium cation and the NTf_2^- anion make the molecular stacking relatively loose. The loose molecular stacking is more susceptible to tiny slip by mechanical stimuli. It is worth noting that **1 · NTf_2** is one of the smallest fluorophores exhibiting an obvious mechanofluorochromic behavior, which constitutes one of the longest emission wavelength red shifts reported so far.

Their mechanofluorochromic mechanism also suggested the generation of a new aggregate during the grinding process.

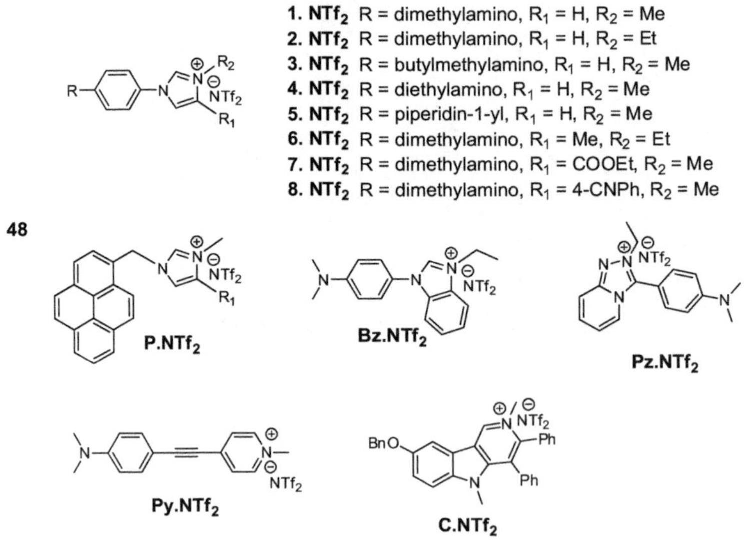

2.3 Conclusion

This chapter provides an exhaustive overview of the recent progress in pure organic mechaoflurochromic systems. These systems have been divided into twenty kinds according to their molecular structures. Even though substantial progress has been made toward understanding the relationship between molecular structure and mechanochromism, the challenge of controlled design and construction of novel organic mechaoflurochromic systems remains unmet.

References

1. P. M. Beaujuge, C. M. Amb and J. R. Reynolds, *Acc. Chem. Res.*, 2010, **43**, 1396–1407.
2. S. R. Forrest, *Nature*, 2004, **428**, 911–918.
3. E. W. Meijer and A. P. H. J. Schenning, *Nature*, 2002, **419**, 353–354.
4. K. Binnemans, *Chem. Rev.*, 2009, **109**, 4283–4374.
5. B.-K. An, J. Gierschner and S. Y. Park, *Acc. Chem. Res.*, 2012, **45**, 544–554.
6. S. Varghese and S. Das, *J. Phys. Chem. Lett.*, 2011, **2**, 863–873.
7. S. J. Yoon, S. Varghese, S. K. Park, R. Wannemacher, J. Gierschner and S. Y. Park, *Adv. Opt. Mater.*, 2013, **1**, 232–237.
8. Y. Sagara, T. Mutai, I. Yoshikawa and K. Araki, *J. Am. Chem. Soc.*, 2007, **129**, 1520–1521.
9. Y. Sagara and T. Kato, *Nat. Chem.*, 2009, **1**, 605–610.
10. M. Sase, S. Yamaguchi, Y. Sagara, I. Yoshikawa, T. Mutai and K. Araki, *J. Mater. Chem.*, 2011, **21**, 8347–8354.

11. S. Yamaguchi, I. Yoshikawa, T. Mutai and K. Araki, *J. Mater. Chem.*, 2012, **22**, 20065–20070.
12. C. Löwe and C. Weder, *Adv. Mater.*, 2002, **14**, 1625–1629.
13. J. Kunzelman, M. Kinami, B. R. Crenshaw, J. D. Protasiewicz and C. Weder, *Adv. Mater.*, 2008, **20**, 119–122.
14. J. Kunzelman, B. R. Crenshaw, M. Kinami and C. Weder, *Macromol. Rapid Commun.*, 2006, **27**, 1981–1987.
15. T. Nakanishi, *Supramolecular Soft Matter*, Wiley-VCH, Weinheim, 2011.
16. S. Yamane, K. Tanabe, Y. Sagara and T. Kato, *Top. Curr. Chem.*, 2012, **318**, 395–405.
17. V. N. Kozhevnikov, B. Donnio and D. W. Bruce, *Angew. Chem., Int. Ed.*, 2008, **47**, 6286–6289.
18. Y. Sagara and T. Kato, *Supramol. Chem.*, 2011, **23**, 310–314.
19. Y. Sagara and T. Kato, *Angew. Chem., Int. Ed.*, 2011, **50**, 9128–9132.
20. F. Würthner, *Chem. Commun.*, 2004, 1564–1579.
21. F. Camerel, L. Bonardi, M. Schmutz and R. Ziessel, *J. Am. Chem. Soc.*, 2006, **128**, 4548–4549.
22. S.-J. Yoon, J. H. Kim, K. S. Kim, J. W. Chung, B. Heinrich, F. Mathevet, P. Kim, B. Donnio, A.-J. Attias, D. Kim and S. Y. Park, *Adv. Funct. Mater.*, 2012, **22**, 61–69.
23. K. Tanabe, Y. Suzui, M. Hasegawa and T. Kato, *J. Am. Chem. Soc.*, 2012, **134**, 5652–5661.
24. S. Yamane, Y. Sagara and T. Kato, *Chem. Commun.*, 2009, 3597–3599.
25. M. O'Neill and S. M. Kelly, *Adv. Mater.*, 2003, **15**, 1135–1146.
26. K. Ichimura, *Chem. Rev.*, 2000, **100**, 1847–1874.
27. A. Natansohn and P. Rochon, *Chem. Rev.*, 2002, **102**, 4139–4175.
28. T. Ikeda, J. Mamiya and Y. Yu, *Angew. Chem., Int. Ed.*, 2007, **46**, 506–528.
29. M. Moriyama, N. Mizoshita, T. Yokota, K. Kishimoto and T. Kato, *Adv. Mater.*, 2003, **15**, 1335–1338.
30. A. Priimagi, K. Ogawa, M. Virkki, J. Mamiya, M. Kauranen and A. Shishido, *Adv. Mater.*, 2012, **24**, 6410–6415.
31. B.-H. Tan, M. Yoshio and T. Kato, *Chem. - Asian J*, 2008, **3**, 534–541.
32. S. Yamane, Y. Sagara, T. Mutai, K. Araki and T. Kato, *J. Mater. Chem. C*, 2013, **1**, 2648–2656.
33. Y. Sagara and T. Kato, *Angew. Chem., Int. Ed.*, 2008, **47**, 5175–5178.
34. Y. Sagara, S. Yamane, T. Mutai, K. Araki and T. Kato, *Adv. Funct. Mater.*, 2009, **19**, 1869–1875.
35. Y. Sagara and T. Kato, *Supramol. Chem.*, 2011, **23**, 310–314.
36. Y. Sagara and T. Kato, *Angew. Chem. Int. Ed.*, 2011, **50**, 9128–9132.
37. Y. Sagara, T. Komatsu, T. Ueno, K. Hanaoka, T. Kato and T. Nagano, *Adv. Funct. Mater.*, 2013, **23**, 5277–5284.
38. Y. Ooyama, G. Ito, H. Fukuoka, T. Nagano, Y. Kagawa, I. Imae, K. Komaguchi and Y. Harima, *Tetrahedron*, 2010, **66**, 7268–7271.
39. Y. Ooyama and Y. Harima, *J. Mater. Chem.*, 2011, **21**, 8372–8380.

40. Y. Ooyama, N. Yamaguchi, S. Inoue, T. Nagano, E. Miyazaki, H. Fukuoka, I. Imae, K. Komaguchi, J. Ohshita and Y. Harima, *Tetrahedron*, 2012, **68**, 529–533.
41. Y. Ooyama, T. Sugiyama, Y. Oda, Y. Hagiwara, N. Yamaguchi, E. Miyazaki, H. Fukuoka, T. Mizumo, Y. Harima and J. Ohshita, *Eur. J. Org. Chem.*, 2012, **25**, 4853–4859.
42. Y. Ooyama and Y. Harima, *Chem. Lett.*, 2006, **35**, 902–903.
43. Y. Ooyama, Y. Kagawa and Y. Harima, *Eur. J. Org. Chem.*, 2007, **22**, 3613–3621.
44. Y. Ooyama, Y. Kagawa, H. Fukuoka, G. Ito and Y. Harima, *Eur. J. Org. Chem.*, 2009, **31**, 5321–5326.
45. Y. Ooyama, G. Ito, K. Kushimoto, K. Komaguchi, I. Imae and Y. Harima, *Org. Biomol. Chem.*, 2010, **8**, 2756–2770.
46. Y. Ooyama, Y. Oda, Y. Hagiwara, H. Fukuoka, E. Miyazaki, T. Mizumo and J. Ohshita, *Tetrahedron*, 2013, **69**, 5818–5822.
47. G. Zhang, J. P. Singer, S. E. Kooi, R. E. Evans, E. L. Thomas and C. L. Fraser, *J. Mater. Chem.*, 2011, **21**, 8295–8299.
48. G. Zhang, J. Lu, M. Sabat and C. L. Fraser, *J. Am. Chem. Soc.*, 2010, **132**, 2160–2162.
49. G. R. Krishna, M. S. R. N. Kiran, C. L. Fraser, U. Ramamurty and C. M. Reddy, *Adv. Funct. Mater.*, 2013, **23**, 1422–1430.
50. N. D. Nguyen, G. Zhang, J. Lu, A. E. Sherman and C. L. Fraser, *J. Mater. Chem.*, 2011, **21**, 8409–8415.
51. G. Zhang, J. Lu and C. L Fraser, *Inorg. Chem.*, 2010, **49**, 10747–10749.
52. T. Liu, A. D. Chien, J. Lu, G. Zhang and C. L. Fraser, *J. Mater. Chem.*, 2011, **21**, 8401–8408.
53. X. Sun, X. Zhang, X. Li, S. Liu and G. Zhang, *J. Mater. Chem.*, 2012, **22**, 17332–17339.
54. F. Ito and T. Sagawa, *RSC Advances*, 2013, **3**, 19785–19788.
55. M. Teng, X. Jia, X. Chen, Z. Ma and Y. Wei, *Chem. Commun.*, 2011, **47**, 6078–6080.
56. S. J. Yoon, J. W. Chung, J. Gierschner, K. S. Kim, M. G. Choi, D. Kim and S. Y. Park, *J. Am. Chem. Soc.*, 2010, **132**, 13675–13683.
57. M. Teng, X. Jia, S. Yang, X. Chen and Y. Wei, *Adv. Mater.*, 2012, **24**, 1255–1261.
58. M. Teng, X. Jia, X. Chen and Y. Wei, *Angew. Chem. Int. Ed.*, 2012, **51**, 6398–6401.
59. Z. Y. Ma, M. J. Teng, Z. J. Wang, S. Yang and X. R. Jia, *Angew. Chem. Int. Ed.*, 2013, **52**, 12268–12272.
60. K. Mizuguchi, H. Kageyama and H. Nakano, *Mater. Lett.*, 2011, **65**, 2658–2661.
61. K. Mizuguchi and H. Nakano, *Dyes Pigments*, 2013, **96**, 76–80.
62. H. Tsuji, G. M. O. Favier, C. Mitsui, S. Lee, D. Hashizume and E. Nakamura, *Chem. Lett.*, 2011, **40**, 576–578.
63. H. Zhang, Z. Zhang, K. Ye, J. Zhang and Y. Wang, *Adv. Mater.*, 2006, **18**, 2369–2372.

64. Z. Zhang, Y. Zhang, D. Yao, H. Bi, I. Javed, Y. Fan, H. Zhang and Y. Wang, *Cryst. Growth Des.*, 2009, **9**, 5069–5076.
65. E. Zhang, D. Yao, T. Zhou, H. Zhang and Y. Wang, *Chem. Commun.*, 2011, **47**, 7782–7784.
66. C. Dou, D. Chen, J. Iqbal, Y. Yuan, H. Zhang and Y. Wang, *Langmuir*, 2011, **27**, 6323–6329.
67. X. Chen, R. Wei, Y. Xiang, Z. Zhou, K. Li, P. Song and A. Tong, *J. Phys. Chem. C*, 2011, **115**, 14353–14359.
68. C. Dou, D. Li, H. Zhang, H. Gao, J. Zhang and Y. Wang, *Sci. China Chem.*, 2011, **54**, 641–650.
69. N. Mizoshita, T. Tani and S. Inagaki, *Adv. Mater.*, 2012, **24**, 3350–3355.
70. S. Mizukami, H. Houjou, K. Sugaya, E. Koyama, H. Tokuhisa, T. Sasaki and M. Kanesato, *Chem. Mater.*, 2005, **17**, 50–56.
71. T. Lasanta, M. E. Olmos, A. Laguna, J. M. López-de-Luzuriaga and P. Naumov, *J. Am. Chem. Soc.*, 2011, **133**, 16358–16361.
72. S. J. Choi, J. Kuwabara, Y. Nishimura, T. Arai and T. Kambara, *Chem. Lett.*, 2012, **41**, 65–70.
73. A. Schönberg, M. Elkaschef, M. Nosseir and M. M. Sidky, *J. Am. Chem. Soc.*, 1958, **80**, 6312–6315.
74. M. Häckel, L. Kador, D. Kropp and H. W. Schmidt, *Adv. Mater.*, 2007, **19**, 227–231.
75. R. Klajn, P. J. Wesson, K. J. M. Bishop and B. A. Grzybowski, *Angew. Chem. Int. Ed.*, 2009, **48**, 7035–7039.
76. F. Chen, J. Zhang and X. Wan, *Chem. Eur. J.*, 2012, **18**, 4558–4567.
77. J. W. Chung, Y. You, H. S. Huh, B. K. An, S. J. Yoon, S. H. Kim, S. W. Lee and S. Y. Park, *J. Am. Chem. Soc.*, 2009, **131**, 8163–8172.
78. X. Luo, J. Li, C. Li, L. Heng, Y. Dong, Z. Liu, Z. Bo and B. Z. Tang, *Adv. Mater.*, 2011, **23**, 3261–3265.
79. Y. Wang, M, J. Li, Y. M. Zhang, J. Yang, S. Y. Zhu, L. Sheng, X. D. Wang, B. Yang and S. X.-A. Zhang, *Chem. Commun.*, 2013, **49**, 6587–6589.
80. J. Luo, L. Li, Y. Song and J. Pei, *Chem. Eur. J.*, 2011, **17**, 10515–10519.
81. M. S. Kwon, J. Gierschner, S. J. Yoon and S. Y. Park, *Adv. Mater.*, 2012, **24**, 5487–5492.
82. P. L. Gentili, M. Nocchetti, C. Miliani and G. Favaro, *New J. Chem.*, 2004, **28**, 379–386.
83. Y. J. Dong, B. Xu, J. B. Zhang, X. Tan, L. J. Wang, J. L. Chen, H. G. Lv, S. P. Wen, B. Li, L. Ye, B. Zou and W. J. Tian, *Angew. Chem. Int. Ed.*, 2012, **51**, 10782–10785.
84. K. Nagura, S. Saito, H. Yusa, H. Yamawaki, H. Fujihisa, H. Sato, Y. Shimoikeda and S. Yamaguchi, *J. Am. Chem. Soc.*, 2013, **135**, 10322–10325.
85. C. G. Wang, S. Y. Chen, K. Wang, S. S. Zhao, J. Y. Zhang and Y. Wang, *J. Phys. Chem. C*, 2012, **116**, 17796–17806.
86. G. C. Li, F. J. Song, D. Wu, J. B. Lan, X. Y. Liu, J. Wu, S. J. Yang, D. Xiao and J. S. You, *Adv. Funct. Mater.*, 2013, **24**, 747–753.

CHAPTER 3

Organic-inorganic Complex Mechanofluorochromic Materials

XIQI ZHANG,[a] YEN WEI,*[a] ZHENGUO CHI[b] AND JIARUI XU[b]

[a] Department of Chemistry, Tsinghua University, Beijing, China; [b] PCFM Lab, State Key Laboratory of Optoelectronic Materials and Technologies, School of Chemistry and Chemical Engineering, Sun Yat-sen University, Guangzhou, China
*Email: weiyen@tsinghua.edu.cn

3.1 Introduction

Based on the content of the previous chapter, a number of organic molecules have been known to change their absorption and emission characteristics in response to mechanical stimuli. Despite the limited number of relevant studies to date on the mechanochromic luminescent metal complexes,[1,2] these materials remain fairly attractive because organometallic or coordination compounds consisting of metal ions and organic ligands are expected to exhibit more intriguing and richer luminescence mechanochromism. A number of cases are now known, where mechanical treatment of certain metal complexes can result in a change in the color of luminescence, and though there is no common mechanism, some sort of structural reorganisation is usually found to accompany this transition. However, the need to grind a crystalline sample to reveal the mechanochromism means that it would be difficult to characterise the structural change involved, as single

RSC Smart Materials No. 8
Mechanochromic Fluorescent Materials: Phenomena, Materials and Applications
Edited by Jiarui Xu and Zhenguo Chi

Published by the Royal Society of Chemistry, www.rsc.org

crystal diffraction can not be applied to the ground form.[3] Meanwhile, this type of mechanochromic luminescence is widely considered to originate from intramolecular conformational folding or twisting, as well as from the variations in intermolecular π-π, metal-metal, or hydrogen-bonding interactions. Therefore, the relevant literature on the mechanochromic luminescent metal complexes has significantly increased in recent years. The aggregation-induced emission (AIE) complexes reported recently may be an important alternative source of mechanochromic luminescent complexes. Our recent article reviews progress in the mechanochromic luminescent complexes in detail and this chapter is based on that review.[4]

3.2 Metal Complexes

3.2.1 Zn(II) Complexes

Mizukami *et al.* (2005)[5] prepared a zinc(II)-salen helical complex **1**. This complex displayed a strong fluorescence both in the solution and in the solid state. When the crystals of **1** were crushed to a powder, the photoluminescence shifted to a shorter wavelength, *i.e.* from green (488 nm) to blue (474 nm) (Figure 3.1), and after the crystal of **1** reacted with certain organic vapors, such as THF or MeOH, it was demonstrated to be vapochromism. The mechanofluorochromic property is assumed to be the result of crystal phase transition, which weakens the intermolecular π-π interactions. The authors supposed that the lifetime and π-π interaction of the chromophore are correlated. This weakening of π-π interaction then results in longer fluorescence lifetime.

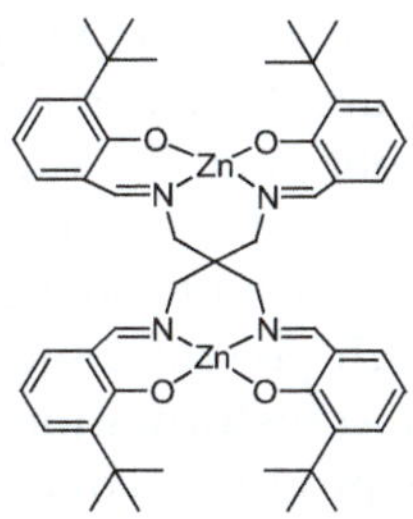

1

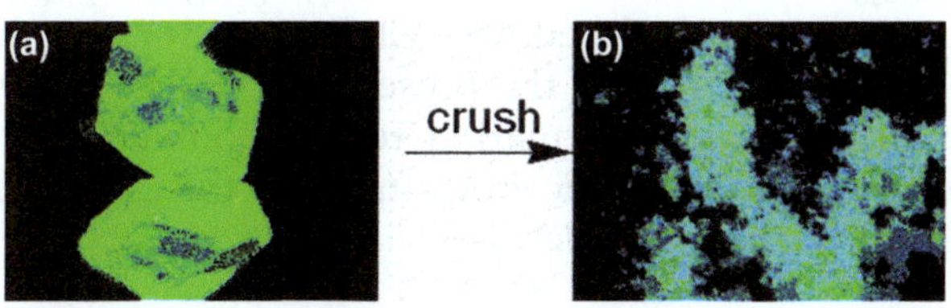

Figure 3.1 Fluorescence microscope images of (a) single crystals, (b) crushed crystals (powder).
Reproduced with permission from reference 5. Copyright 2005, American Chemical Society.

Tzeng *et al.* (2010)[6] reported on the luminescent mechanochromism of two Zn(II)-based coordination frameworks containing a new N,N′-bis(pyridyl carbonyl)-4,4′-diaminodiphenyl thioether (paps) ligand. The complex can assemble into double-zigzag or polyrotaxane frameworks depending on the presence and absence of water. The 1D double-zigzag framework, $\{[Zn(paps)_2(H_2O)_2](ClO_4)_2\}_n$ (**2a**), was synthesized through the reaction of $Zn(ClO_4)_2$ with the paps using a layer method. But the same reaction using dry solvents led to the formation of a 2D polyrotaxane framework, $[Zn(paps)_2(ClO_4)_2]_n$ (**2b**) (Figure 3.2A). Complex **2a** is strongly emissive, with an emission maximum of approximately 495 nm at room temperature and 77 K, but complex **2b** is poorly emissive in a similar spectral region. The poor emission of **2b** may be due to the intermolecular stacking interactions (*i.e.*, π-π interactions). The significant luminescent mechanochromism can be observed from **2b** to **2a**: when the solid samples of **2b** were placed in a mortar, they showed dark and no emission under an ultraviolet (UV) lamp; however, a bright blue emission followed by a pale-yellow emission was immediately observed while grinding the solid samples mixed with one drop of water (manual grinding for 1 min and drying for 30 min) (Figure 3.2B). It was considered that the emission change originated from the phase transformation from the 2D polyrotaxane framework to the 1D double-zigzag one. The phase transformation could be reversibly interconverted by heating or grinding in the presence of moisture.

Tzeng *et al.* (2012)[7] used a similar ligand N,N′-bis-(pyridylcarbonyl)-4,4′-diaminodiphenyl ether (papo), which has O atom instead of S to synthesize similar Zn(II)-based coordination frameworks. Under different conditions, they obtained three complexes, $[Zn(papo)(CH_3OH)_4](ClO_4)_2$ (**3a**), $[Zn(papo)_2(ClO_4)_2]_n$ (**3b**) and $\{[Zn(papo)_2(H_2O)_4](ClO_4)_2\}_n$ (**3c**), with different supramolecular structures (Figure 3.3). The pure form of the polyrotaxane framework of **3b** cannot be obtained in the reaction. Instead, a major product **3a** and a minor product **3b** were obtained. However, grinding a mixture of **3b** and **3a** in the presence of moisture yielded the double-zigzag framework of **3c**. Therefore, both polyrotaxane and the mononuclear structures can undergo structural transformation to yield the double-zigzag framework of **3c** as the final product. This indicates that the structural transformation processes for **3b** to **3c** are reversible. A very similar complex $[Zn(papc)_2(ClO_4)_2]_n$ was also obtained by changing ligand paps and papo to papc (*i.e.* S or O atom was replaced with CH_2). However, grinding the solid samples of $[Zn(papc)_2(ClO_4)_2]_n$ in the presence of moisture has proven unsuccessful for structural transformation to the double-zigzag framework, which may be ascribed to the steric crowding from the CH_2 group relative to the S and O atoms in the crystal lattice.

3.2.2 Au(I) Complexes

Au(I) complexes have the tendency to aggregate through the closed shell "aurophilic" interactions, which cause unique luminescence properties.

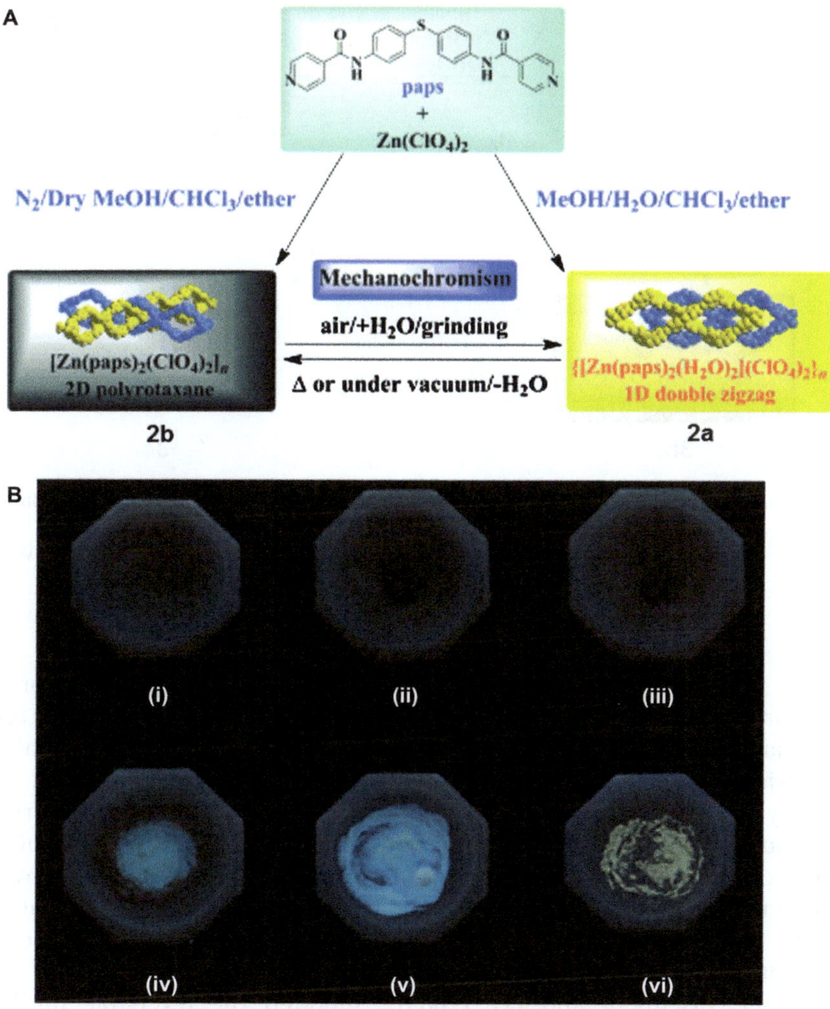

Figure 3.2 (A) Schematic of the synthesis of a 1D double-zigzag framework (**2a**) and a 2D polyrotaxane framework (**2b**) and the luminescent mechanochromism in their transformation process. (B) Luminescent images of 2 in the mortar: (i) no solid samples, (ii) with solid samples, (iii) with solid samples ground, (iv) with solid samples ground and addition of one drop of water, (v) grinding solid samples in the presence of one drop of water, (vi) grinding in the absence of solvent. A UV lamp at the excitation wavelength 365 nm was used.

The aurophilic interactions also play a key role in determining the solid-state structures that many Au(I) compounds exhibit.[8–13]

However, the mechanochromism mechanism of Au(I) complexes upon grinding is not well understood. For example, Fackler and coworkers

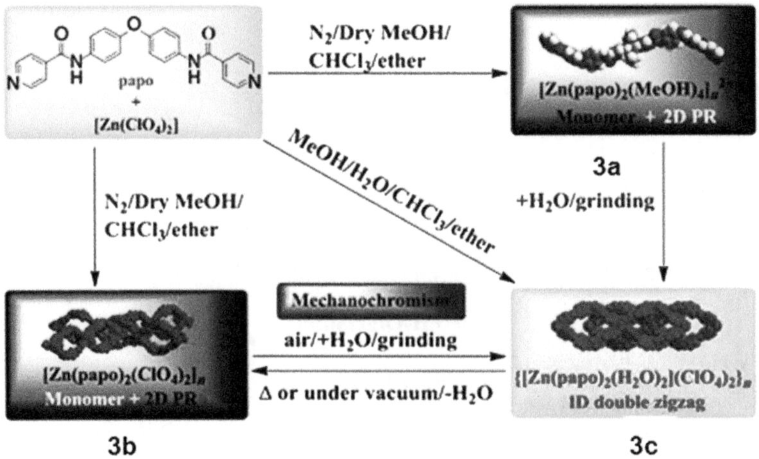

Figure 3.3 Schematic representation of the structural transformation of **3a**, **3b**, and **3c**.
Reproduced with permission from reference 7. Copyright 2012, Wiley-VCH.

(2002)[14] reported a change in emission intensity upon grinding for the linear one-dimensional chain compound $[(TPA)_2Au][Au(CN)_2]$, where TPA is 1,3,5-triaza-7-phosphaadamantine. The single crystals of this material are non-emissive, but become strongly photoluminescent after grinding. This property was attributed to the generation of lattice defects near the surface. Catalano and Horner[15] have reported that the colorless, binuclear gold (I) complex $[Au_2(dpim)_2](ClO_4)_2$.2MeCN (dpim = 2-(diphenylphosphanyl)-1-methylimidazole) exhibited an orange emission, however, the material showed a stronger blue emission after grinding, and thus supported the above lattice defect mechanism.

However, Eisenberg and Lee (2003)[16] reported the reversible mechanochromic luminescence from an Au(I) thiouracilate complexes (**4** and **5**), whose unique emissive properties can be related to two different structural motifs interconverted by protonation of the uracilate ligand. The complexes are either non-emissive or weakly photoluminescent. For **4**, the weak emission is white in color, which is indicative of multiple emissions. When the solid samples of **4** are gently crushed under a spatula, a dramatic conversion occurs to give samples bright blue or cyan photoluminescence at the ambient temperature. The powder X-ray diffraction patterns of the polycrystalline and crushed samples of **4** reveal that no gross phase change transpires upon crushing. The authors proposed that the effect could be attributed in part to (1) the disruption of weak Au···Au intermolecular interactions accompanied by rearrangement to dimers having stronger intermolecular contacts; and (2) the evolution of volatile acid upon crushing. In the non-emissive state with its helical structure, the intermolecular Au···Au

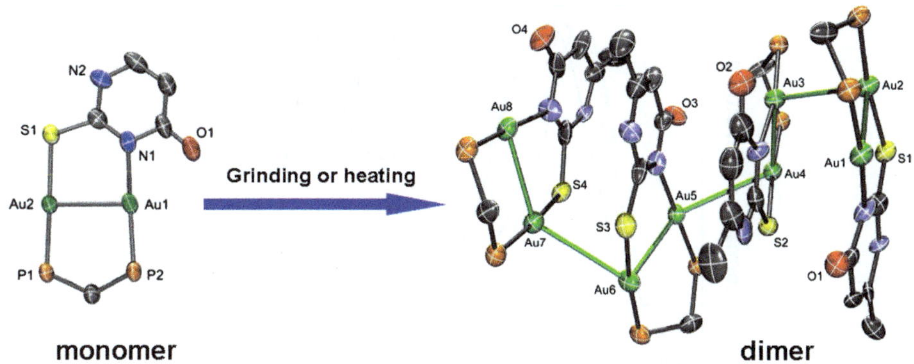

Figure 3.4 Aggregated state change between monomer and dimer of **4** under grinding or heating.
Adapted with permission from reference 16. Copyright 2003, American Chemical Society.

interactions are weak, and the extended Au···Au interactions are severely kinked, whereas in the emissive state with its dimeric structure (Figure 3.4), the intermolecular interaction is strong, and the Au···Au···Au···Au arrangement more linear.

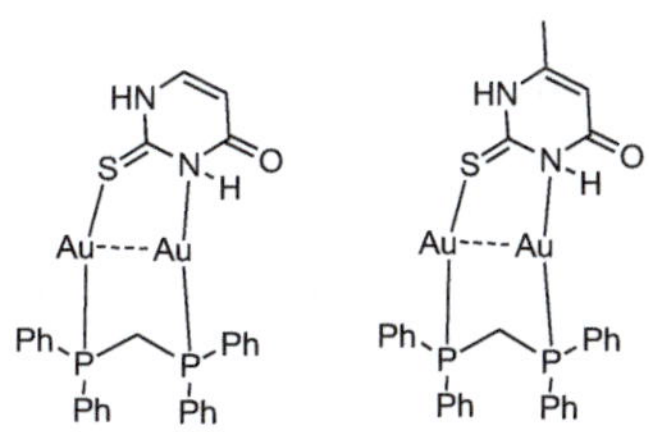

4 **5**

Eisenberg *et al.* (2008)[17] studied the structural and photophysical properties of a series of cationic Au(I) dinuclear compounds (**6**) bridged by(diphenylphosphino)methane (dppm) and substituted benzimidazolethiolate (X-BIT) ligands, where X = H (**6a**), Me (**6b**), OMe (**6c**) and Cl (**6d**). A unique solid-state emission property observed with **6a, 6b** and **6d** is of mechanochromism luminescence. When the crystals of **6a, 6b** and **6d** are irradiated with 365 nm light, they show emission maxima of 484, 481 and 478 nm, respectively. Upon gentle crushing of the crystals with a spatula, the emission maxima shift to higher energies: 459, 466 and 459 nm for **6a, 6b** and **6d**, respectively (Figure 3.5). For **6c**, this unique mechanical effect is not observed to the same extent as the other three cationic complexes. In contrast with the **4** and **5** complexes, the cationic complexes **6a, 6b** and **6d** exhibit no evolution of volatile CF_3CO_2H after gentle heating or sonication. In the report, the authors gave few details about the mechanochromic mechanism.

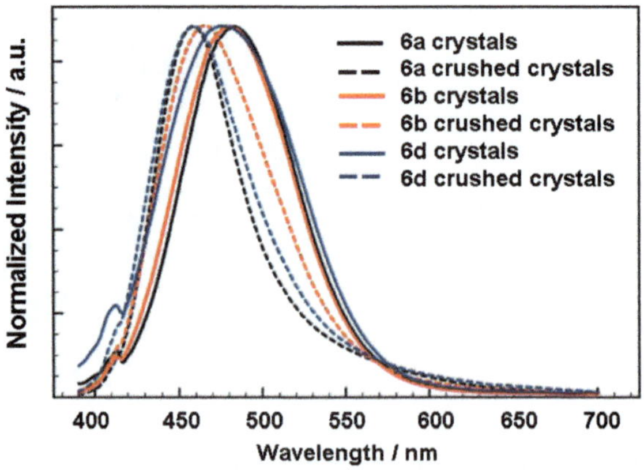

Figure 3.5 Solid-state emission ($\lambda_{ex} = 365$ nm) spectra showing luminescence tribochromism of **6a–b** and **6d** in a finely ground matrix of KBr at room temperature.
Reproduced with permission from reference 17. Copyright 2008, American Chemical Society.

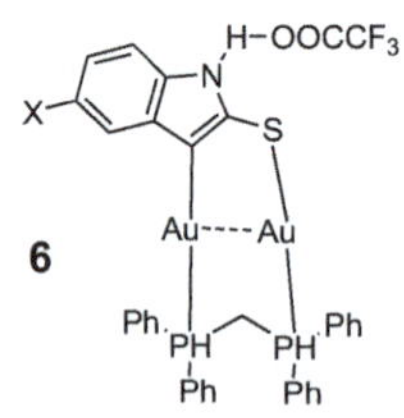

6

X = H (**6a**), Me (**6b**), OMe (**6c**), Cl (**6d**)

Ito *et al.* (2008)[18] reported an Au(I) complex [($C_6F_5Au)_2$(μ-1,4-diisocyanobenzene)] (7), which has a solid-state photoemission that can be reversibly switched by external stimuli (Figure 3.6). Upon gentle grinding of 7 using a spatula or a pestle, the blue luminescence ($\lambda_{em} = 415$ nm) changed to an intense yellow luminescence ($\lambda_{em} = 533$ nm) with a 118 nm red shift under 365 nm UV light. However, the color changes were not observed under the ambient light. Upon treatment of the ground powder with drops of dichloromethane, the yellow luminescence reverted to the original blue color. Evaporation of the solvent produced the original blue luminescent solid. The blue-to-yellow conversion achieved by subsequent regrinding can be repeated several times without causing any degradation in the luminescence. The solvent treatment is achieved by the addition of other solvents as well as by exposure to dichloromethane vapor. The authors proposed a mechanism for the mechanochromic process of 7 that involves two phases:[18] the first phase transforms the microcrystalline powder into a metastable amorphous phase by grinding. This phase has aurophilic interactions that are responsible for the red-shifted emission. The shortest Au···Au distance in the single crystal was 5.19 Å, which is beyond the range of significant

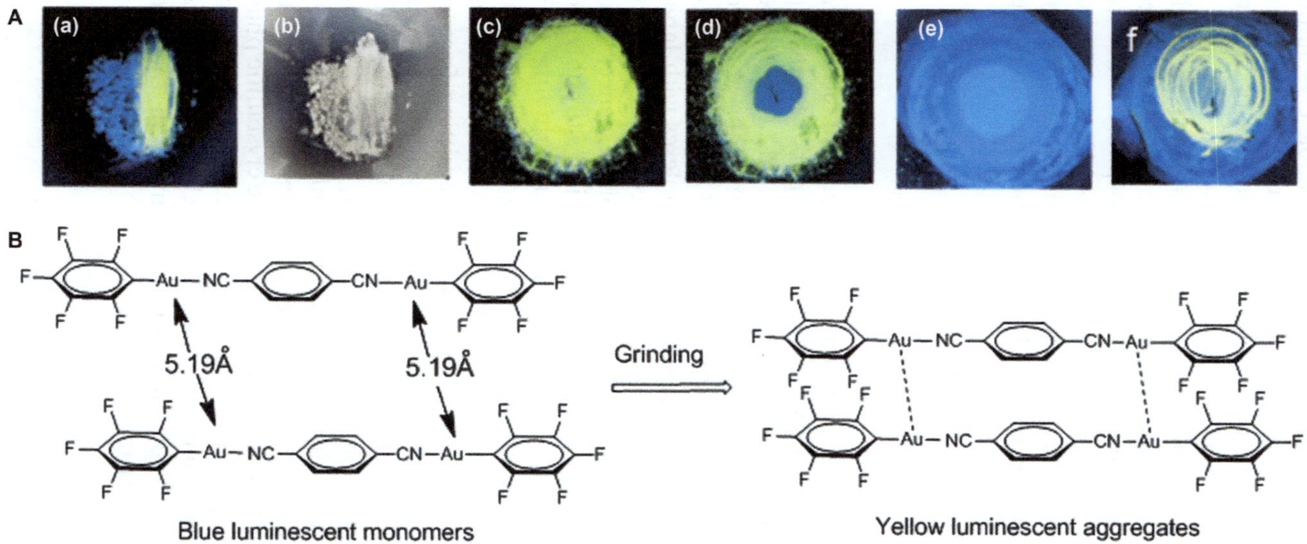

Figure 3.6 (A) Photographs showing complex 7 on an agate mortar under UV irradiation with black light (365 nm): (a) after grinding the right-half with a pestle, (b) the same sample under ambient light, (c) entirely ground powder of 7, (d) partial reversion to the blue luminescence by dropwise treatment using dichloromethane onto the center of the ground powder, (e) powder after treatment with dichloromethane, and (f) repetition of the yellow emission by scratching the powder with a pestle. (B) Structural changes accompanying the luminescence changes of 7.
Reproduced with permission from reference 18. Copyright 2008, American Chemical Society.

aurophilic bonding (2.7 Å to 3.3 Å). Thus, the aurophilic interactions were absent within the crystal (Figure 3.7). The second phase involves the rearrangement of the amorphous phase into the more stable crystalline phase *via* a partial dissolution and recrystallization process upon solvent treatment of the ground sample. Within the crystals of 7, the lack of strong intermolecular forces (such as C–H···π interactions) and the planarity of the molecular structure allow the slipping of the molecular stacks by mechanical stimuli. Thus, the solid-state blue emission from the sample occurred from single molecules, which can be attributed to the phosphorescence from the intraligand-localized π-π* excited state; and the red-shifted emission from the ground sample appears to arise from an amorphous material in which aurophilic interactions may be responsible for the new emissive state.

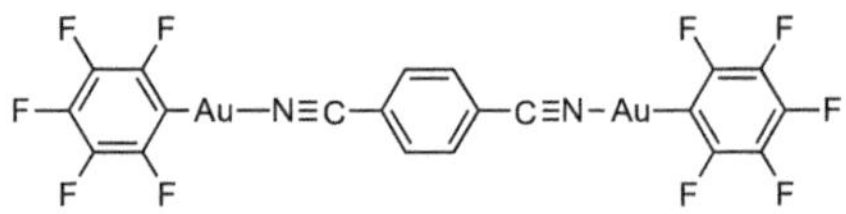

7

Osawa *et al.* (2010)[19] designed and synthesized a series of tetrahedral Au(I) complexes (**8**) containing the diphosphane ligand [dppb = 1,2-bis(diphenylphosphino) benzene], [Au(dppb)$_2$]X [X = Cl, Br, I, NO_3, BF_4, PF_6, B(C_6H_4F-4)$_4$]. The complexes, which have relatively small counteranions and two dppb ligands, are symmetrically coordinated to the central Au(I) atom and display intense blue phosphorescence. Meanwhile, the complexes with large counteranions and two dppb ligands are asymmetrically coordinated to the Au(I) atom and display yellow or yellow-orange phosphorescence. The latter complexes exhibit clear responsiveness to mechanical stimuli or organic solvents. As shown in Figure 3.7, after continuous grinding of sample [Au(dppb)$_2$]BF_4, the phosphorescence eventually changed from blue (λ_{max} = 494 nm) to yellow orange (λ_{max} = 575 nm). A reversible transformation from a yellow-orange-emitting form to a blue-emitting form was observed by treating the sample with diethyl ether. This yellow-orange-emitting form was found to show vapochromism. Upon exposure of this form to the vapors of volatile organic solvents, such as ether, acetone, acetonitrile, dichloromethane and tetrahydrofuran, under the ambient temperature, the phosphorescence turned from yellow orange to intense blue within 0.5 min to 5 min. The XRD patterns revealed the change of the crystalline phase to an amorphous phase after grinding. Almost no regular intermolecular interactions occur in the amorphous phase. The researchers proposed that the symmetry reductions are responsible for the changes in the phosphorescence color caused by external stimuli.

8

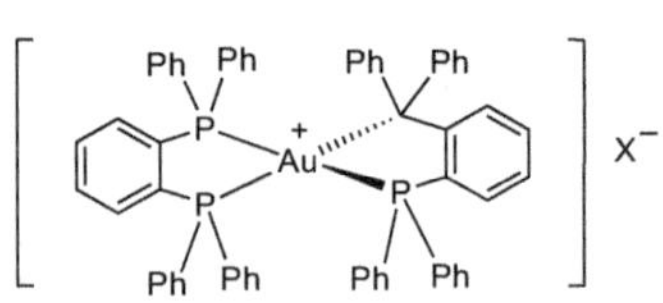

X=Cl, Br, I, NO_3, BF_4, PF_6, B(C_6H_4F-4)$_4$

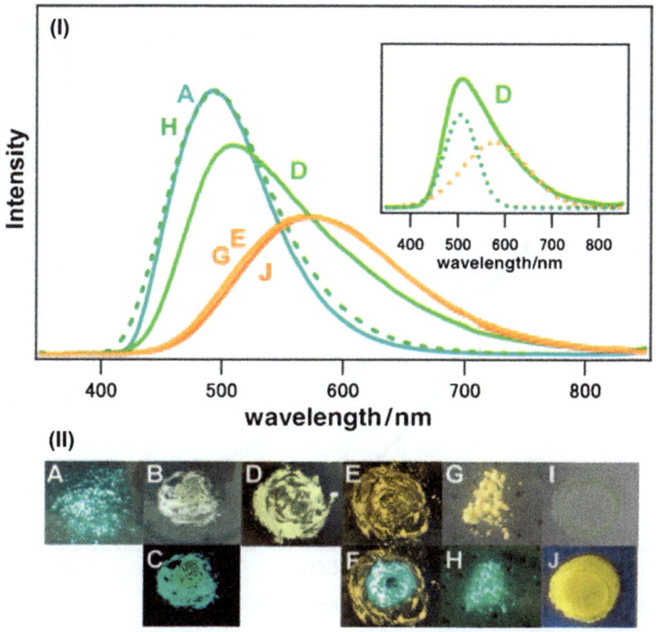

Figure 3.7 (I) Corrected emission spectra ($\lambda_{ex} = 350$ nm) and (II) images upon UV irradiation (except for B and I) in various states of $[Au(dppb)_2]PF_6$ (A–J, see reference). The inset shows two segmentalized spectra in state D.
Reproduced with permission from reference 19. Copyright 2010, Wiley-VCH.

Table 3.1 Solid-state absorption and luminescence data of complexes **9a–9e** at ambient temperature.[20]

		λ_{em}/nm	
Compound	λ_{abs}/nm	*Before grinding*	*After grinding*
9a	335	432,459	531
9b	332	435,469	528
9c	374	477	502
9d	332	405	405
9e	330	488	488

Yin *et al.* (2012)[20] designed and synthesized a series of diisocyano-based Au(I) complexes (**9**) with various aromatic bridges instead of phenyl linkages. The mechanochromic properties of these complexes indicated that the ligand structure significantly affected the mechanochromic activity. The gold complexes with methyl-substituted phenyl bridges (**9a** and **9b**) exhibited mechanochromism. In addition, an approximately 100 nm red-shift in the fluorescence spectrum was observed after grinding (Table 3.1, Figure 3.8).

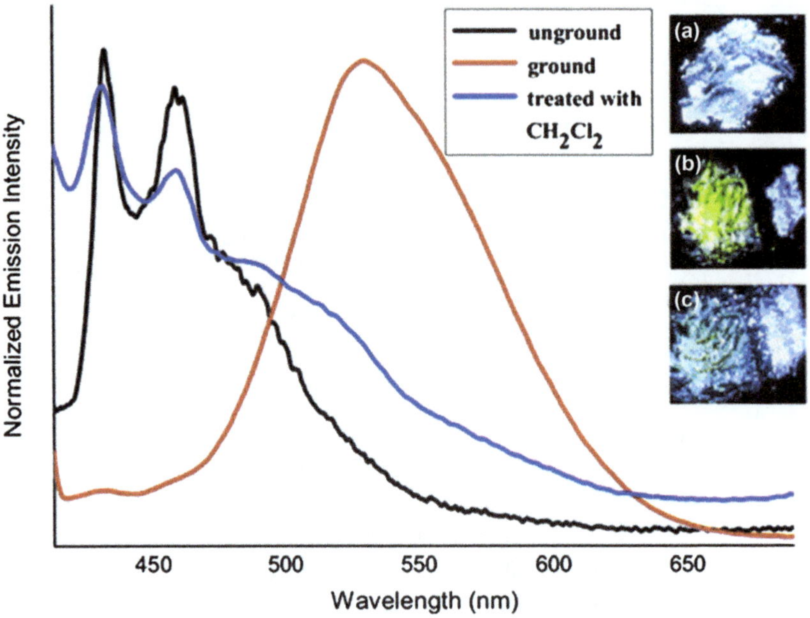

Figure 3.8 Emission spectra of **9a** in various states. Insets show the fluorescence images of Au (I) complex **9a** powder under 365 nm UV light: (a) before grinding, (b) after grinding, (c) after treatment with dichloromethane. Reproduced with permission from reference 20. Copyright 2012, Elsevier.

The complex with a diphenylmethane bridge (**9c**) exhibited mechanochromism through a change in fluorescence from blue to green, as well as a 25 nm red shift after grinding. No mechanochromism was observed for either the biphenyl- or diphenylethane-containing complexes (**9d**, **9e**). The ground complexes with mechanochromic activities reverted to their original states upon CH_2Cl_2 treatment. The single crystals of complexes **9a,** **9c** and **9e** were then examined using single crystal X-ray analysis. The results confirmed the different crystalline structures of these complexes. The molecular packing of complex **9a** exhibited a head-to-head arrangement. Moreover, the intermolecular distance of the gold atoms was 5.64 Å, which indicates no obvious intermolecular Au···Au interactions. The molecular packing of complex **9c** was arranged in a head-to-tail pattern. The distance between the gold atoms was 3.23 Å, which is indicative of intermolecular Au···Au interactions. Furthermore, XRD was used to study the mechanochromic behavior. The results showed that the unground samples **9a** to **9c** exhibited clear reflection peaks. However, grinding caused a crystalline-to-metastable phase conversion. By contrast, the non-mechanochromic species **9d** and **9e** were thoroughly ground and their XRD patterns were essentially unchanged, which implied that no crystalline to non-crystalline transition occurred.

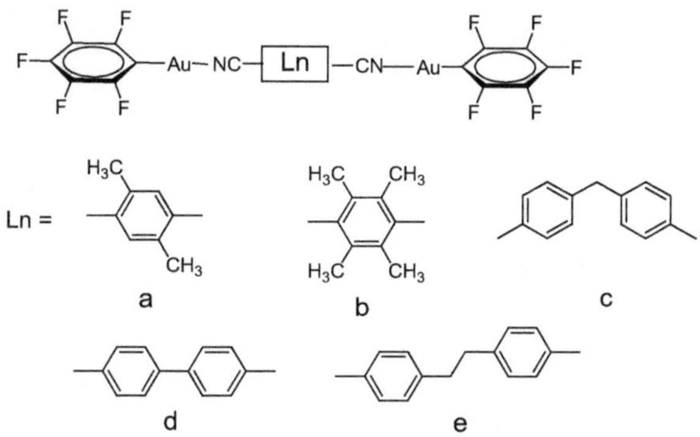

9

3.2.3 Pt(II) Complexes

Recently, square-planar Pt(II) coordination complexes have attracted attention, largely on account of their optoelectronic properties.[21–23] The optoelectronic properties are directly related to the intermolecular interactions between the adjacent Pt atoms. The Pt-Pt interactions can be strongly influenced by the coordinating ligand and its substituents, as well as the counter-ion used, which can allow for tuning of the spectroscopic properties.[24,25] Thus, the absorption and emission properties of these complexes are quite sensitive to external stimuli in solution and as solids. For example, some square-planar platinum complexes have a sensitive response in absorption or photoluminescence spectra when exposed to volatile organic compounds (VOCs).[26–28] Another type of stimuli-responsiveness is mechanochromic luminescence, and some platinum(II) complexes have shown this behavior. Grinding or rubbing the solid samples of these complexes resulted in strong red-shifted emissions and analogous color changes, which are related to a force-induced structural rearrangement, resulting in the increase of intermolecular Pt-Pt interactions.

Kozhevnikov *et al.* (2008)[29] reported two series of liquid-crystalline Pt(II) complexes **10** and **11**. These complexes exhibited different emission colors at different phase states. For example, it has been observed that the same sample of pure **10b** can produce different emission colors depending on the method of sample preparation, that is, the thermal history. When cooling slowly from the isotropic melt to 170 °C and then rapidly to room temperature. The emission spectrum exhibited a fine structure and two peaks at 575 nm and 624 nm (Figure 3.9a), which were considered as the monomer emission of the complex. However, when the same complex was cooled rapidly from the isotropic melt directly, no monomer emission is observed, rather than excimer-like emission (660 nm) (Figure 3.9b).

It was found that a pure film of **11a** by spin coating displayed only excimer-like emission (660 nm). However, after heating the film to 110 °C, followed by cooling to room temperature, there was a drastic change in emission color from the red of the excimer to yellow, indicating simultaneous emission

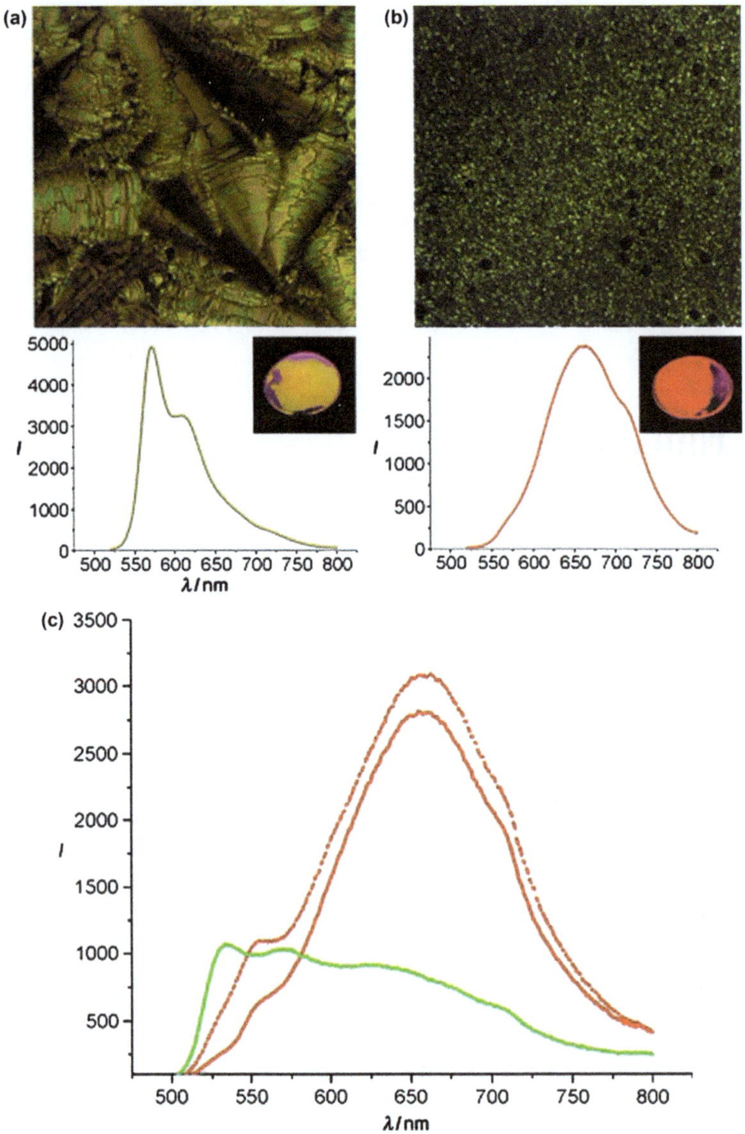

Figure 3.9 Photomicrographs of Complex **10b** (taken between crossed polarizers at room temperature) and corresponding emission spectra ($\lambda_{ex} = 420$ nm): (a) fast-cooled from the liquid crystal (LC) phase after complete development of the texture, and (b) fast-cooled direct from the isotropic phase. Inset: real samples of pure films sandwiched between glass slides. (c) The emission spectra of pure films of **11a** obtained by spincoating on a glass surface, as prepared (red), the same slide but after heating to 110 °C for 5 min (green), the same slide, disrupted mechanically at room temperature (dotted red). $\lambda_{ex} = 410$ nm.
Reproduced with permission from reference 29. Copyright 2008, Wiley-VCH.

from the monomer and excimer. However, if the film is subjected to mechanical disturbance (such as rubbing), the red emission of the excimer is returned (Figure 3.9c). A further heat-cool cycle re-establishes monomer emission. The authors believed that the emission was under tribological control and the initial state could be readily re-set, suggesting re-usable, stimulus-responsive applications.

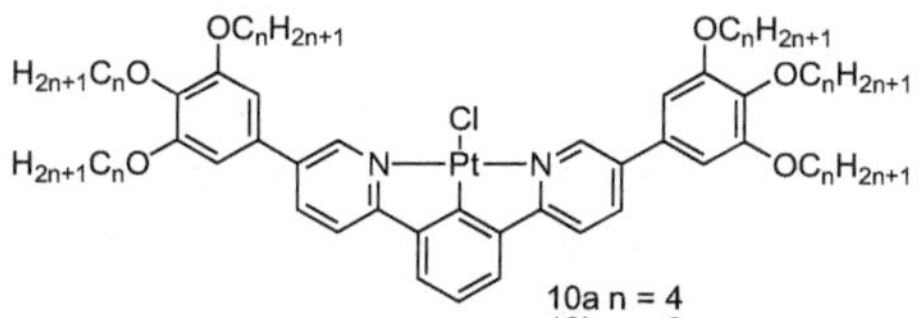

10 10a n = 4
10b n = 6
10c n = 8
10d n = 10
10e n = 12

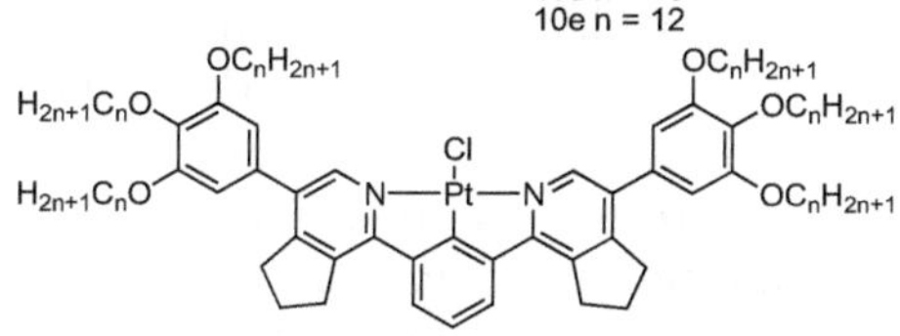

11 11a n = 10
11b n = 12

Abe *et al.* (2009)[30] reported on the mechanochromic behavior of two Pt(II) complexes **12** and **13**. When the crystal of **12** was crushed and ground on a glass substrate with a spatula, the luminescence changed from yellow to bright orange, while the crystal color itself remained yellow. A broad emission band at approximately 670 nm increased in the longer wavelength region as compared with that in the initial spectrum (peaking at 570 nm) and the vibronic structure at 490 and 530 nm observed for the crystal still remained after grinding (Figure 3.10), which was similar to the excimer emission observed in solution. The chemical structure determination and

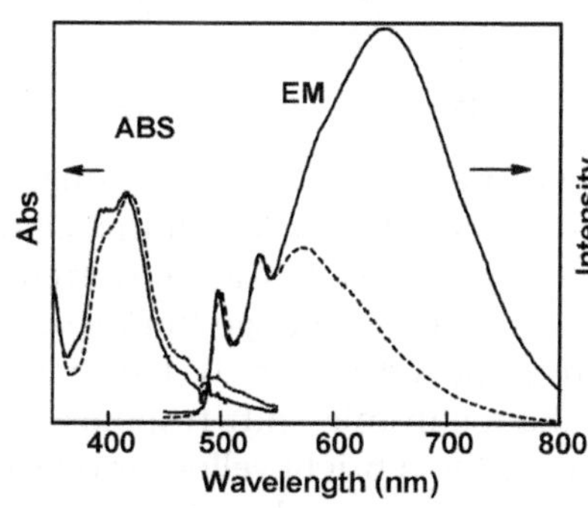

Figure 3.10 Absorption and emission spectra of **12** in the solid state before (dotted line) and after (solid line) grinding. Absorbance and emission intensities are normalized at 430 and 495 nm, respectively.
Reproduced with permission from reference 30. Copyright 2009, The Royal Society of Chemistry.

unchanged color and absorption spectrum after grinding indicated that bond-breaking or isomerization reactions did not occur, *i.e.* no solid-state reaction proceeded during the grinding process. The mechanochromic luminescence in the crystalline structure of the complex solid was investigated with respect to the excimer emission. XRD measurements were performed for complex **12** before and after grinding. The results indicated that the crystalline structure of the ground powder did not change and remained similar to that of the initial crystal before grinding. The luminescence change may be due to a dynamic process in the photoexcited state; this mechanism is in contrast to the static process of vapochromism observed for other Pt(II) complexes. The XRD pattern for the powder is in good agreement with that of the crystal and the simulation calculated from the X-ray crystallographic data of **12** using the Mercury software, indicating that the crystalline structure of the powder was not changed from that of the initial crystal before grinding. The authors suggested that this was a phenomenon arising from a dynamic process in the photoexcited state of **12**, in contrast to the static process of vapochromism known for other Pt(II) complexes. The solid-state emission properties of a related complex, **13**, with no methyl groups, were also studied. Orange coloration and red luminescence were observed upon grinding. The results were apparently different from those of **12**. In the emission spectrum, vibronic bands at around 500 nm before grinding were completely absent. Instead, only a structureless band peaking at 700 nm was observed. The XRD result indicated that mechanical grinding destroyed the regular arrangement of molecules in the crystal, because the powder was almost amorphous. Thus, the authors believed that the red luminescence of **13** powder might be attributed to emission of a dimer or aggregate generated in the amorphous phase.

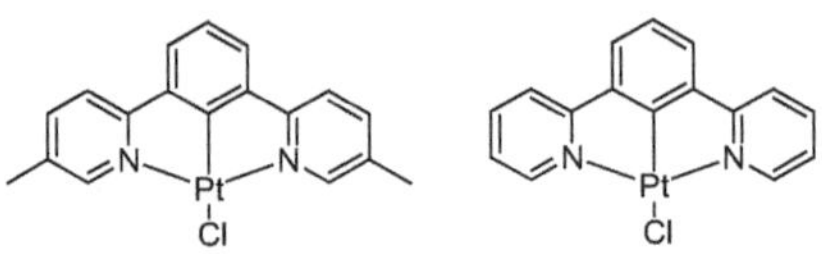

12 **13**

Chen *et al.* (2011)[31] synthesized a series of square-planar bis(s-fluorophenylacetylide) Pt(II) complexes **14–16**. When the solid species were gently crushed or ground, the crystalline state was converted into the amorphous phase. Meanwhile, the bright yellow-orange luminescence of the crystalline species was converted to dark red under UV light irradiation, exhibiting a 100 nm to 160 nm mechanochromic shift response (Figure 3.11). This luminescence could be restored to a bright yellow or yellow-orange emission by adding a drop of acetone to the ground sample. Complex **14** was found to display selective luminescence vapochromic properties to CH_2Cl_2 and $CHCl_3$ vapors, with a luminescence vapochromic shift response at approximately 150 nm to 200 nm. The author suggested that the highly reversible vapochromic or mechanochromic responses triggered by halohydrocarbon vapor

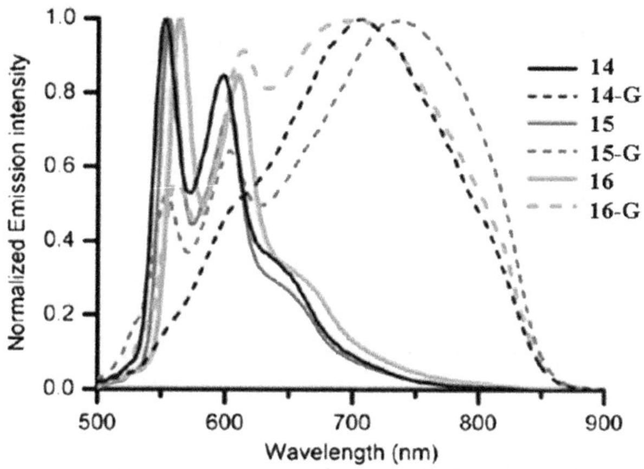

Figure 3.11 Emission spectra of desolvated crystalline species **14–16** before (solid lines) and after (dashed lines) mechanically ground, showing the significant emission spectral changes in response to mechanical grinding, corresponding to about 100–180 nm of mechanochromic luminescence response shift.
Reproduced with permission from reference 31. Copyright 2011, Wiley-VCH.

or mechanical grinding arose mostly from the increased contribution of intermolecular Pt-Pt contact, and consequently an excited state conversion from metal-to-ligand charge transfer (MLCT)/ligand-to-ligand charge transfer (LLCT) to metal-metal-to-ligand charge transfer (MMLCT)/LLCT character.

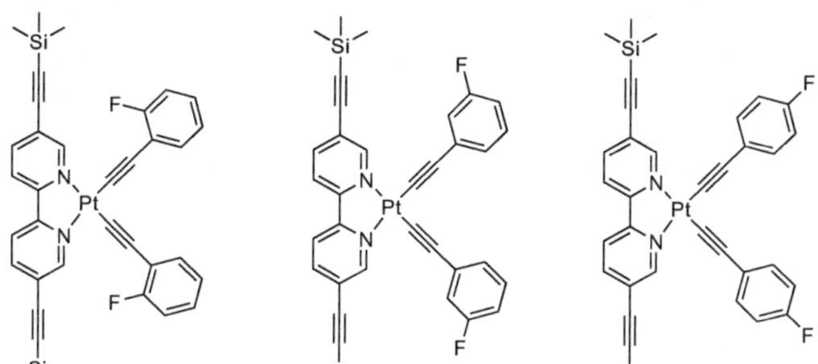

14 **15** **16**

Two planar 5-trimethylsilylethynyl-2,2′-bipyridine Pt(II) complexes **17** and **18** have been studied by Zhang *et al.* (2011).[32] The complexes exhibited reversible and reproducible mechanical stimuli-responsive color and luminescence switch. When the crystalline samples were ground, the bright yellow-green emission with well-resolved vibronic-structured emission was immediately converted to red luminescence with a red-shift emission of 121–155 nm for **17** and 53–89 nm for **18** (Figure 3.12). The results indicated that the corresponding response shift in **18** was significantly smaller because the

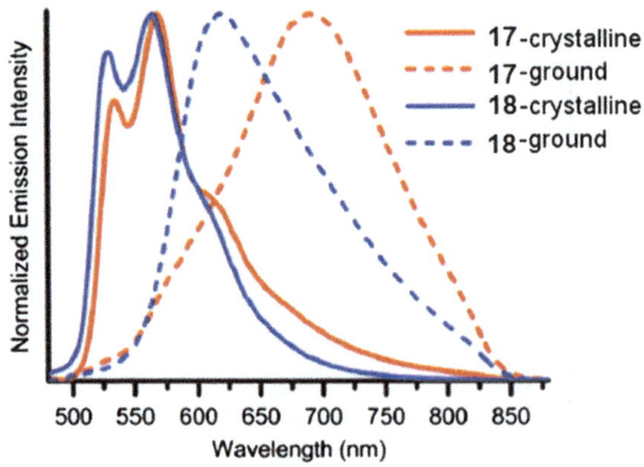

Figure 3.12 Normalized emission spectra of **17** (red) and **18** (blue) in the crystalline state (solid) and ground state (dash), showing a mechanical grinding-triggered emission red shift of 121–155 nm for **17** and 53–89 nm for **18**.
Reproduced with permission from reference 32. Copyright 2011, American Chemical Society.

bulky *tert*-butyl group induced the planar Pt(II) molecules to stack through a longer Pt-Pt distance with less intermetallic contact as compared with that in **17**. The XRD results also revealed that the crystalline state was transformed to the amorphous phase that could be reverted to the original crystalline state by organic vapor adsorbing or heating, along with red luminescence turning back to yellow-green emitting. In addition, the vibronic-structured emission bands entirely disappeared after grinding. The authors proposed that the drastic grinding-triggered emission red shift was likely involved in the formation of a dimer or an aggregate through Pt-Pt interaction, resulting in the conversion of the ^{3}MLCT/^{3}LLCT emissive state in the crystalline state into the ^{3}MMLCT triplet state in the amorphous phase.

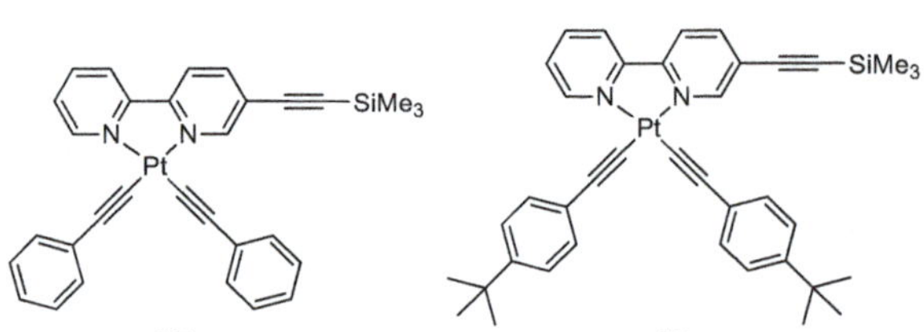

17 **18**

Zhang *et al.* (2012)[33] synthesized a Pt(II) complex containing 4-trifluoromethylphenylacetylide (**19**). It was found that the complex exhibited remarkable luminescence vapochromic and mechanochromic properties and a thermo-triggered luminescence change. The solid state sample was selectively sensitive to the vapors of oxygen-containing volatile compounds such as THF, dioxane and tetrahydropyrane (THP), exhibiting vapochromic phosphorescent response. The emission wavelength red shifted from 561 and 608 nm to 698 nm (THF), 689 nm (dioxane) and 715 nm (THP). The photographic images of **19** in the poly(methyl methacrylate) films showed remarkable luminescence color switches under UV light (365 nm) irradiation in response to vapor or mechanical stimuli. When the compounds were heated, a thermo-triggered luminescence change occurred, in which bright yellow luminescence at 561 and 608 nm turned to red luminescence at 667 nm with a drastic red shift. The vapo-, thermo- and mechano-triggered luminescence switches have been monitored by the changes in emission spectra and X-ray diffraction patterns, which revealed that the multi-stimulus-responsive luminescence changes were mostly reversible, with the conversion from one state to another by vapor adsorption or recrystallization (Figure 3.13a–c). Both X-ray crystallographic and DFT studies suggested that the variation in the intermolecular Pt-Pt interaction was the key factor in inducing intriguing luminescence switches. The photographic images of **19** in the PMMA films are shown in Figure 3.13d, showing remarkable luminescence color switches under UV light (365 nm) irradiation in response to vapor or mechanical stimuli, exhibiting good recording performance.

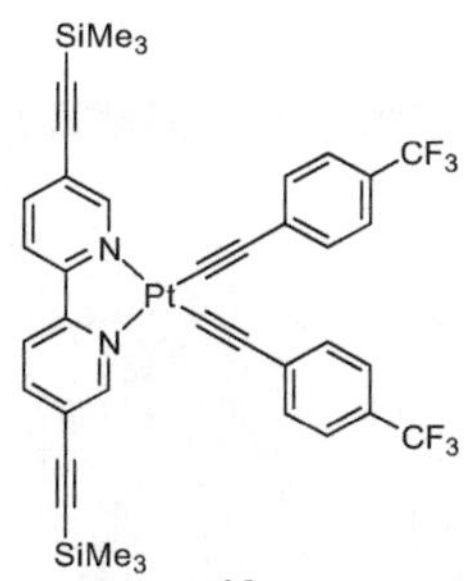

19

Kanbara *et al.* (2012)[34] studied the photophysical properties of three pincer Pt(II) complexes containing amide groups (**20**–**22**). Complex **20** exhibited a weak luminescence at 518 nm (Φ_F <0.01) due to the carbamoyl group, whereas complexes **21** and **22** exhibited strong luminescence in solid state at a wavelength of 480 nm ($\Phi_F = 0.39$ and 0.40, respectively). The results of luminescence decay time determinations (both $\tau = 1.59$ μs) revealed that **21** and **22** were phosphorescent compounds. It was found that in this series of complexes, only complex **22** had mechanochromic luminescence activity and complexes **20** and **21** exhibited no mechanochromic luminescence. Green luminescence was observed from the crystals obtained by recrystallization from DMF (**22** · DMF) under UV light. The green luminescent crystals changed to orange (**22** · Powder) when grinding in a ceramic mortar. Exposure of the orange luminescent **22** · Powder to methanol vapor induced a change to yellow luminescence (**22** · MeOH). The orange luminescence was recovered from the yellow luminescence by mechanical grinding or by heating at 200 °C. The XRD data showed that the green luminescent crystals of **22** · DMF exhibited a sharp diffraction at 5.90°. By contrast, the ground

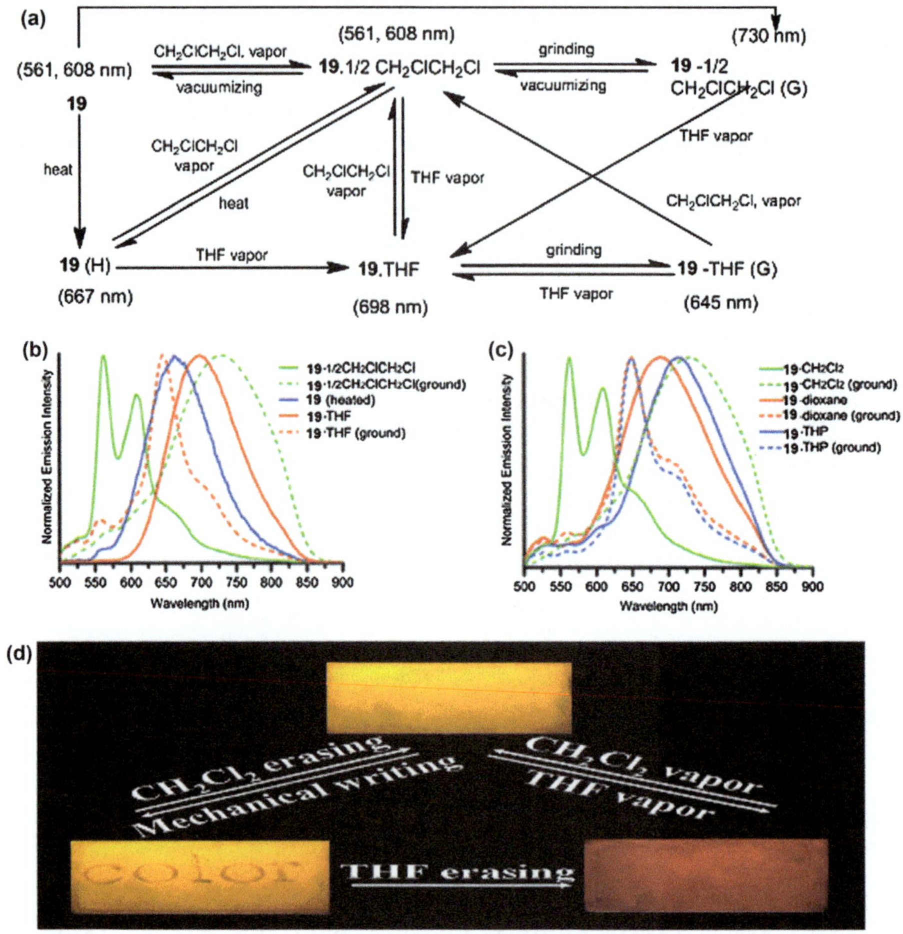

Figure 3.13 (a) Interconversion processes for solid materials based on complex **19**. (b) and (c) Solid-state emission spectra of crystalline and corresponding ground materials based on complex **19** containing organic solvents at ambient temperature. (d) Photographic images of **19** in a PMMA film in response to CH_2Cl_2 vapor or mechanical stimuli under UV light (365 nm) irradiation.

22·Powder did not exhibit clear diffraction. This result indicates that mechanical grinding scatters the regular arrangements of **22**·DMF. A new sharp diffraction peak appeared at 4.94° upon exposure of **22**·Powder to methanol vapor for several minutes. The molecular structures of the square-planar Pt(II) complex in **22**·DMF and **22**·MeOH are similar, whereas their molecular orientations and hydrogen-bonding networks are different. One of the N-H moieties in the amide groups interacts with the chloro ligand in the neighboring Pt(II) complex. Another N-H moiety has hydrogen-bonding interaction with the solvated DMF molecule. In the case of **22** · MeOH, three types of hydrogen bonds can be observed. In addition to N–H···Cl and N–H···O (MeOH), one of the N–H moieties shows hydrogen-bonding interaction with the carbonyl group in the neighboring complex. The hydrogen-bonding network affects the Pt-Pt distance in the crystals. The packing diagram of **22** · DMF reveals a Pt-Pt distance of 4.854 Å, whereas the packing diagram of **22** · MeOH reveals a shorter Pt-Pt distance of 3.385 Å. The shorter Pt-Pt distance indicates the metal-metal interaction, which induces excimer luminescence. Since the solvent molecules affect the Pt-Pt distances in the solid state *via* hydrogen bonding, the hydrogen-bonding capability of the luminescent material is expected to be a key factor for tuning luminescence color in the solid state. Thus, the authors considered that the mechanical grinding and heating induced the change in the luminescence color of **22** · DMF to orange which caused by excimer formation through Pt-Pt interactions.

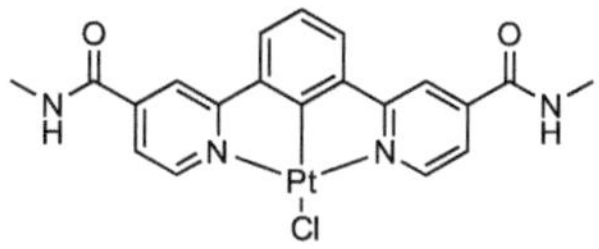

20

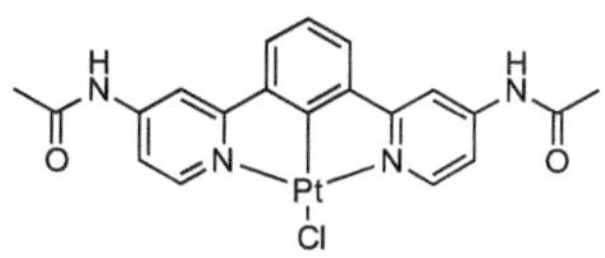

21

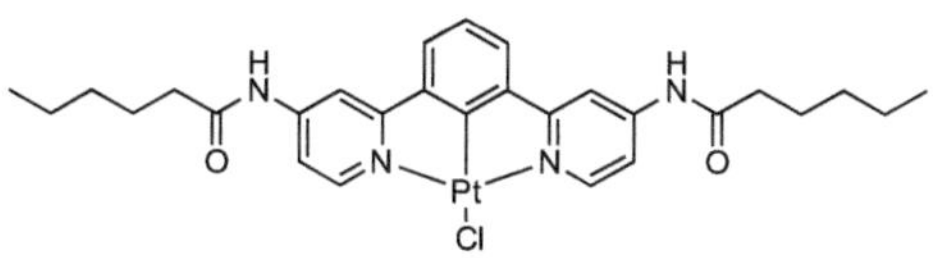

22

3.2.4 Cu(I) Complex

Copper iodide clusters are well known for their thermochromic luminescence properties.[35] For example, $Cu_4I_4(PPh_2(CH_2)_2Si(OCH_2CH_3)_3)_4$, was found to be thermochromic luminescence with bright yellow luminescence at room temperature and purple emission at 77 K. It was suggested that as a result of weak Cu-Cu interactions, the two emissive states appeared as highly coupled with a low energy barrier (2 kJ · mol^{-1}), leading to a controlled thermochromism in a large temperature range.[36] Copper iodide clusters are chosen as inorganic moieties, not only for their thermochromic performances but also for the lower price and larger abundance properties of Cu, which make it possible for wider applications and larger-scale productions.

Perruchas *et al.* (2010)[37] reported a molecular Cu(I) iodide cluster **23** (Figure 3.14A) with mechanochromic and thermochromic luminescence properties (Figure 3.14B and C). This molecular cluster underwent a significant change in its solid-state emission properties from weak green to intense yellow upon grinding. This change was reversible upon exposure to solvent or heating. By contrast, the crushed compound had a yellow emission at room temperature, which was replaced by a purple one at 77 K. When

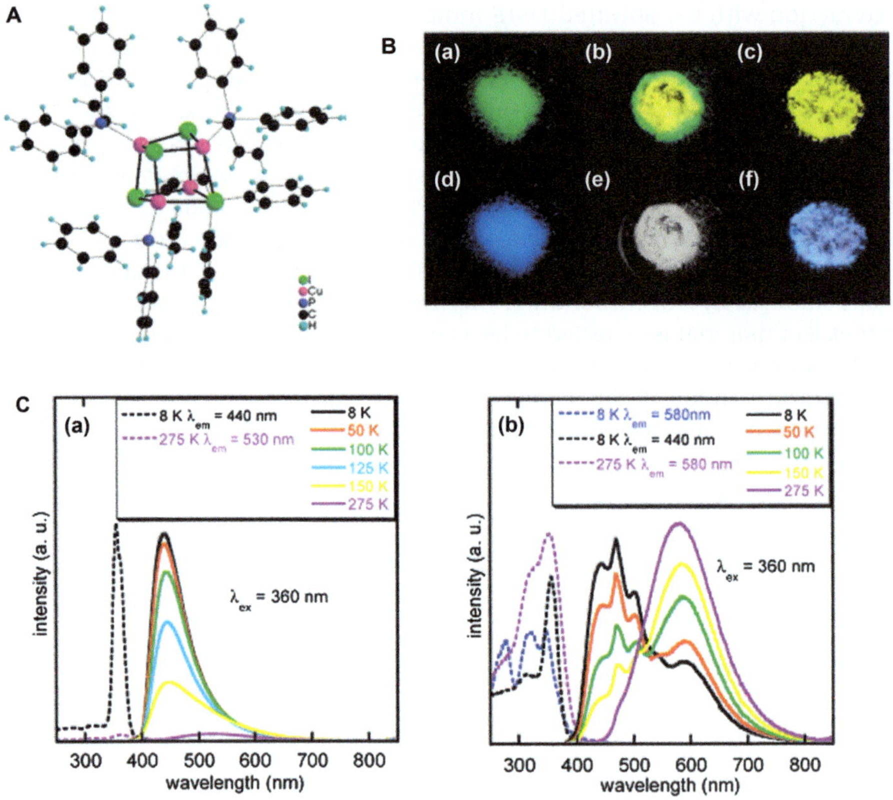

Figure 3.14 (A) Molecular structure of **23**. (B) Photos showing luminescence changes of **23** upon grinding and at different temperatures. (a) Cluster before grinding and (b–c) cluster upon grinding under 365 nm (UV lamp) at room temperature. Cluster (d) before and (f) after grinding under 365 nm (UV lamp) at 77 K. (e) Corresponding photo of (b) under ambient light at room temperature. (C) Solid-state luminescence spectra of (a) uncrushed and (b) crushed **23** recorded between 275 and 8 K. Reproduced with permission from reference 37. Copyright 2010, American Chemical Society.

both samples were progressively warmed, the room temperature emission could be recovered, thus indicating a completely reversible thermochromic luminescence mechanism. Prior to grinding, an intense emission band (30 times higher compared with that at 275 K) at $\lambda_{max} = 440$ nm was observed for the cluster at 8 K. By increasing the temperature, the intensity of this blue band decreased and the emission red shifted to 530 nm at 275 K. For the crushed cluster, two emission bands at 470 and 580 nm were observed at 8 K. By increasing the temperature from 8 K to 275 K, the intensity of the 470 nm band progressively decreased, whereas that of the 580 nm increased, without exhibiting significant wavelength shifts. Through the crystalline structure analysis, it was found that no intermolecular interactions directly involved the Cu(I) iodide core, and no typical π-π stacking interactions existed despite the presence of numerous phenyl groups in the crystalline structure. Only several short CH···H contacts involving the allyl groups of the ligands were detected. The authors considered that the mechanical constraints induced by the grinding modified these inter-ligand interactions leading to a different cluster packing in the crushed compound. These local distortions allowed the relaxation of the $[Cu_4I_4]$ core, which was probably sterically constrained in the crystalline phase, and drastically modified the luminescence properties. Thus, this reversible phenomenon was attributed to the distortions in the crystal packing, leading to modifications of the intermolecular interactions and thus of the $[Cu_4I_4]$ cluster core geometry.

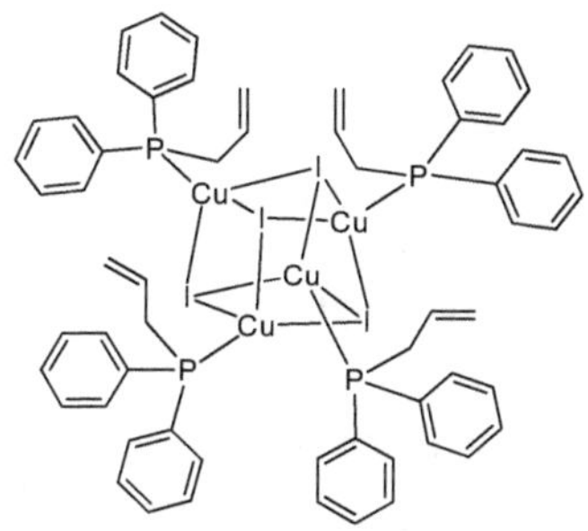

23

Hong *et al.* (2013)[38] successfully prepared a tri-stimuli-responsive luminescent material, $[Cu_4I_4(\text{4-dpda})_4]$ (**24**) (4-dpda = 4-(diphenylphosphino)-N,N-dimethylaniline) with reversible and independent thermochromism, mechanochromism and vapochromism. The X-ray crystal analyses revealed that only CH_3CN molecules were incorporated into the final product of **24**·CH_3CN, and the formation of this compound showed strongly dependent on the solvents. Upon grinding of **24**·CH_3CN using a mortar, the yellow-green luminescence changed to an intense orange luminescence observed under UV irradiation at 365 nm. While under ambient light, the color changes of the sample were not detected. After treatment with drops of CH_3CN and drying, the emission color could be reverted to the original yellow-green emission. The dynamic recovery of the luminescent color from yellow-green and orange was further proved with the repeating cycles of grinding and treating the sample with solvents.

The UV-vis absorption spectrum of the ground sample of **24**·CH_3CN showed a broad band ranging from 270 to 340 nm, with no obvious shift as compared to the unground sample. Upon excitation at 355 nm at ambient temperature, the ground sample showed a broad emission band centered at 590 nm with an obvious red shift as compared to the 540 nm emission of the unground one. Then, by treating it with CH_3CN, the spectra of absorption and emission recovered and appeared to be almost the same as those before grinding (Figure 3.15A). The spectroscopic investigations also showed no obvious change in the sample which was heated at 120 °C in a vacuum for

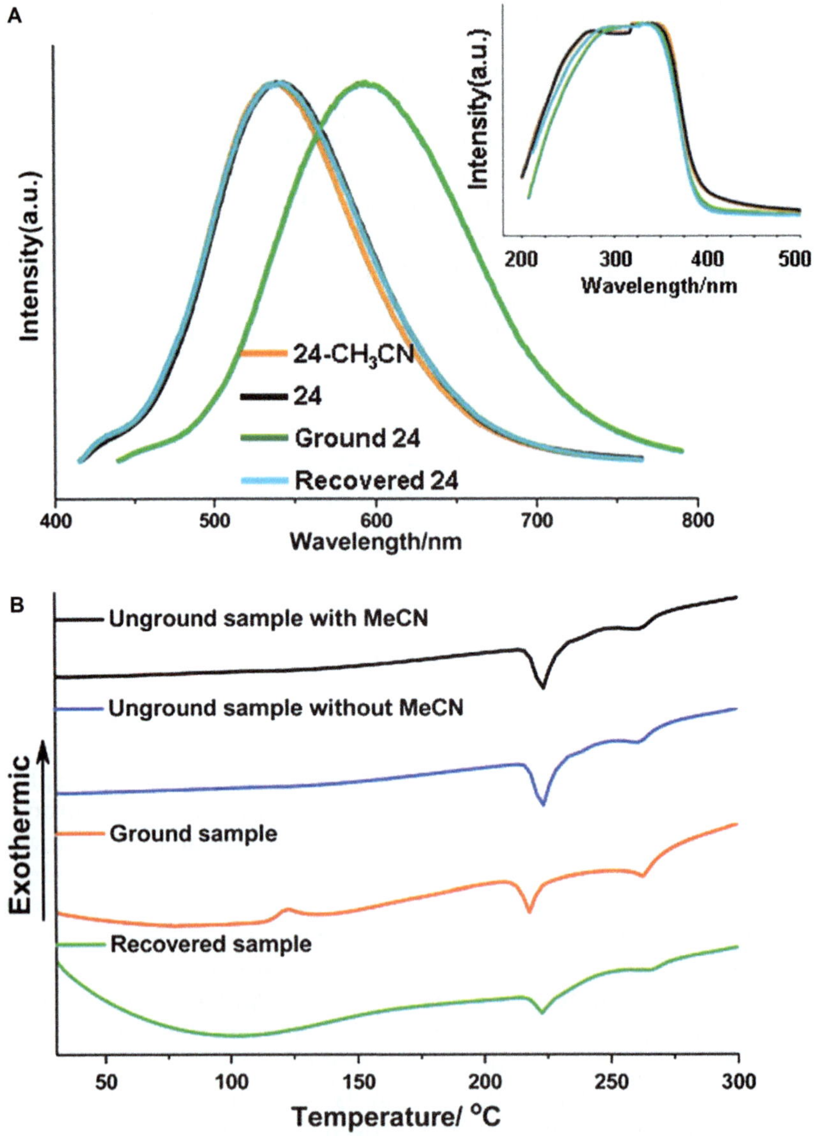

Figure 3.15 (A) The emission spectra ($\lambda_{ex} = 355$ nm) of compound **24**·CH_3CN in different states. Inset: their corresponding UV-vis absorption spectra. (B) DSC measurements of compound **24**·CH_3CN in different states. Reproduced with permission from reference 38. Copyright 2013, The Royal Society of Chemistry.

12 hours to remove all the solvents, and it exhibited the same mechanochromic performance as before heating, and the PXRD pattern of the solvent-removed sample was coincident with the previous one, both indicating that the solvents were not responsible for the mechanochromism.

By treating with CH_3CN, the recovered ground sample returned to its original structural arrangement and luminescent properties even without CH_3CN in it. It was concluded that the guest CH_3CN determines the molecular packing modes, but losing it had no influence on either the crystal structures or the photoluminescent properties.

The XRD pattern of **24**·CH_3CN exhibited diffraction peaks ascribed to well-defined microcrystalline structures, and in contrast, the ground sample showed only weak and ambiguous reflections, indicating that the crystal lattice was significantly disrupted and there was a crystal-to-amorphous phase conversion caused by grinding. Upon treating the ground sample with CH_3CN, reversion from the amorphous to the crystalline phase occurred as the reflection peaks were restored. The DSC measurements showed that the unground **24**·CH_3CN and recovered samples exhibited a similar melting point at around 224 °C. However, the ground sample showed another exothermic process at 122 °C besides a lower melting point at 218 °C (Figure 3.15B). The newly observed exothermic peak should be ascribed to a phase-to-phase transition that converts the thermodynamically stable crystalline phase to a metastable one, which seems a common feature for most of the mechanochromic compounds. Thus, the authors believed that the mechanochromism transitions which represent switching between the crystalline and amorphous phases were likely to be driven by stacking mode changes by the change of external stimuli-induced cuprophilic interactions.

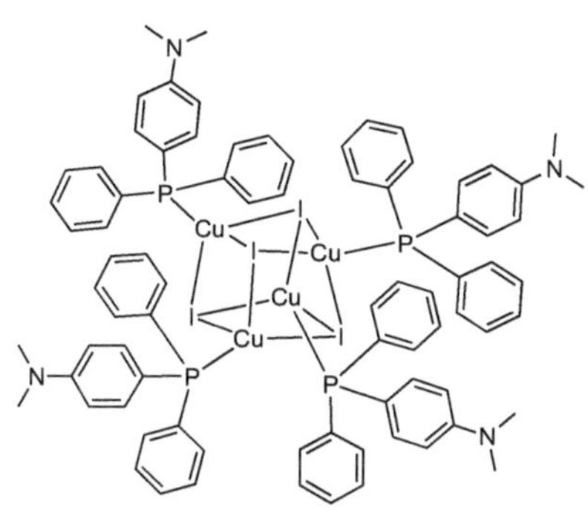

24

Jiang *et al.* (2013)[39] synthesized a copper complex cluster $[Cu_8I_8(4\text{-dpda})_6]_n$ (**25**, Figure 3.16A), which crystallized in the ethanol solvent of mixed CuI and 4-(diphenylphosphino)-N,N-dimethylaniline (4-dpda). When the solid samples of **25** were ground in a mortar, the white powder obtained exhibited orange luminescence (~630 nm) which could be reverted to the original green luminescence (530 nm) after treating it with drops of ethanol and drying (Figure 3.16B). The dynamic recovery of the emission color between green and orange was attested with repeating cycles of grinding and treating the sample with solvent. The XRD results indicated that after grinding the crystal lattice was significantly disrupted and there was a

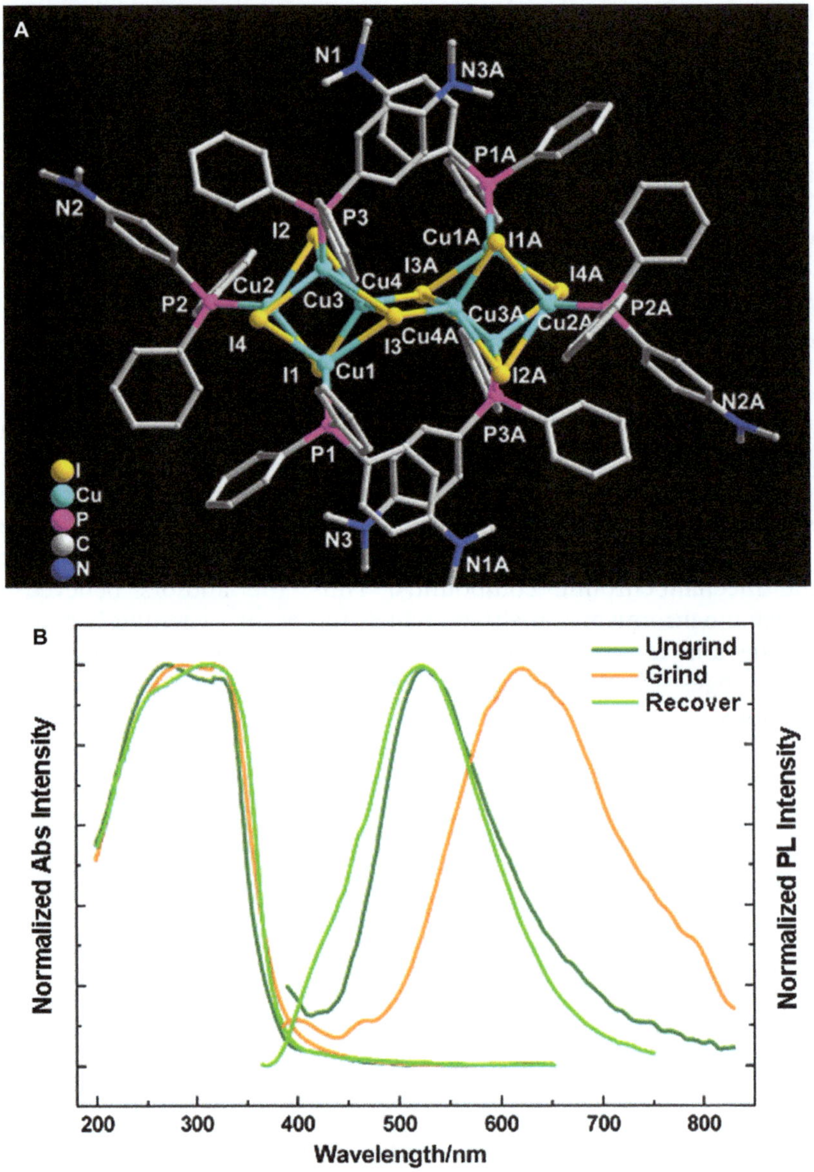

Figure 3.16 (A) The monomer structure of **25** with a Cu_8I_8 cluster. (B) UV-vis absorption spectra of the unground, ground, and recovered solid samples (left) and their emission spectra (right).
Reproduced with permission from reference 39. Copyright 2013, American Chemical Society.

crystal-to-amorphous phase conversion. According to the single-crystal X-ray analysis, the authors thought that when **25** was in the crystalline state, the monomer clusters were likely to stack and form a planar "molecular sheet" with the help of π-π interactions. After the crystalline **25** was ground, the regular structure was disrupted by pressure, which led to the arrangement shifts and the molecules moving disorderly. As the molecular packing changes, the molecules got more random and closer to each other so as to extrude the monomer clusters and make the Cu-Cu distance shorter. As a result, the Cu-Cu interactions inside the Cu_8I_8 monomer increased, which led to a ~100 nm bathochromic shift in the amorphous state as compared with the crystalline state.

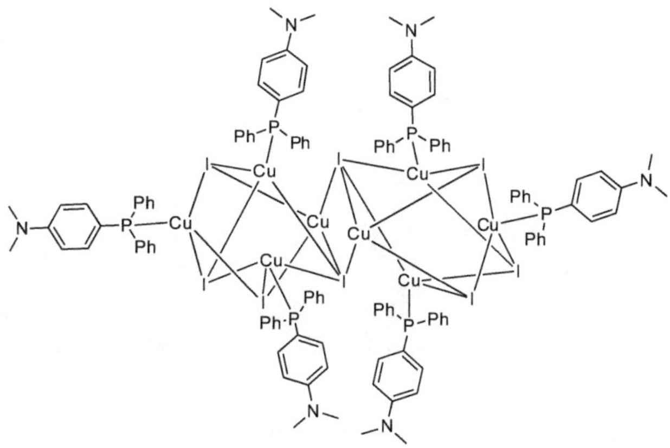

25

Zhang *et al.* (2013)[40] reported a multifunctional, one-dimensional Cu(I) coordination polymer $[Cu(dm\text{-}bim)]_n$ (**26**, dm-bim = 5,6-dimethylbenzimidazolate, Figure 3.17A) with unusual mechanochromic, sensing of nitrobenzene and photocatalytic properties for degradation of organic dyes. Obvious mechanical-grinding triggered-luminescence change from bright green (λ_{em} = 539 nm) to wine red (λ_{em} = 700 nm) was observed for this complex (Figure 3.17B and C). The emission of **26** could be tentatively assigned to the [Cu-π*(dm-bim)] metal-to-ligand charge transfer (MLCT). The powder X-ray diffraction pattern revealed that there were only very weak and ambiguous reflections, indicating that the crystal lattice was significantly disrupted and there was a crystal-to-amorphous phase conversion caused by the strong grinding. Since compound **26** has a helical chain structure, the disorder slide between chains, accompanying the alteration of Cu-Cu interactions and even π-π interactions in the structure, may respond to this luminescent change. As described above, many ground mechanochromic metal complexes can recover their original crystalline phase and emission properties upon exposure to solvent or heating. However, the ground sample **26** was found to be very difficult to recover to the crystalline state in several days by exposure to some organic solvents, such as toluene, ethanol or chloroform *etc.* The authors believed that this was probably due to the fact that the polymer could not dissolve in organic solvents, and the recrystallization was hardly carried out as other metal complexes did. However, after one month, the crystalline sample of **26** was found to successfully recover

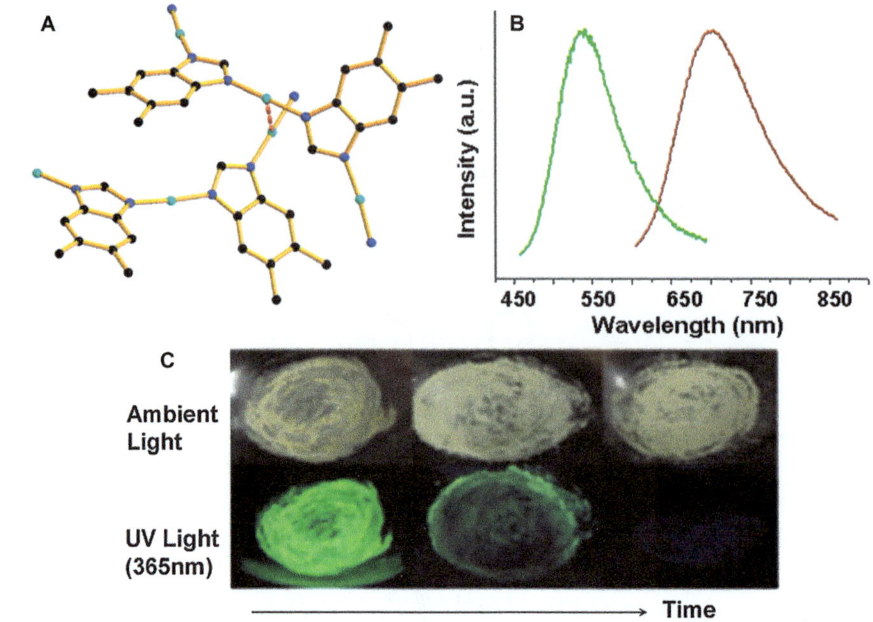

Figure 3.17 (A) The coordination environment in **26**, showing Cu···Cu interaction (red dashed line). (B) Normalized solid-state emission spectra ($\lambda_{ex} = 365$ nm) for the unground (green line) and ground (red line) samples of **26**. (C) photographs of sample **26** show luminescence changes with grinding time at room temperature under ambient light and 365 nm UV lamp, respectively.

from a mixed chloroform and benzene solvent, also demonstrating the inherent self-restore function of coordination polymers.

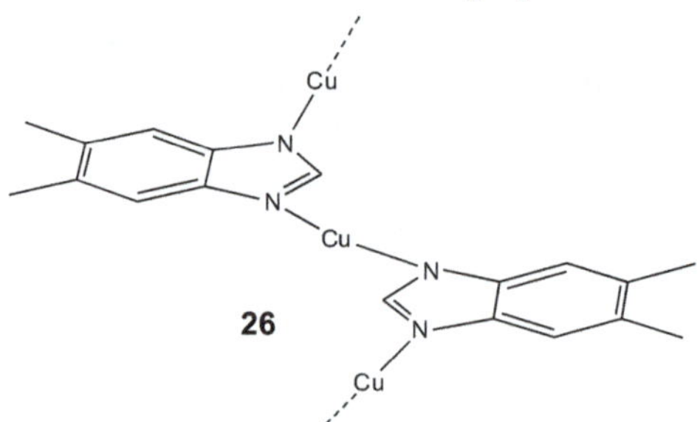

3.2.5 Ag(I) Complexes

Tsukuda *et al.* (2010)[41] obtained an Ag(I) complex **27**, which contained *o*-bis(diphenylphosphino)benzene. This complex showed a reversible luminescent mechanochromism. The colorless crystals obtained by

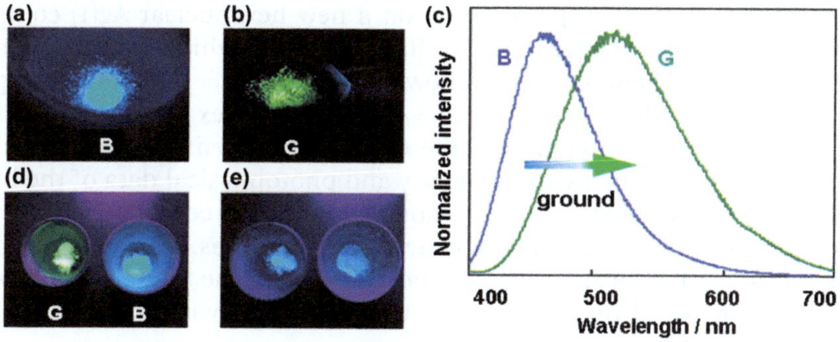

Figure 3.18 Emitting color on irradiation of UV-light (365 nm) of B-form (a) and G-form (b) and luminescence spectra of B-form and G-form (c); emitting color of G-form and B-form before heating (d) and after heating at 200 °C for 10 min (e).
Reproduced with permission from reference 41. Copyright 2010, The Royal Society of Chemistry.

recrystallization from $CHCl_3$-ether showed blue emission (B-form) upon UV light irradiation. When ground in a ceramic mortar, the obtained white powder emitted green luminescence (G-form). The maximum emission of the G-form occurred at 518 nm, whereas that of the B-form occurred at 458 nm. Upon treatment of the ground sample with drops of $CHCl_3$-hexane (1 : 2) or upon heating at 200 °C for 10 min, the G-form reverted to the B-form (Figure 3.18). The XRD results showed that reversible phase conversion occurred from crystalline to amorphous upon grinding and heating. The synthesis of the complex in CH_3CN instead of in $CHCl_3$ yielded needle-like crystals with a green emission (which has the same formula as the B-form). From the crystal data of the B-form, the distances between the carbon atoms of the phenylene rings in the adjacent molecules are all 3.55 Å, suggesting that noticeable intermolecular interactions occur between the phenylene rings of the ligand. Such interactions were not observed in the G-form because the shortest distance between phenylene rings was longer (7 Å). Thus, this research proposes that mechanical grinding leads to a disruption of the intermolecular interactions between the phenylene rings of adjacent molecules in the G-form solid, which results in the green emission. The π-π interactions between the aromatic rings lead to a blue-shift of the emission wavelength, which was considered uncommon.

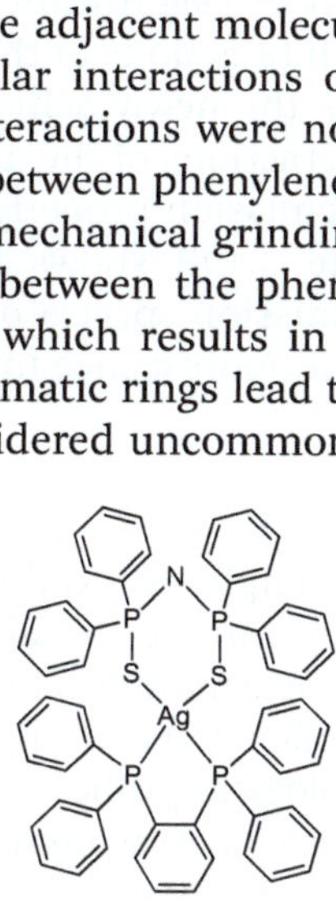

27

Babashkina *et al.* (2011)[42] reported on a new hexanuclear Ag(I) complex [{Ag_3(iPrNHC(S)NP(S)(OiPr)$_2$-S,S')$_3$}$_2$] ([(Ag_3L_3)$_2$], **28**), which exhibited reversible conversion between the yellow-emitting ($\lambda_{em} = 573$ nm) complex [(Ag_3L_3)$_2$] and the blue-emitting ($\lambda_{em} = 469$ nm) complex upon grinding and recrystallization. A comparison of the elemental analysis, infrared, nuclear magnetic resonance, mass spectrometry and photophysical data of the blue-emitting complex powder and **28** showed that these compounds had the same composition and exhibited structural similarities. Thus, the authors proposed a [Ag_3L_3] formulation for the complex of the white powder with blue emission based on less nuclearity as compared with the hexanuclear complex [(Ag_3L_3)$_2$].

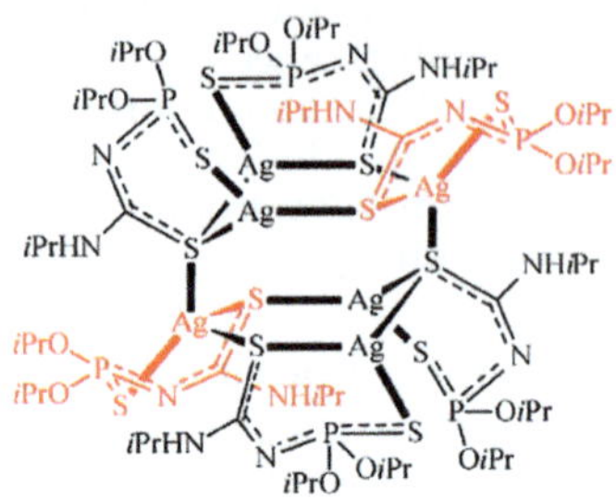

[(Ag_3L_3)$_2$] **28**

3.2.6 Al(III) Complexes

Zhang *et al.* (2011)[43] reported a green emissive ($\lambda_{em} = 510$ nm) Alq_3 solid with a facial isomeric, which was formed by grinding the typical blue luminescent ($\lambda_{em} = 470$ nm) fac-Alq_3 (**29**) crystalline powder. This powder showed clear mechanofluorochromic behaviors (Figure 3.19A). However, the ground mer-Alq_3 (**30**) did not show a significant color change under UV light. The enhanced intermolecular interactions of the quinoline ligands could be the key factor that determines the newly achieved luminescence of **29**. Although the intermolecular interactions of the amorphous **29** powder cannot be accurately determined, its emission property may be strongly dependent on molecular packing. Figure 3.19B shows the pattern in the fac-Alq_3 nanostructure film and the emission spectra of the patterned fac-Alq_3 film examined by the laser confocal fluorescence microscope.

3.2.7 Ir(III) Complexes

Organometallic iridium(III) complexes exhibiting high phosphorescence efficiency and rich photophysical properties as well as superior optical stability have emerged as promising candidates for high-tech applications such as organic light-emitting diodes (OLED), light-emitting electrochemical cells, chemical sensor and biological probes, *etc.* Recently, a series of Ir(III) complexes have been found to possess remarkable reversible mechanochromic luminescent property, and new opportunities for Ir(III) complexes in optical recording could be well expected.[44–48]

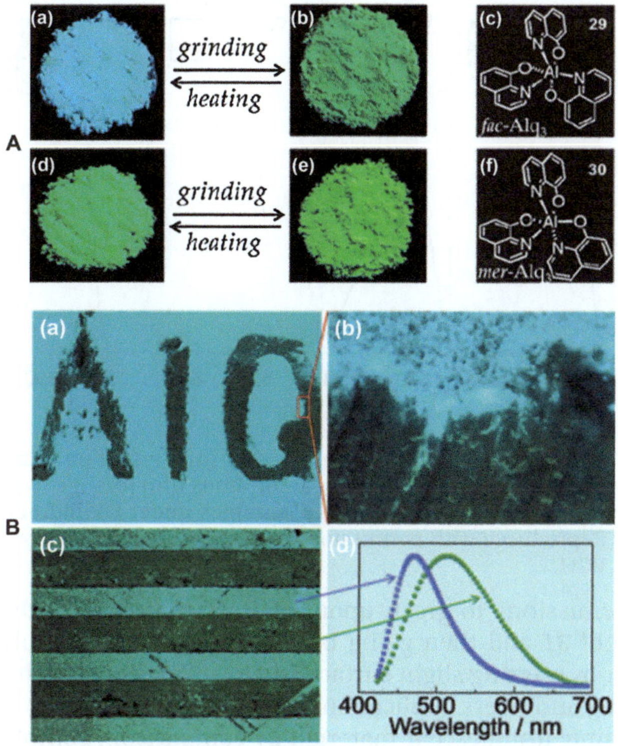

Figure 3.19 (A) Mechanofluorochromic behaviors and molecular conformations of fac-Alq_3 (**29**) and mer-Alq_3 (**30**). (B) Fluorescence microscope images of (a) fac-Alq_3 film with A, L, Q patterned characters; (b) joint area of the fac-Alq_3 film and patterned area; (c) laser confocal fluorescence microscope images and (d) emission spectra of ground and unground areas in the micro-scale patterned thin film.
Reproduced with permission from reference 43. Copyright 2011, The Royal Society of Chemistry.

Ghedini *et al.* (2010)[49] reported two Ir(III)-based complexes (**31** and **32**) showing high phosphorescence luminescence in their solid state. The introduction of long alkoxy chains on the bipyridine ligand in **31** induced mesomorphism. Thus, a dynamic functional material with properties modulated by external stimuli was obtained as the first Ir mechanochromic complex reported. The kinetically favored mesophase of **31** was preferentially formed from the isotropic phase under fast cooling and was characterized as a columnar hexagonal phase at room temperature. Yellow phosphorescence at 560 nm was observed. Starting from the isotropic phase, a crystalline phase characterized by a bright green emission of 520 nm was obtained upon slow cooling. Moreover, an amorphous thin film could be obtained by spin coating solutions of **31**. The photophysical properties of this thin film showed that the high molecular disorder achieved in this condensed phase resulted in a further shift in the emission wavelength to 580 nm, turning it into orange-red (Figure 3.20). These thin films could

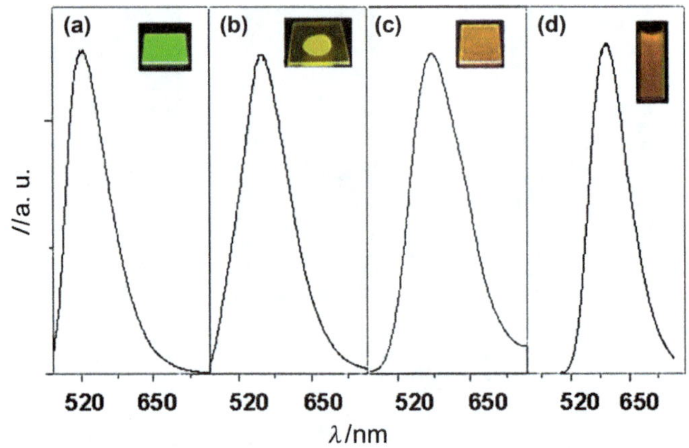

Figure 3.20 Emission spectra of complex **31**: (a) crystalline film (green), (b) mesophase film (yellow), (c) amorphous film (orange-red), (d) dichloromethane solution; insets: real samples under UV illumination. Reproduced with permission from reference 49. Copyright 2010, Wiley-VCH.

switch their emissions to green upon heating, returning to the solid crystalline phase of **31** and then going back to 580 nm by just rubbing with a velvet cloth or by exerting slight surface friction. This color-tuning process by surface heating and stress application is fully reversible and is indicative of a mechanochromic luminescent material. By comparison, complex **32** has no mechanochromic property. The XRD measurement of **32** revealed the absence of strong interchromophoric interactions such as π-π stacking in the crystal packing, thus illustrating that the Ir(III) cations behaved in a manner similar to that of isolated molecules. The presence of long alkoxy chains on the bipyridine ligand in complex **31** causes further aggregation with respect to complex **32**, which results in the differences in emissions.

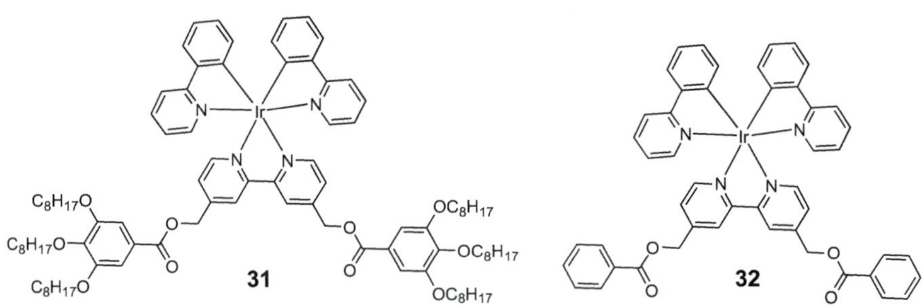

Su *et al.* (2012)[50] reported the synthesis of two new cationic iridium complexes **33** and **34** with mechanochromic luminescent properties. Upon grinding, the emission colors of these complexes showed reversible interconversion from blue (λ_{em} = 461 nm) to blue-green (λ_{em} = 493 nm) for **33** and from green (λ_{em} = 515 nm) to orange (λ_{em} = 560 nm) for **34** (Figure 3.21). In addition, the crystalline structure analyses showed weak C-H···π

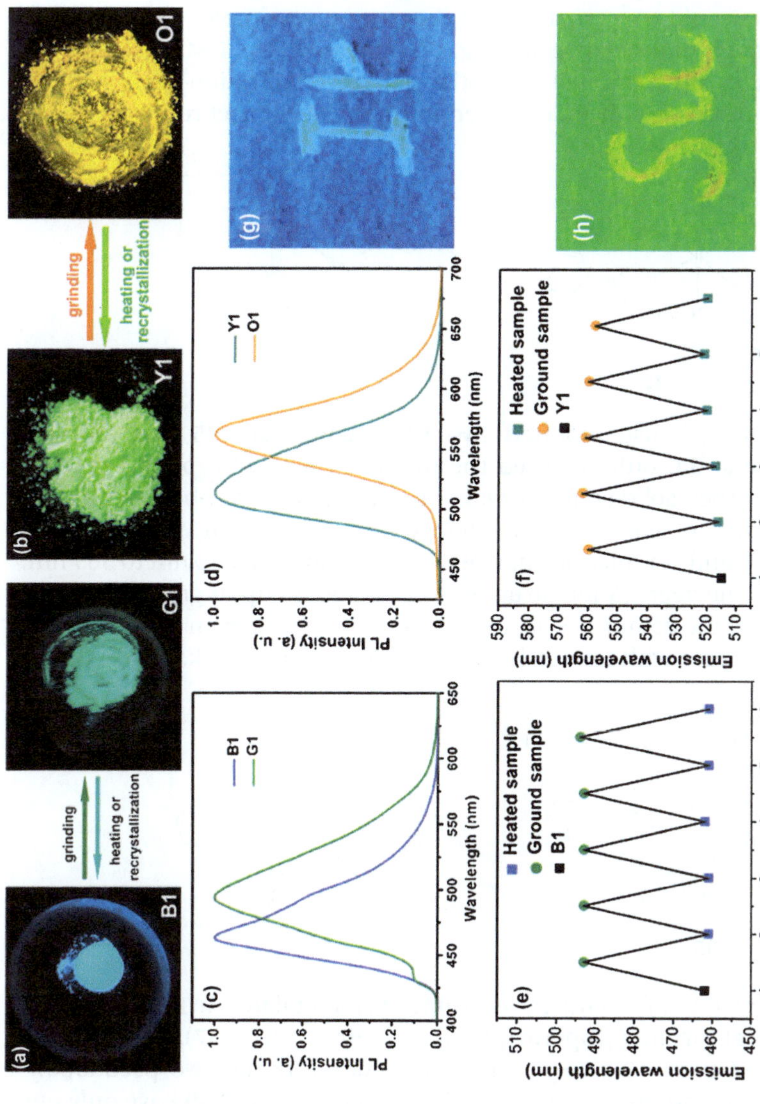

Figure 3.21 Emitting color upon irradiation of UV-light of complexes **33** (a) and **34** (b); the luminescence spectra of complexes **33** (c) and **34** (d); repeated cycles of the mechanochromism of **33** (e) and **34** (f); **33** (g) and **34** (h) were cast on the filter paper and the letters "Ir" and "Su" were written with a spatula under UV-light at room temperature. Reproduced with permission from reference 50. Copyright 2012, The Royal Society of Chemistry.

intermolecular interactions, as well as dimer-like structures in the packing diagram. It was considered that the molecular packing was relatively loose because of the weak intermolecular interactions. Thus, the ordered crystalline structures can be easily collapsed by the external pressure. The XRD results showed that the original sample consisted of highly ordered crystalline aggregates, with the amorphous states forming only after grinding. Heating or recrystallization could convert the amorphous ground samples to the crystalline states through molecular repacking. Upon heating, the DSC curves of the amorphous states also exhibited an exothermic recrystallization peak, such an exothermic transition has been observed in many mechanochromic cases.

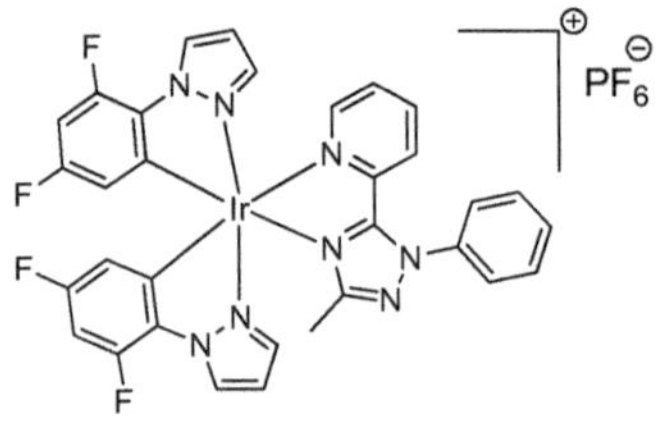

33

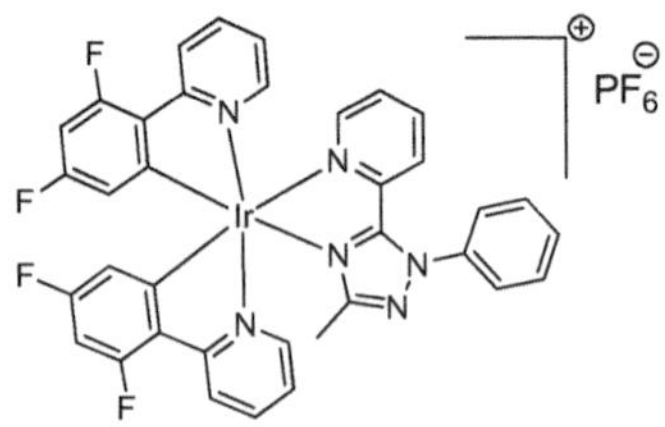

34

Su *et al.* (2012)[51] also reported two Ir(III) complexes **35** and **36**, very similar to **33** and **34**, only replaced the methyl groups with phenyl groups. After ground the solid-state powder samples, the sky-blue emission ($\lambda_{em} = 471$ nm) of complex **35** changes to blue-green luminescence ($\lambda_{em} = 499$ nm); and complex **36** changes its emission from 542 nm to 563 nm. The change in the aggregation structures of **35** and **36** upon grinding is also similar to that of the complexes **33** and **34**, *i.e.*, conversion from the crystalline to amorphous states with an exothermic recrystallization peak.

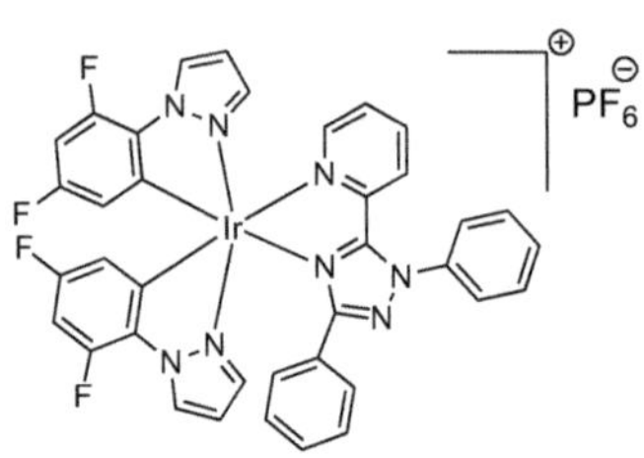

35

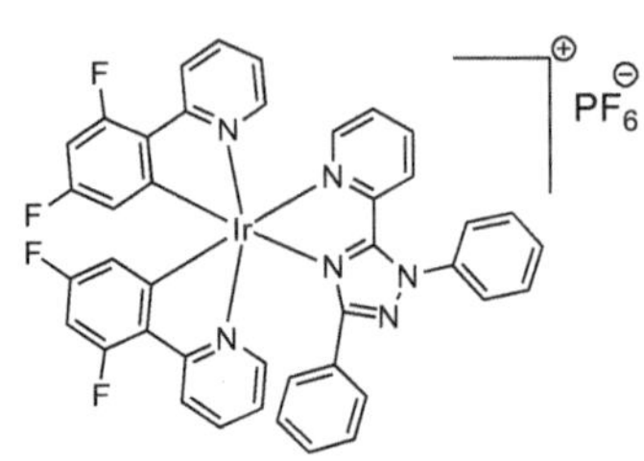

36

Talarico *et al.* (2012)[52] synthesized three cyclometalated Ir(III) complexes with the general formula $[(ppy)_2Ir(pam)]X$, where $X = Cl^-$ (**37**), PF_6^- (**38**), or ClO_4^- (**39**), and pam = 2-picolylamine. Compound **37** has two pseudopolymorphs (solvated and non-solvated species), whereas **38** and **39** have only one form. For the non-solvated species of **37**, the initial multiple peaks luminescence ($\lambda_{em} = 475$, 508 and 605 nm) could be almost completely converted into green ($\lambda_{em} = 508$ nm) after grinding. Although the initial luminescence colors of the crystals **38** and **39** are almost the same, a green emission at ~605 nm, they exhibit a dual emission, at 508 and 605 nm for **38**, and 508 and 590 nm for **39** (Figure 3.22). The authors suggested that the dual

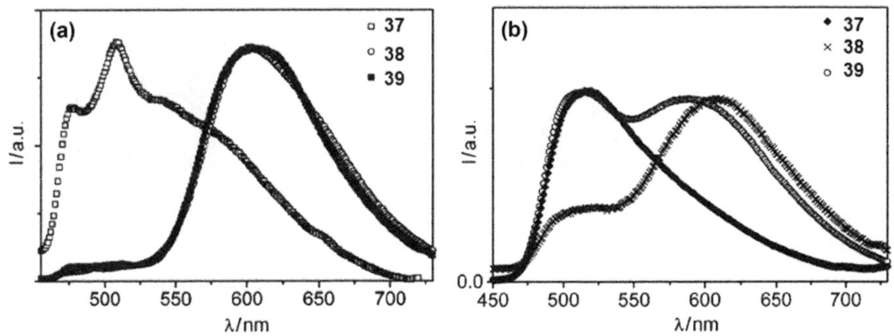

Figure 3.22 (a) Emission spectra of the initial (a) and ground (b) samples. Sample **37** is the non-solvated species.
Adapted with permission from reference 52. Copyright 2012, The Royal Society of Chemistry.

emission spectra were most likely attributed to the presence of crystallites of two different emission centers, one on the surface and the other in the crystallite bulk. The continuous grinding enhanced the surface area, which increased the surface site contribution to the light emission without completely disrupting the initial supramolecular organization.

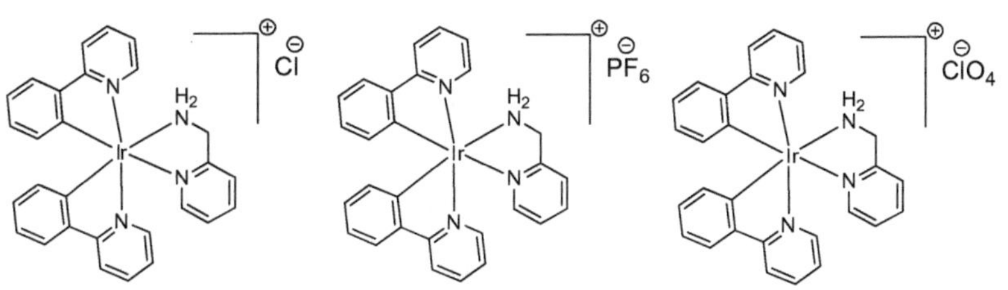

37 **38** **39**

In some organic mechanofluorochromic compounds, it has been demonstrated that grafts of different long alkyl chains on their molecular structures can not only construct a series of new mechanofluorochromic materials, but also provide an efficient way to study the relationship between the molecular structures and mechanofluorochromic behaviors.[53–56]

Su *et al.* (2013)[57] designed and prepared a series of iridium(III) complexes with different length of alkyls (**40a–e**). All the complexes exhibit strong sky-blue emission with peak wavelength of 461–472 nm. When these solid state powders, obtained through column chromatography (namely, as-synthesized powders), were ground in a ceramic mortar, naked-eye visible luminescent color change could be clearly observed (Figure 3.23A, Table 3.2). The grinding-induced spectral shifts (Δλ) of samples **40a–40e** are alkyl length-dependent, although they all exhibit obvious mechanochormic luminescence behavior. The Δλ for **40a, 40b, 40c, 40d** and **40e** are 17, 20, 21, 10 and 8 nm, respectively. The sample **40e** with a longest alkyl chain among these samples exhibits the smallest red shift in emission. The result also suggests that alkyl chains play a functional role in turning the

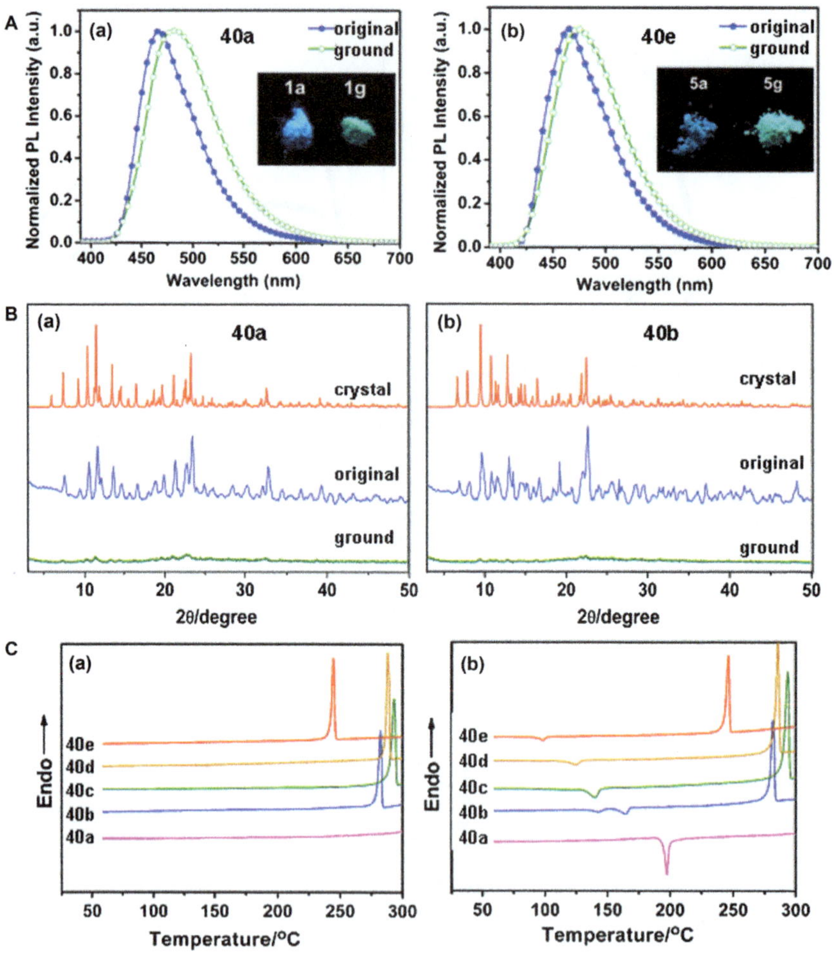

Figure 3.23 (A) Emission spectra of complexes **40a** and **40e** in the solid states. (B) XRD patterns of **40a** and **40b** under different states. (C) DSC curves of as-synthesized samples (a) and ground samples (b).
Adapted with permission from reference 57. Copyright 2013, Elsevier.

Table 3.2 Peak emission wavelengths (λ, in nm) of **40** series samples upon grinding.

Compounds	$\lambda_{pristine}$	λ_{ground}	$\Delta\lambda^{a}$
40a	466	483	17
40b	463	483	20
40c	461	483	22
40d	472	482	10
40e	465	473	8

[a] $\Delta\lambda = \lambda_{ground} - \lambda_{pristine}$

mechanochromic luminescent behavior and the order of color changes induced by grinding is similar to the previously reported organic fluorescent mechanochromic compounds.

The powder X-ray diffraction experiments were performed on both the as-synthesized and ground samples and revealed that the as-synthesized samples with sharp and intense reflections should be crystalline and the ground samples with weak and broad XRD pattern should be amorphous (Figure 3.23B). The differential scanning calorimetry studies further confirm the formation of the amorphous powders upon grinding, that is, for the ground samples, each DSC curves firstly appeared an exothermic recrystallization peak (Figure 3.23C), which is a sign for many mechanochromic luminescent compounds. For the as-synthesized samples, there are no such transitions, because they are crystalline, demonstrating that the amorphous ground samples are all in the metastable states. Therefore, if the ground samples are heated above their recrystallization temperatures, they will restore to thermodynamically stable crystals through the exothermic recrystallization process, showing reversibility upon grinding and heating.

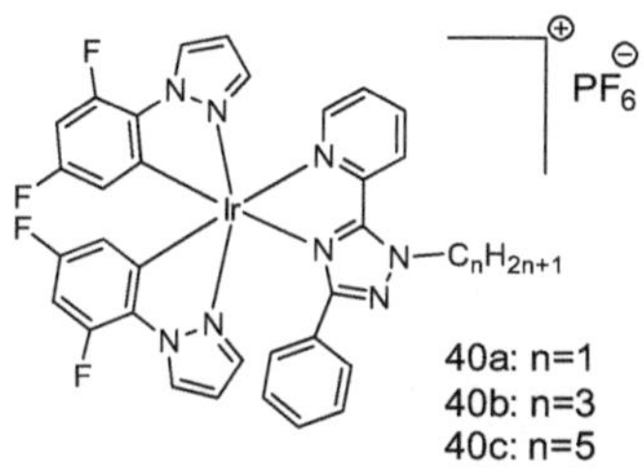

40

3.2.8 Heterodinuclear Complexes

All the complexes described above are homometallic dinuclear or multinuclear. How about the electronic effects of metal-metal interactions on the mechanochromic behavior if different metal ions are introduced into the same motif? The heterodinuclear complexes (**41**), *syn*-[MPt(μ-pyt)2(bpy)2]$^{n+}$ (*syn*-[MPt], M = Pd^{2+}, Au^{3+}, Hpyt = pyridine-2-thiol, bpy = 2,2′-bipyridine), were synthesized by Kato *et al.* (2013)[58] using stepwise complexation. In addition to the vapochromic response, the heterodinuclear complexes were also found to exhibit mechanochromic behavior. When they were ground in an agate mortar, distinct color changes occurred even in the absence of vapor, from orange to red for *syn*-[PdPt], and from light-red to dark-red for syn-[PtPt] (Figure 3.24A). After grinding, the samples did not exhibit any diffraction peaks, indicating that the desorbed crystalline forms of *syn*-[PtPt] and *syn*-[PdPt] were transformed to amorphous forms. The amorphous states were also obtained by grinding the vapor-included forms (Figure 3.24B). This indicates that the exhibiting of mechanochromic behaviors is also resulted through the crystal-to-amorphous transformation by grinding. The metal-metal interactions often occur in the amorphous state by grinding, forming short, local metal-metal contacts. The vapor-included forms were reconstructed on exposure the ground samples to CH_3CN vapor, also showing a good reversibility. It is noteworthy that structural

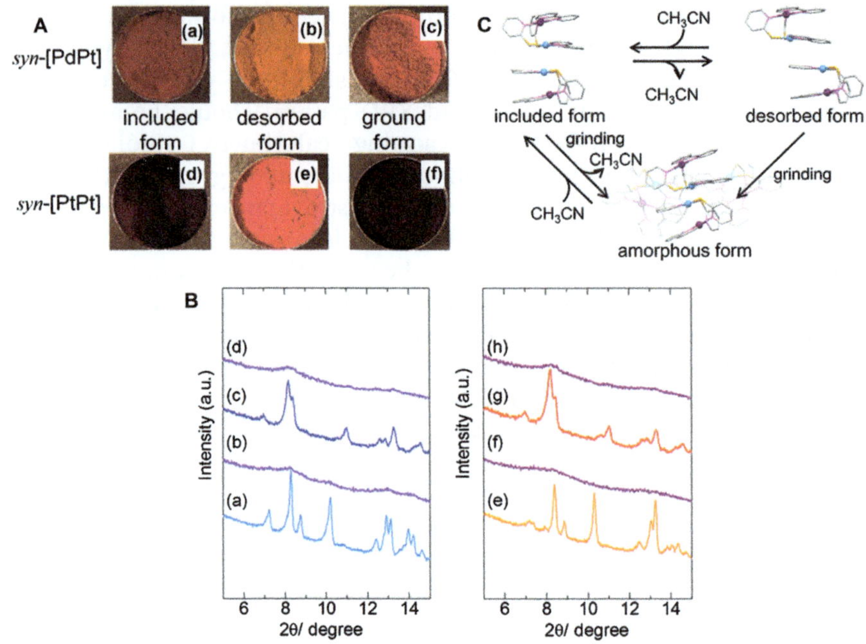

Figure 3.24 (A) Photographs of powder samples: (a and d) vapor-included forms; (b and e) desorbed forms; and (c and f) ground forms, for *syn*-[PdPt] and *syn*-[PtPt], respectively. (B) Changes in PXRD patterns by grinding the samples for *syn*-[**PtPt**] (a–d) and *syn*-[**PdPt**] (e–h): (a) and (e) desorbed forms (starting materials), (b) and (f) after grinding, (c) and (g) after exposure of acetonitrile vapor to ground samples, (d) and (h) re-grinding. (C) Structural transformations of the dimer-of-dimer motif of *syn*-[MPt] (M = Pd^{2+} and Pt^{2+}) induced by vapor and grinding. Adapted with permission from reference 58. Copyright 2013, The Royal Society of Chemistry.

reconstruction was not observed by heating to 50 °C, at which the structural transformations occurred from the included to the desorbed forms. The authors considered that the vapor molecules played an essential role in controlling the structural construction and transformation. The overall chromic behaviors of these binuclear complexes are summarized in Figure 3.24C.

41

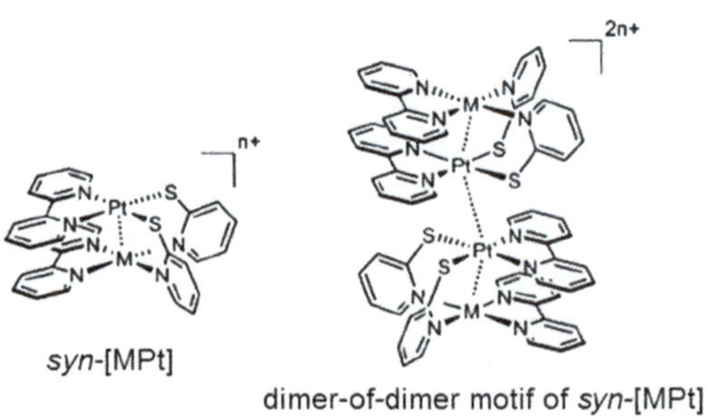

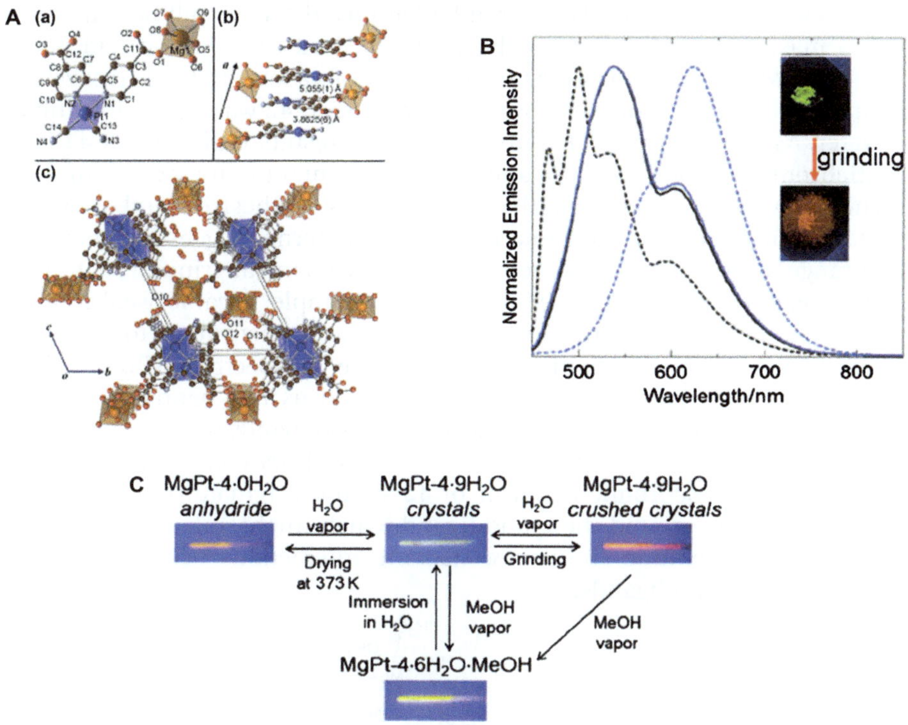

Figure 3.25 (A) (a) Molecular structure, (b) 1-D stacked structure, and (c) packing diagram viewed down along the *a* axis of **42**. The coordination spheres of the Pt(II) and Mg(II) ions are shown as blue planes and orange octahedrons, respectively. H atoms are omitted for clarity. Solvated water molecules are omitted in (a) and (b) for clarity. The brown, light-blue, and red spheres represent C, N and O atoms, respectively. (B) Luminescence spectral changes in **42** by mechanical grinding and exposure to MeOH vapor ($\lambda_{ex} = 400$ nm). The black and blue lines represent the spectra of as-synthesized and crushed crystals, respectively, whereas the dotted and solid lines represent the spectra before and after exposure to MeOH vapor for 1 day at room temperature, respectively. Inset: luminescence images of as-synthesized and ground samples. (C) Schematic diagram summarizing the multichromic behavior of **42**. Reproduced with permission from reference 59. Copyright 2012, The Royal Society of Chemistry.

Kato *et al.* (2012)[59] reported a heterodinuclear coordination polymer $[Mg(H_2O)_5]$-$[Pt(CN)_2(4,4'\text{-dcbpy})] \cdot 4H_2O$ (**42**, Figure 3.25) prepared by reactions of a Pt(II)-diimine-based metalloligand $Na_2[Pt(CN)_2(4,4'\text{-dcbpy})]$ ($4,4'\text{-}H_2dcbpy = 4,4'$-dicarboxy-2,2′-bipyridine) with $Mg(NO_3)_2 \cdot 6H_2O$ in aqueous solution. Complex **42** exhibited vapochromic and mechanochromic behaviors. The luminescence spectrum of the as-synthesized crystals changed from a greenish yellow to a yellow emission centered at 537 nm, with a shoulder at 607 nm after exposure to MeOH vapor. The color and emission also changed to a reddish orange emission centered at 624 nm after mechanical grinding of the crystals, as shown in Figure 3.25B. After exposing the crushed sample to MeOH vapor, it had almost the same yellow emission, with a maximum at 537 nm, as the as-synthesized crystals that had been exposed to MeOH vapor. The original emission of **42** was gradually recovered by immersing these samples in water, even though **42** was nearly insoluble in water. As summarized in Figure 3.25C, these results indicate that **42** is a multichromic (vapo-, mechano- and thermochromic) luminescent complex.

The XRD analysis revealed that both the as-synthesized and crushed crystals showed almost the same diffraction patterns, suggesting that the long-range structure of **42** was retained after mechanical grinding. But, the peak intensities and widths in the crushed sample became weaker and broader, indicating that the crystalline **42** was transformed to a partly amorphous solid by mechanical grinding. The IR spectrum showed that ν(C≡N) mode of the cyano ligand was observed as two bands in the as-synthesized crystals, whereas it was observed as two bands with a shoulder at 2138 cm^{-1} in the crushed sample. Thus, the authors considered that the observed mechanochromic behavior of **42** should originate from the local structural change around the cyano ligand which might enhance the intermolecular metallophilic and/or π-π stacking interactions between the adjacent Pt(II) metallo ligands.

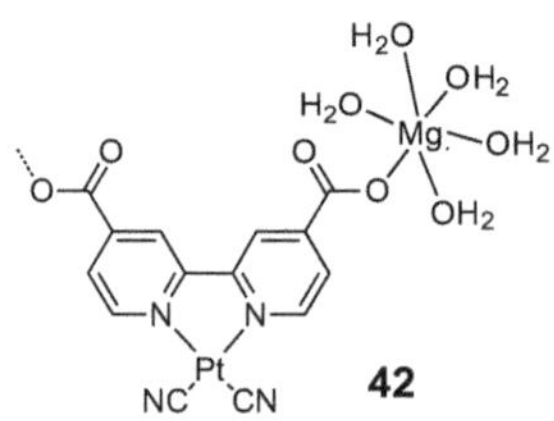

42

3.2.9 AIE Complexes

As the number of reported mechanochromic luminescent AIE complexes to date are still limited, and some of the AIE complexes (**43, 44** and **45** in Chapter 4) will be described in Chapter 6 about AIE mechanochromic luminescent compounds, we would not describe them here.[60–62] Through the reported instances, it is our profound belief that when the AIE unit(s) is/are introduced into a ligand to synthesize an AIE complex with metal ions, the AIE complex can be expected to possess mechanochromic luminescent properties. Due to the attractive properties and potential applications of organometallic or coordination compounds containing AIE moieties, it can be expected that more and more mechanochromic luminescent AIE complexes are to be exploited.

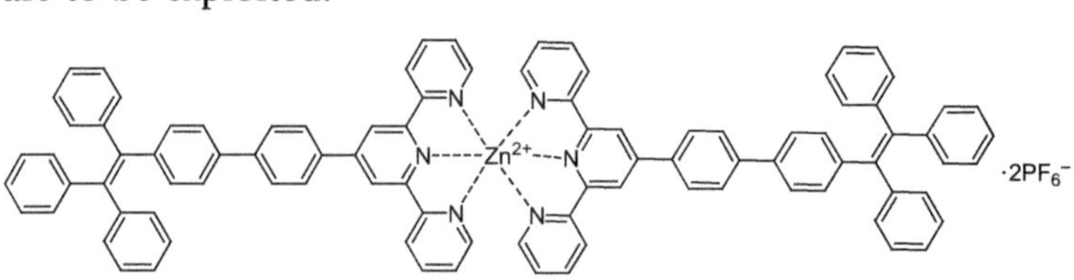

43

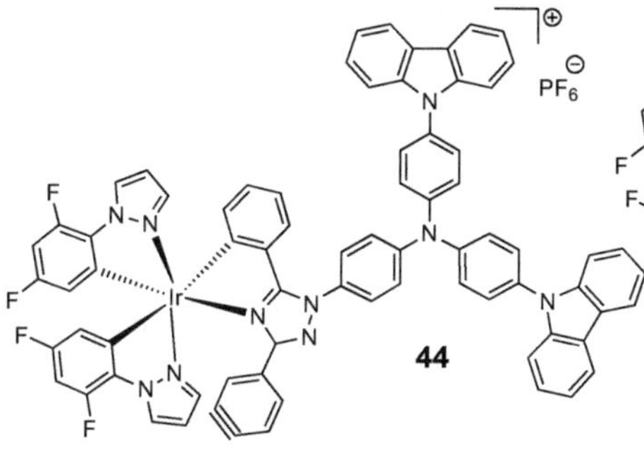

44

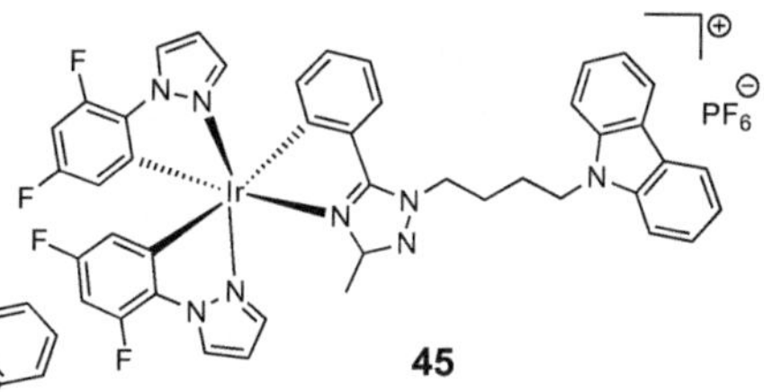

45

3.3 Conclusion

In this chapter, recent research on mechanochromic luminescent metal complex materials has been presented. The reported mechanochromic luminescent metal complex systems include Zn(II), Au(I), Pt(II), Ag(I), Cu(I), Al(III), Ir(III) and heterodinuclear complexes. The structure-property relationships, as well as the mechanisms proposed for the mechanochromic luminescent metal complexes have also been described. Finally, we point out that AIE complexes are expected to become a rich source of mechanochromic luminescent complexes.

References

1. A. L. Balch, *Angew. Chem. Int. Ed.*, 2009, **48**, 2641–2644.
2. Y. Sagara and T. Kato, *Nat. Chem.*, 2009, **1**, 605–610.
3. K. W. Galloway, S. A. Moggach, P. Parois, A. R. Lennie, J. E. Warren, E. K. Brechin, R. D. Peacock, R. Valiente, J. González, F. Rodríguez, S. Parsons and M. Murrie, *CrystEngComm*, 2010, **12**, 2516–2519.
4. X. Q. Zhang, Z. G. Chi, Y. Zhang, S. W. Liu and J. R. Xu, *J. Mater. Chem. C*, 2013, **1**, 3376–3390.
5. S. Mizukami, H. Houjou, K. Sugaya, E. Koyama, H. Tokuhisa, T. Sasaki and M. Kanesato, *Chem. Mater.*, 2005, **17**, 50–56.
6. B. Tzeng, T. Chang and H. Sheu, *Chem. Eur. J.*, 2010, **16**, 9990–9993.
7. B. Tzeng, T. Chang, S. Wei and H. Sheu, *Chem. Eur. J.*, 2012, **18**, 5105–5112.
8. P. Pyykkö, *Chem. Rev.*, 1997, **97**, 597–636.
9. V. W. W. Yam, E. C. C. Cheng and K. K. Cheung, *Angew. Chem., Int. Ed.*, 1999, **38**, 197–199.
10. H. Schmidbaur, *Nature*, 2001, **413**, 31–33.
11. B.-C. Tzeng, A. Schier and H. Schmidbaur, *Inorg. Chem.*, 1999, **38**, 3978–3984.
12. D. B. Leznoff, B.-Y. Xue, B. O. Patrick, V. Sanchez and R. C. Thompson, *Chem. Commun.*, 2001, 259–260.
13. Y.-A Lee and R. Eisenberg, *J. Am. Chem. Soc.*, 2003, **125**, 7778–7779.
14. Z. Assefa, M. A. Omary, B. G. McBurnett, A. A. Mohamed, H. H. Patterson, R. J. Staples and J. P. Jr. Fackler, *Inorg. Chem.*, 2002, **41**, 6274–6280.

15. V. J. Catalano and S. J. Horner, *Inorg. Chem.*, 2003, **42**, 8430–8438.
16. Y.-A. Lee and R. Eisenberg, *J. Am. Chem. Soc.*, 2003, **125**, 7778–7779.
17. J. Schneider, Y.-A Lee, J. Pérez, W. W. Brennessel, C. Flaschenriem and R. Eisenberg, *Inorg. Chem.*, 2008, **47**, 957–968.
18. H. Ito, T. Saito, N. Oshima, N. Kitamura, S. Ishizaka, Y. Hinatsu, M. Wakeshima, M. Kato, K. Tsuge and M. Sawamura, *J. Am. Chem. Soc.*, 2008, **130**, 10044–10045.
19. M. Osawa, I. Kawata, S. Igawa, M. Hoshino, T. Fukunaga and D. Hashizume, *Chem. Eur. J.*, 2010, **16**, 12114–12126.
20. J. Liang, F. Hu, X. Lv, Z. Chen, Z. Chen, J. Yin, G. Yu and S. Liu, *Dyes Pigments*, 2012, **95**, 485–490.
21. I. Eryazici, C. N. Moorefield and G. R. Newkome, *Chem. Rev.*, 2008, **108**, 1834–1895.
22. W. Paw, S. D. Cummings, M. A. Mansour, W. B. Connick, D. K. Geiger and R. Eisenberg, *Coord. Chem. Rev.*, 1998, **171**, 125–150.
23. J. A. G. Williams, S. Develay, D. L. Rochester and L. Murphy, *Coord. Chem. Rev.*, 2008, **252**, 2596–2611.
24. Z. M. Hudson, C. Sun, K. J. Harris, B. E. G. Lucier, R. W. Schurko and S. Wang, *Inorg. Chem.*, 2011, **50**, 3447–3457.
25. W. Lu, B. X. Mi, M. C. W. Chan, Z. Hui, C. M. Che, N. Zhu and S. T. Lee, *J. Am. Chem. Soc.*, 2004, **126**, 4958–4971.
26. M. L. Muro, C. A. Daws and F. N. Castellano, *Chem. Commun.*, 2008, 6134–6136.
27. J. S. Field, C. D. Grimmer, O. Q. Munro and B. P. Waldron, *Dalton Trans.*, 2010, **39**, 1558–1567.
28. T. J. Wadas, Q. M. Wang, Y. J. Kim, C. Flaschenreim, T. N. Blanton and R. Eisenberg, *J. Am. Chem. Soc.*, 2004, **126**, 16841–16849.
29. V. Kozhevnikov, B. Donnio and D. Bruce, *Angew. Chem. Int. Ed.*, 2008, **47**, 6286–6289.
30. T. Abe, T. Itakura, N. Ikeda and K. Shinozaki, *Dalton. Trans.*, 2009, 711–715.
31. J. Ni, X. Zhang, Y. Wu, L. Zhang and Z. Chen, *Chem. Eur. J.*, 2011, **17**, 1171–1183.
32. J. Ni, X. Zhang, N. Qiu, Y. Wu, L. Zhang, J. Zhang and Z. Chen, *Inorg. Chem.*, 2011, **50**, 9090–9096.
33. X. Zhang, J. Wang, J. Ni, L. Zhang and Z. Chen, *Inorg. Chem.*, 2012, **51**, 5569–5579.
34. S. Choi, J. Kuwabara, Y. Nishimura, T. Arai and T. Kanbara, *Chem. Lett.*, 2012, **41**, 65–67.
35. H. D. Hardt and A. Z. Pierre, *Anorg. Allg. Chem.*, 1973, **402**, 107–112.
36. C. Tard, S. Perruchas, S. Maron, X. F. Le Goff, F. Guillen, A. Garcia, J. Vigneron, A. Etcheberry, T. Gacoin and J.-P. Boilot, *Chem. Mater.*, 2008, **20**, 7010–7016.
37. S. Perruchas, X. Le Goff, S. Maron, I. Maurin, F. Guillen, A. Garcia, T. Gacoin and J. Boilot, *J. Am. Chem. Soc.*, 2010, **132**, 10967–10969.
38. X. C. Shan, F. L. Jiang, L. Chen, M. Y. Wu, J. Pan, X. Y. Wan and M. C. Hong, *J. Mater. Chem. C*, 2013, **1**, 4339–4349.
39. X. C. Shan, H. B. Zhang, L. Chen, M. Y. Wu, F. L. Jiang and M. C. Hong, *Cryst. Growth Des.*, 2013, **13**, 1377–1381.

40. T. Wen, D. X. Zhang, J. Liu, R. Lin and J. Zhang, *Chem. Commun.*, 2013, **49**, 5660–5662.
41. T. Tsukuda, M. Kawase, A. Dairiki, K. Matsumoto and T. Tsubomura, *Chem. Commun.*, 2010, **46**, 1905–1907.
42. M. Babashkina, D. Safin, M. Bolte and Y. Garcia, *Dalton Trans.*, 2011, **40**, 8523–8526.
43. H. Bi, D. Chen, D. Li, Y. Yuan, D. Xia, Z. Zhang, H. Zhang and Y. Wang, *Chem. Commun.*, 2011, **47**, 4135–4137.
44. S. Ladouceur and E. Zysman-Colman, *Eur. J. Inorg. Chem.*, 2013, **2013**, 2985–3007.
45. Y. You and W. Nam, *Chem. Soc. Rev.*, 2012, **41**, 7061–7084.
46. R. D. Costa, E. Ortí, H. J. Bolink, F. Monti, G. Accorsi and N. Armaroli, *Angew.Chem. Int. Ed.*, 2012, **51**, 8178–8211.
47. C. Ulbricht, B. Beyer, C. Friebe, A. Winter and U. S. Schubert, *Adv. Mater.*, 2009, **21**, 4418–4441.
48. H. Xu, D. H. Yu, L. L. Liu, P. F. Yan, L. W. Jia, G. M. Li and Z. Y. Yue, *J. Phys. Chem. B*, 2010, **114**, 141–150.
49. E. Szerb, A. Talarico, I. Aiello, A. Crispini, N. Godbert, D. Pucci, T. Pugliese and M. Ghedini, *Eur. J. Inorg. Chem.*, 2010, **2010**, 3270–3277.
50. G. Shan, H. Li, H. Cao, D. Zhu, P. Li, Z. Su and Y. Liao, *Chem. Commun.*, 2012, **48**, 2000–2002.
51. G. Shan, H. Li, D. Zhu, Z. Su and Y. Liao, *J. Mater. Chem.*, 2012, **22**, 12736–12744.
52. T. Mastropietro, Y. Yadav, E. Szerb, A. Talarico, M. Ghedini and A. Crispini, *Dalton Trans.*, 2012, **41**, 8899–8907.
53. N. D. Nguyen, G. Zhang, J. Lu, A. E. Sherman and C. L. Fraser, *J. Mater. Chem.*, 2011, **21**, 8409–8415.
54. L. Y. Bu, M. X. Sun, D. T. Zhang, W. Liu, Y. L. Wang, M. Zheng, S. F. Xue and W. J. Yang, *J. Mater. Chem. C*, 2013, **1**, 2028–2035.
55. Y. L. Wang, W. Liu, L. Y. Bu, J. F. Li, M. Zheng, D. T. Zhang, M. X. Sun, Y. Tao, S. F. Xue and W. J. Yang, *J. Mater. Chem. C*, 2013, **1**, 856–862.
56. X. Q. Zhang, Z. G. Chi, B. J. Xu, J. Jiang, X. Zhou, Y. Zhang, S. W. Liu and J. R. Xu, *Chem. Commun.*, 2012, **48**, 10895–10897.
57. G. G. Shan, H. B. Li, H. T. Cao, H. Z. Sun, D. X. Zhu and Z. M. Su, *Dyes Pigments*, 2013, **99**, 1082–1090.
58. T. Ohba, A. Kobayashi, H. C. Chang and M. Kato, *Dalton Trans.*, 2013, **42**, 5514–5523.
59. A. Kobayashi, H. Hara, T. Yonemura, H. C. Chang and M. Kato, *Dalton Trans.*, 2012, **41**, 1878–1888.
60. B. Xu, Z. Chi, X. Zhang, H. Li, C. Chen, S. Liu, Y. Zhang and J. Xu, *Chem. Commun.*, 2011, **47**, 11080–11082.
61. G. G. Shan, H. B. Li, J. S. Qin, D. X. Zhu, Y. Liao and Z. M. Su, *Dalton Trans.*, 2012, **41**, 9590–9593.
62. G. G. Shan, H. B. Li, H. Z. Sun, D. X. Zhu, H. T. Cao and Z. M. Su, *J. Mater. Chem. C*, 2013, **1**, 1440–1449.

CHAPTER 4

Polymer Mechanofluorochromic Materials

YI ZHANG AND ZHENGUO CHI*

PCFM Lab, State Key Laboratory of Optoelectronic Materials and Technologies, School of Chemistry and Chemical Engineering, Sun Yat-sen University, Guangzhou, China
*Email: chizhg@mail.sysu.edu.cn

4.1 Introduction

In terms of practical uses, film-/fiber-forming materials such as polymers have many advantages, such as better processability and wider applications. Thus, mechanofluorochromic polymers, either dye-doped polymers or pure polymeric dyes, should be superior to other mechanoflurochromic materials in actual applications.

However, thermoplastic polymers are made of long, flexible molecules with single covalent bonds. Hence, in such chemical bonds electrons reside in low-level orbitals, with a very large energy gap between bonding and antibonding orbitals. For this reason, natural and man-made polymers are usually colorless, with the exception of highly conjugated polymers, in which the extended conjugation reduces the energy gap for at least part of the backbone-binding electrons. The structural rigidity and free electron mobility render conjugation, and then color the polymer. So these polymers are easily converted into semiconductors and even electroconductive materials, which are known as "organic metals". However, these materials are characterized by a rigid backbone that adversely affects the typical viscoelastic behavior of the thermoplastic polymers, which may be related to the high

RSC Smart Materials No. 8
Mechanochromic Fluorescent Materials: Phenomena, Materials and Applications
Edited by Jiarui Xu and Zhenguo Chi

Published by the Royal Society of Chemistry, www.rsc.org

entropy gain provided by the easy coiling of the flexible chains. Two possible procedures must be followed in order to prepare materials showing both the typical thermoplastic polymer properties (viscoelasticity) and optical response to visible light:[1]

The first approach is based on the principle that a colored polymeric material can be obtained by dispersing an appropriate dye in the bulk of the pristine colorless polymer matrix. In this case, the macromolecules remain structurally unaltered, and the system is generally biphasic unless the resulting dyes are fully soluble in the matrix polymer. The dispersion of a dye into a polymer is largely applied to the commodity plastics and the procedure has been used for pigmented materials since the early days of the plastic era. The dye is simply a filler which is able to provide more properties than simply color for aesthetic applications. Moreover, the specific interactions occurring between the macromolecule and the chromophore tune the optical behavior of the system, by means of driven geometrical assemblies to exciton coupling and self-organisation. The resulting optical behavior is then generated by the synergic interaction with the polymer structure and system dynamics, providing also innovative characteristics and responsive character upon external stimuli.

The second approach consists of chromophoric units covalently inserted into the macromolecular chains, allowing the backbone to remain stable and flexible. The modified macromolecules form a polymeric material with color and intensity determined, respectively, by the selected chromophore and the extent of chemical modification in terms of chromophore concentration. The resulting colored macromolecules are similar in the structure of a copolymer, with random or block distribution of colorless and colored monomer residues. However, the synthetic possibility is often limited by the need for appropriate functional groups on both macromolecules and chromophores. The preparative method can, in some cases, be facilitated by direct grafting of a dye onto the preformed macromolecule, which is much easier to perform than a chemical condensation/addition reaction between complementary functional groups or a copolymerization between the colorless monomer and the dye modified monomer.

In the literature, many examples are reported on mechanochromic chromophores covalently incorporated into the polymer backbones. Some interesting examples come from polydiacetylenes, poly(urethane-diacetylene) segmented copolymers and, as recently reported, poly(methyl acrylate) or polycaprolactone with the spiropyran motifs covalently linked in the middle of the polymeric chains. Other substances that also displayed significant color changes upon mechanical solicitations include azobenzenes and cyano-substituted oligo(*p*-phenylene vinylene) chromophores incorporated into polyurethanes or polyesters. Due to the large number of available organic dyes and suitable semicrystalline polymer matrices, significative and intriguing examples are recently reported on the mechanochromic materials prepared by means of blending.[2]

4.2 Synthetic Method

In their review article, Pucci *et al.*[1] have comprehensively summarized the synthetic methods for covalently linked dye-polymer systems, as described below.

For a covalently linked dye-polymer system to implement the function of mechanochromic luminescence, the mechanical stimuli often induce selective chemical transformations of the mechanophore at the atomic level and the consequent color change. The mechanophore unit should be appropriately positioned within a polymer chain to experience mechanical perturbation in a controlled manner. And furthermore, the mechanophore must possess mechanically labile bonds that change through isomerisations or precise bond scission events, thus producing distinct optical variations.[3,4]

The introduction of mechanophore moieties into polymers through covalent bonding is generally achieved by different pathways (Scheme 4.1),[1] depending on the type of polymers to be obtained.

For example, in the case of polystyrene (PS) and poly(methyl methacrylate) (PMMA), the polymer labeled with the generic dye D is usually obtained by radical copolymerization of the corresponding monomer with a methacryloyl or styryl derivative of D. The amount of dye covalently linked to the polymer is modulated by adjusting the content of the labeled co-monomer in the feed (Scheme 4.1a).

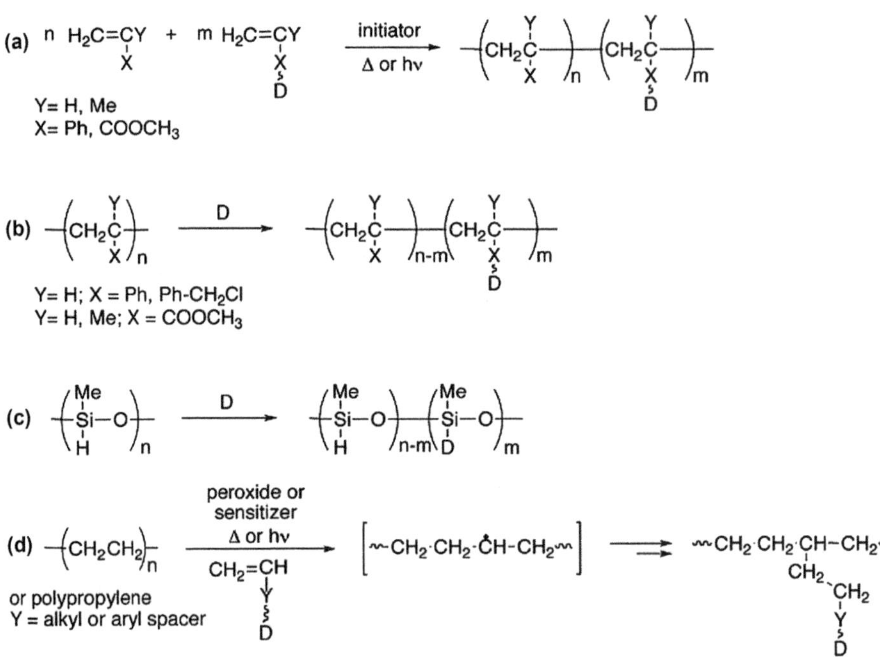

Scheme 4.1 Common synthetic routes for the preparation of covalently linked dye-polymer materials (D, dye).
Reproduced with permission from reference 1. Copyright 2013, Royal Society of Chemistry.

In the case of polydimethylsiloxane (PDMS) elastomers, the covalent bonding of dye molecules is generally realized by the catalytic functionalization of polyhydromethylsiloxane (Scheme 4.1c).

The (a) and (b) synthetic pathways cannot be applied in the case of polyolefins using Ziegler–Natta catalysts, as their syntheses do not usually tolerate the olefinic monomers with polar moieties or heteroatoms, which are commonly found in the chemical structure of dyes.[5] In this case, the dye can be directly grafted to the macromolecule chain through a radical process (Scheme 4.1d), initiated by a thermally or photochemically induced hydrogen abstraction from the polyolefin backbone. The macro-radical thus formed is then added to a vinyl monomer, which contains the dye functionality or a precursor in its structure.

In the case that mechanophores must be localized at a specific site, such as the chain-end or the junction between blocks, dye-polymer systems can be effectively prepared using living/controlled polymerization techniques.[6] The dye can be introduced at three different steps in the polymerization:[7] (i) at the very beginning *via* the initiator or a chain transfer agent, (ii) during polymerization *via* a functional monomer or (iii) after polymerization *via* covalent binding of a dye-derivative (Figure 4.1).

In this context, Moore and Sottos[8,9] proposed for the first time the artificial polymers in which mechanical stress provides the activation energy for specific chemical reactions. They realized the mechanoresponsive materials

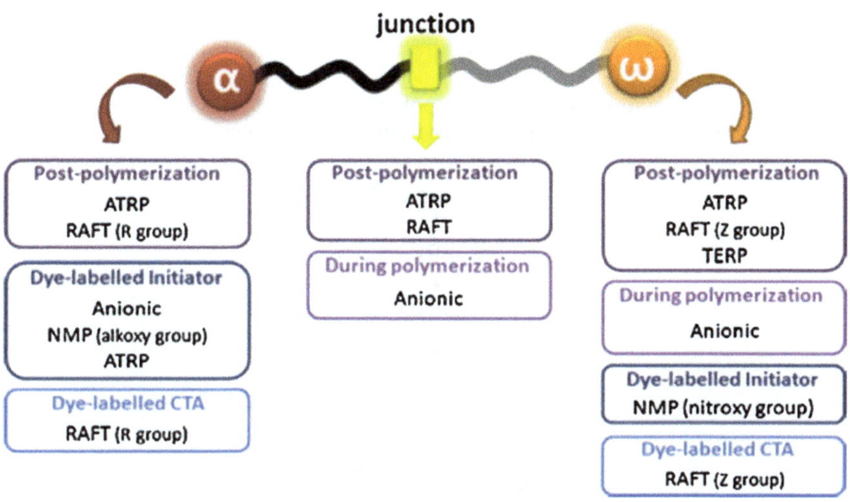

Figure 4.1 Summary of the different controlled synthetic methodologies that have been used for the synthesis of α-, junction- and ω-dye-labeled (co)polymers. ATRP = atom transfer radical polymerization; RAFT = reversible addition-fragmentation chain-transfer polymerization; TERP = organotellurium-mediated living radical polymerization; NMP = nitroxide-mediated radical polymerization.
Reprinted from reference 7. Copyright (2011), with permission from Elsevier.

by directly linking spiropyran (SP) dyes into the polymer chains of elastomeric poly(methyl acrylate) (PMA) or on glassy cross-linked poly(methyl methacrylate) (PMMA) by single electron transfer living radical polymerization (SET-LRP)[10] (Scheme 4.2). The synthetic method has been used in the synthesis of PMA, incorporating bis(adamantyl)-1,2-dioxetane mechanophore units.[11]

In the case of proprietary polymers such as polyesters, polyamides or polyurethanes, the covalent incorporation of dyes into the polymer backbone involves the preparation of di-functional monomers (diacids, diamines or glycols) bearing mechanophore units to be polymerized by stepwise processes. Cyano-OPVs were covalently incorporated into the main chain of polyesters or polyurethanes in order to prepare self-assessing photoluminescent elastomers.[12,13] In the latter case, the covalent incorporation of a 1,4-bis(α-cyano-4-(12-hydroxydodecyloxy)-styryl)-2,5-dimethoxybenzene into the thermoplastic polyurethanes (TPU) based on poly(tetramethylene glycol), butanediol, and 4,4′-methylene-bis(phenyl isocyanate) was illustrated (Scheme 4.3).[13]

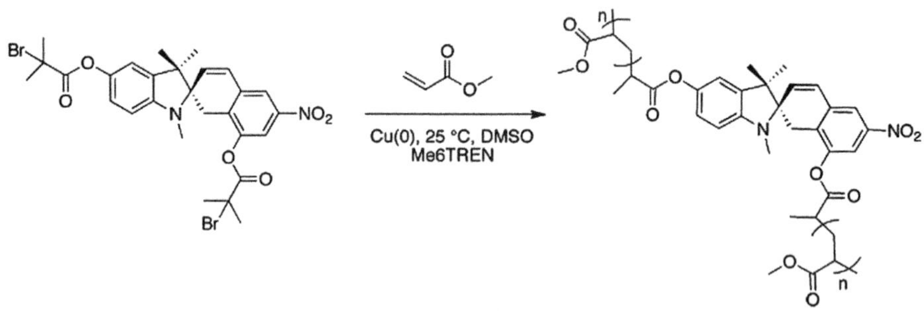

Scheme 4.2 Spiropyran-functionalized poly(methyl acrylate) chain synthesized by the single electron transfer living radical polymerization.

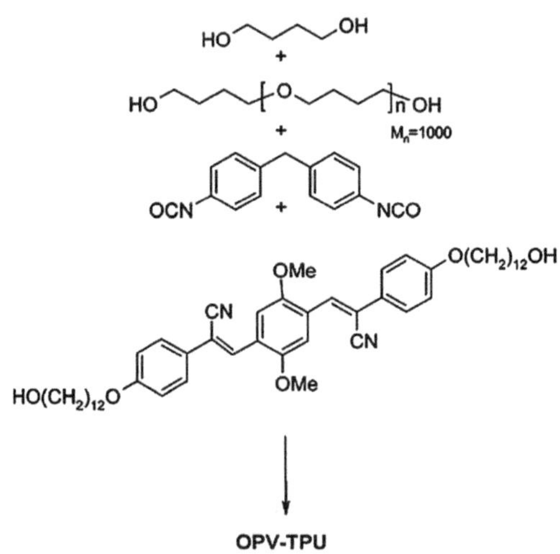

Scheme 4.3 Covalent incorporation of Cyano-OPVs into TPUs (9.2 wt% of OPV in the monomer feed).

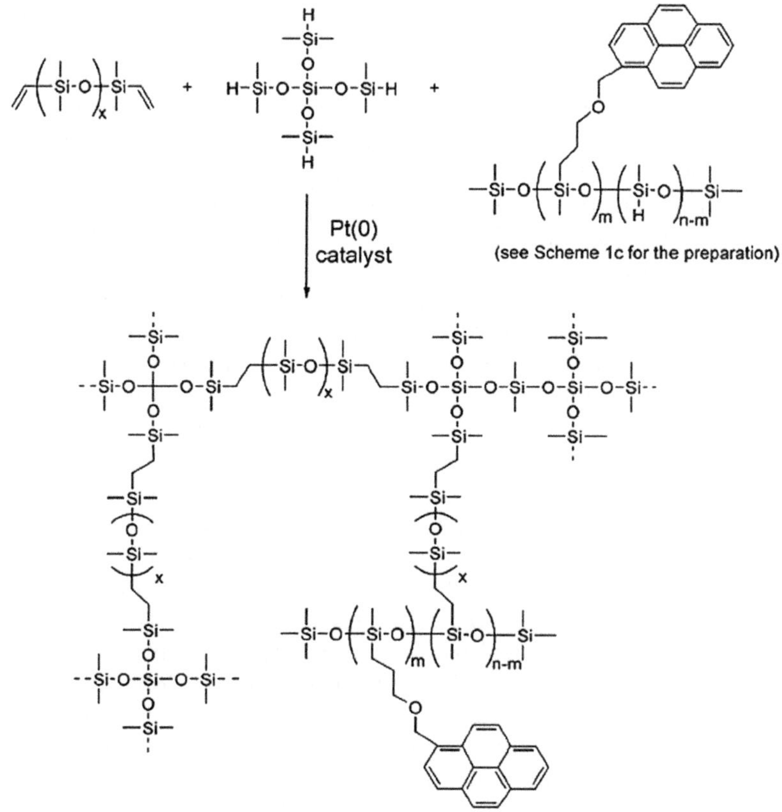

Scheme 4.4 Synthesis of pyrene-substituted PDMS elastomers. Reproduced with permission from reference 1. Copyright 2013, Royal Society of Chemistry.

As far as mechanochromic elastomers are concerned, pyrene-substituted polyhydromethylsiloxanes (PHMS-Pyx) were prepared by the hydrosilylation reaction of prop-3-enyloxymethylpyrene with polyhydromethylsiloxane.[14] The ratio of pyrene substituent to Si-H unit was modulated to afford a range of pyrene-functionalized polysiloxanes. The elastomeric response was achieved by crosslinking with divinyl-terminated PDMS (Scheme 4.4).[1]

4.3 Mechanofluorochromic Polymer Systems

4.3.1 Based on Chemical Reaction of Mechanophore-polymer

4.3.1.1 Spiropyran Mechanophore

As mentioned above, a common polymer chain includes numberless co-valent bonds. To obtain its mechanofluorochromism, one of the most common variations of molecular structure involves cleavages of bonds. However, the cleavage of covalent bonds is difficult to detect, specify and analyze. In order to overcome this difficulty, it is necessary to introduce some reactive structures as a sensitive moiety for detection of mechanical stimuli into the tested materials. These structures are sometimes called mechanophores. In 2009, the research teams at the University of Illinois used a spiropyran (SP) mechanophore as the molecular force sensor (Figure 4.6).[9] This mechanophore enabled visible detection and mapping of mechanical stresses within bulk polymeric materials, because the cleaved product merocyanine (MC) form exhibits a slight red color. The spiropyran units are expected to show a reversible ring-opening reaction under tensile stress, realizing locally visualization of the mechanochemical reaction. Mechanoresponsive polymeric materials were synthesized by linking spiropyran moieties into the chains of bulk polymers or by using spiropyran as the crosslinkers. Applying mechanical forces to the spiropyran-linked poly(methyl acrylate) induced reversible transformation between the colorless spiropyran form and a highly colored merocyanine structure through rupture of the spiro group carbon-oxygen (C–O) bond, because the increased conjugation of the merocyanine form shifts the absorption to longer wavelengths, thus giving rise to visible color and fluorescence. The mechanism is completely based on chemical reaction. Here, it is very important that the mechanophore motif must efficiently transfer external force to a relatively small number of specific bonds in the bulk polymer, with this mechanochemical activation depending critically on the molecular structure of the mechanophore and the attachment points to the polymer chains. The spiro C–O bond is the critical point for mechanically induced activation, and there are many possible attachment points on the spiropyran that differ in their ability to transmit force efficiently across the junction (Figure 4.2). The simple mechanical models indicate that attachment on opposing sides of the spiro-junction, specifically, positions 5′ or 6′ of the indole side and positions 7 or 8 of the benzopyran side, preferentially stresses the spiro C–O bond over the spiro C–C bond.

The orientation of force-sensitive chemical species (mechanophores) in the bulk polymers was measured *via* the anisotropy of fluorescence polarization by Sottos *et al.* (2012)[14] Orientation measurements were utilized to investigate the role of mechanophore alignment on mechanically driven chemical reactions. The mechanophore, spiropyran (SP), was covalently bonded into the backbone of poly(methyl acrylate) (PMA) and poly(methyl methacrylate) (PMMA) polymers. Under the UV light or tensile force, SP reacts to a merocyanine (MC) form, which exhibits a strong fluorescence, polarized roughly across the long axis of the MC subspecies (Figure 4.3A). An order parameter was calculated, based on the anisotropy of fluorescence polarization, to characterize the orientation of the MC subspecies relative to tensile force. For the UV-activated SP-linked PMA samples, the order parameter increased with the applied strain, up to an order parameter of approximately 0.5. Significantly higher order parameters were obtained

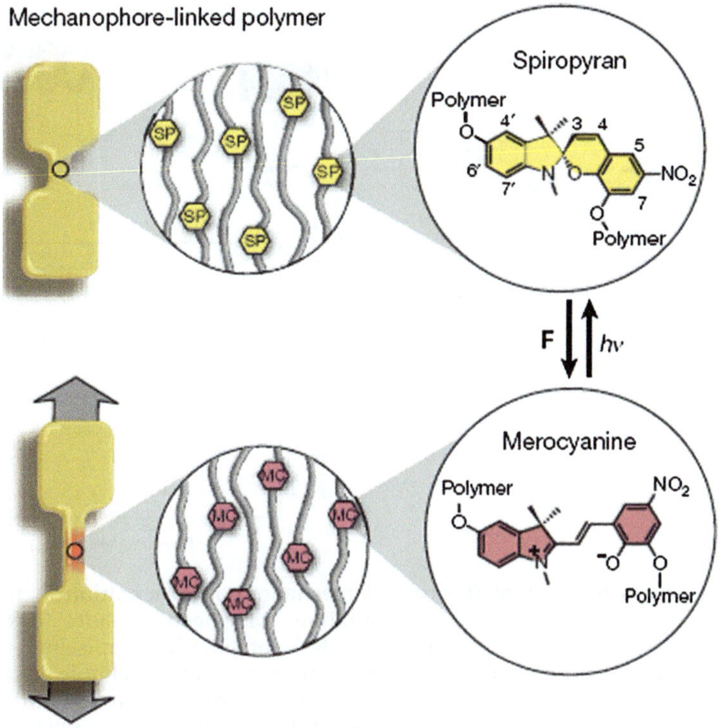

Figure 4.2 Schematic of mechanical stretching of isomerization between spiropyran and merocyanine attached to polymer chains.
Reproduced with permission from reference 9. Copyright 2009, NPG.

for mechanically activated SP-linked PMA samples (Figure 4.3B), indicating preferential mechanochemical activation of species oriented in the tensile direction. The anisotropy of fluorescence polarization in SP-linked PMMA also provided insight into polymer drawing and polymer relaxation at failure.

Based on the above-mentioned spiropyran ring-opening to change emission color mechanism, A photochromic poly(ε-caprolactone) polymer was prepared by O'Bryan *et al.* (2010)[15] using ring-opening polymerization (ROP) with a di-functional spiropyran (Figure 4.4A). The resulting polymer was characterized by standard techniques and found to respond to mechanically induced stress events and UV irradiation due to ring-opening of the spiro C–O bond to form the blue-colored merocyanine (Figure 4.4B). It was found that deformation of the thin polymer films results in bond cleavage at the weak C–O bond of the integrated spiropyran initiator. Light-induced stress relaxation was demonstrated by irradiating the polymer films while holding the samples at a constant strain. Both phenomena are a direct result of incorporating a photochromic spiropyran diol within the polymer backbone, a configuration that is readily attained *via* ROP with the initiating molecule. Through both the experimental and theoretical studies, this work illustrates

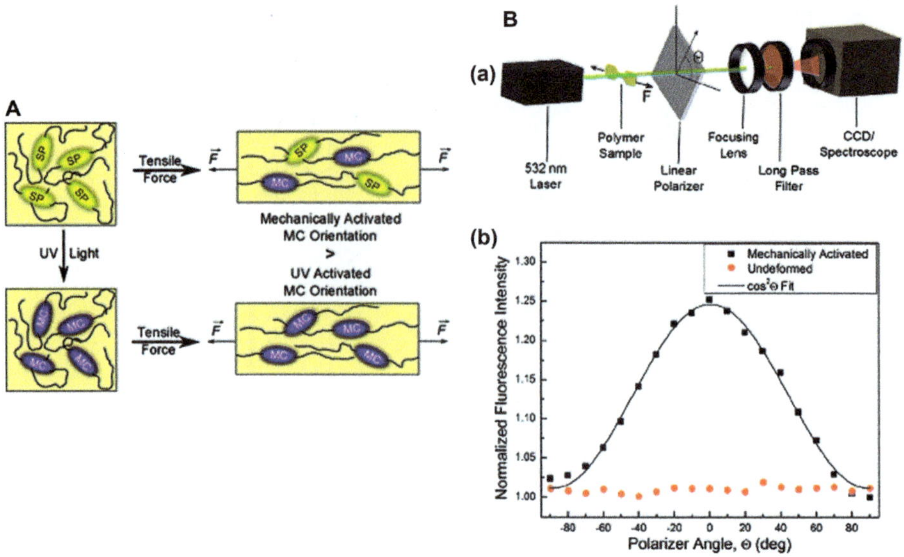

Figure 4.3 (A) Schematic of the role of mechanophore orientation in mechanochemical reactions. (B) Analysis of mechanophore orientation. (a) Experimental setup for analysis of polarized emission. (b) Fluorescence intensity *vs.* polarizer orientation for mechanically activated ($\varepsilon = 35\%$) and undeformed thermally activated SP-linked PMMA.
Reproduced with permission from reference 14. Copyright 2012, American Chemical Society.

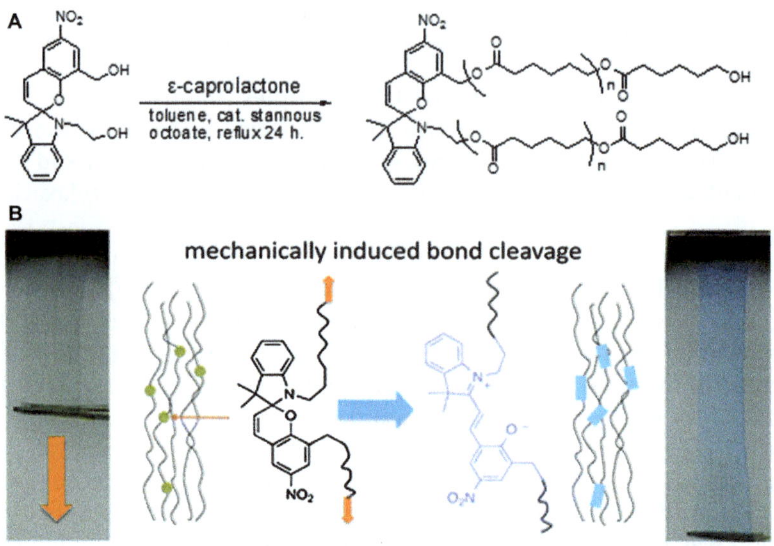

Figure 4.4 (A) The preparation method of the spiropyran-poly(ε-caprolactone) polymer. (B) Mechanchromic luminescence behavior and mechanism of spiropyran- poly(ε-caprolactone) polymer.
Reproduced with permission from reference 15. Copyright 2010, American Chemical Society.

that an induced stress (even for small loads) can lead to a significant mechanochromic response in the bulk polymer materials. A sharp transition between spiropyran and zwitterionic merocyanine forms was detected at 39% elongation length.

This mechanophore motif could also be useful in other systems including those being polymerized, cross-linked or mechanically fused. Recently, the same mechanopore was introduced into a step-growth polyurethane (PU) for more detailed studies.[16] The inherent mechanical toughness, elasticity and low glass transition temperature ($Tg \approx -60$ °C) of PU enable the effects of mechanical force on the SP-MC equilibrium (Figure 4.5) to be studied. The initial polymer systems studied were not amenable to kinetic analysis because they were either too soft [elastomeric poly(methyl acrylate) (PMA)], resulting in mechanical activation only at high strains close to the strain of failure, or too glassy [poly(methyl methacrylate) (PMMA), $Tg = 105$ °C]. In PMMA, because of the high *Tg*, the rate of conversion from MC to SP form could not be investigated at room temperature.[9] The experimental results show that the SP mechanophore is mechanochromic in PU and that the equilibrium between the colored MC and colorless SP form can be directly controlled by mechanical strain. Because of the low *Tg*, equilibrium is reached in experimentally accessible times at room temperature. It was the first time that the kinetics of the mechanically activated SP to MC

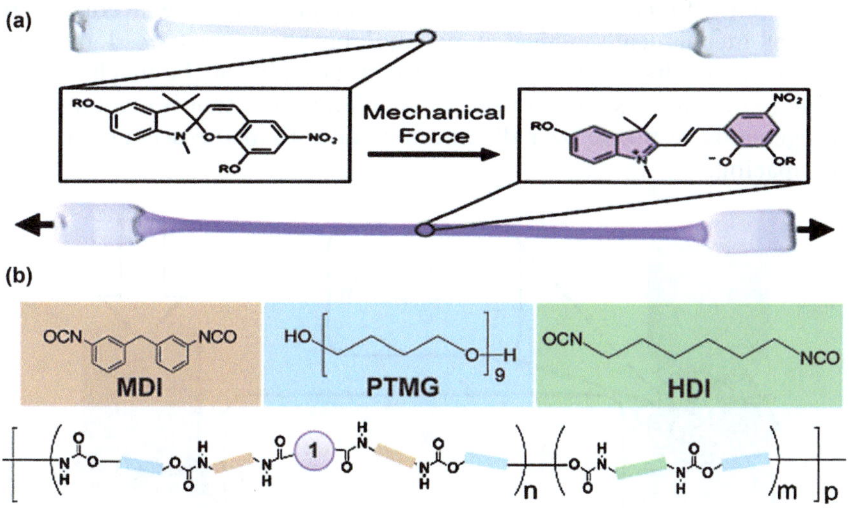

Figure 4.5 (a) Chemical structures of spiropyran (SP) and merocyanine (MC) and the mechanically or optically triggered conversion equilibrium between the colorless SP and colored MC forms. Optical images of prestretched PU-1 "dog bone" containing 0.03 wt %1 before (left) and after (right) being stretched to a stretch ratio (final length/initial length) of 2.0 and released. (b) Schematic of the incorporation of SP mechanophore (**1**) into PU *via* step growth polymerization where $m \gg n$.
Adapted with permission from reference 16. Copyright 2010, American Chemical Society.

conversion, as well as the thermally activated conversion of MC to SP in a bulk polymer were reported. The absorbance due to the ring-opening increased linearly with strain, and the merocyanine form did not revert back to the thermodynamically preferred spiropyran form when held at a constant strain. The latter feature indicates a strain-induced change in the energy landscape of the spiropyran isomerization system. PU is of specific interest not only because it is a ubiquitous engineering polymer but also because it can be synthesized *via* step growth polymerization, so the mechanophore concentration can be modulated independently of molecular weight or cross-linking density. From a practical standpoint, the synthesis of the mechanochromic PU can be scaled to large volumes, simplifying mechanical testing studies.

4.3.1.2 gem-Dihalocyclopropanated Polybutadiene

Sottos *et al.* (2011)[17] have also used *gem*-dihalocyclopropane (gDHC) as the sensitive components in polybutadiene for assessing the size of mechanically activated moieties in polymers (Figure 4.6). This molecular structure undergoes strain-triggered ring opening, providing quantitative information regarding the number of mechanically active monomer units, *i.e.*, the size of the mechanically activated domains along individual polymer backbones within the bulk materials. It was found that upon applying compressive force to the highly functionalized *gem*-dihalocyclopropane, the average block length of mechanically activated polymer backbone was *ca.* 4 monomer units. These results suggest that the critical domain for the mechanical activation in polymers is greater than a single monomer that is responsible for heterogeneous response over distances smaller than the average entanglement spacing.

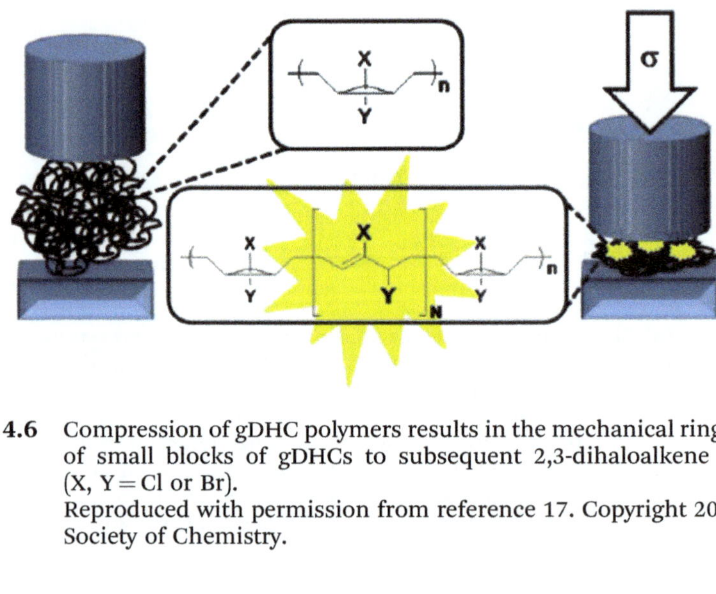

Figure 4.6 Compression of gDHC polymers results in the mechanical ring opening of small blocks of gDHCs to subsequent 2,3-dihaloalkene products (X, Y = Cl or Br).
Reproduced with permission from reference 17. Copyright 2011, Royal Society of Chemistry.

4.3.1.3 Bis(adamantyl)-1,2-dioxetane Mechanophore

Another interesting report that has recently emerged involves the ring opening of 1,2-dioxetanes by Sijbesma's group (2012).[18] The mechanically labile dioxetane ring, covalently linked to linear poly(methyl acrylate) (PMA) by means of living radical polymerization, decomposes into electronically excited carbonyl products upon mechanical activation. The electronic relaxation of these derivatives is often accompanied by the emission of blue light (Figure 4.7a and b). In this case, no chromophoric unit exists in the material composition before solicitation, and the phenomenon resembles the fascinating emission of light by marine organisms when subjected to mechanical stress. To investigate the time course of luminescence during straining, the light intensity was increased by the addition (0.5 wt%) of an appropriate energy-transfer molecule such as thiadiazole or perylene derivatives. Representative images as the sample underwent a mechanical test reveal the emergence of light during the increasing plastic deformation process, with the strongest emission at the position where the film broke (Figure 4.7c). The real-time monitoring of the phenomenon allowed the study of polymeric material failure with remarkable details, thus opening the possibility to develop innovative optical sensors with miniaturized dimensions and elevated sensitivity.

4.3.2 Based on Macromolecular Packing Change

4.3.2.1 Polymers with Conjugated Main Chains

π-Conjugated polymers exhibit interesting optical and electrical properties.[19,20] It has been reported that some π-conjugated polymers change their optical properties upon altering the solvent (solvatochromism),[21–23] temperature (thermochromism)[24–26] and/or external force (mechanochromism).[27–30] However, in comparison with solvatochromism and thermochromism, examples of mechanochromism are still limited.

4.3.2.1.1 Poly(3-alkylthiophene)s. Poly(3-alkylthiophene)s have a strong tendency to self-assemble and to form ordered structures. Formation of such an ordered structure is considered to be crucial for superior mobility in field effect transistors[31–33] and large third-order optical nonlinear susceptibility.[34] Thus, the relationship between the spectra behavior and the molecular packing of the poly(3-alkylthiophene)s under high pressure has attracted strong interest. But, some of the studies have been mainly focused on the mechanochromism based on changes of absorption spectra, not the emission spectra.

The conductive, melting, optical and photoluminescent properties of poly(3-alkylthiophenes) under high pressure have been more extensively studied by Yoshino's group.[35–37] Yoshino *et al.* have recently measured absorption and photoluminescence in poly(3-alkylthiophenes) films and

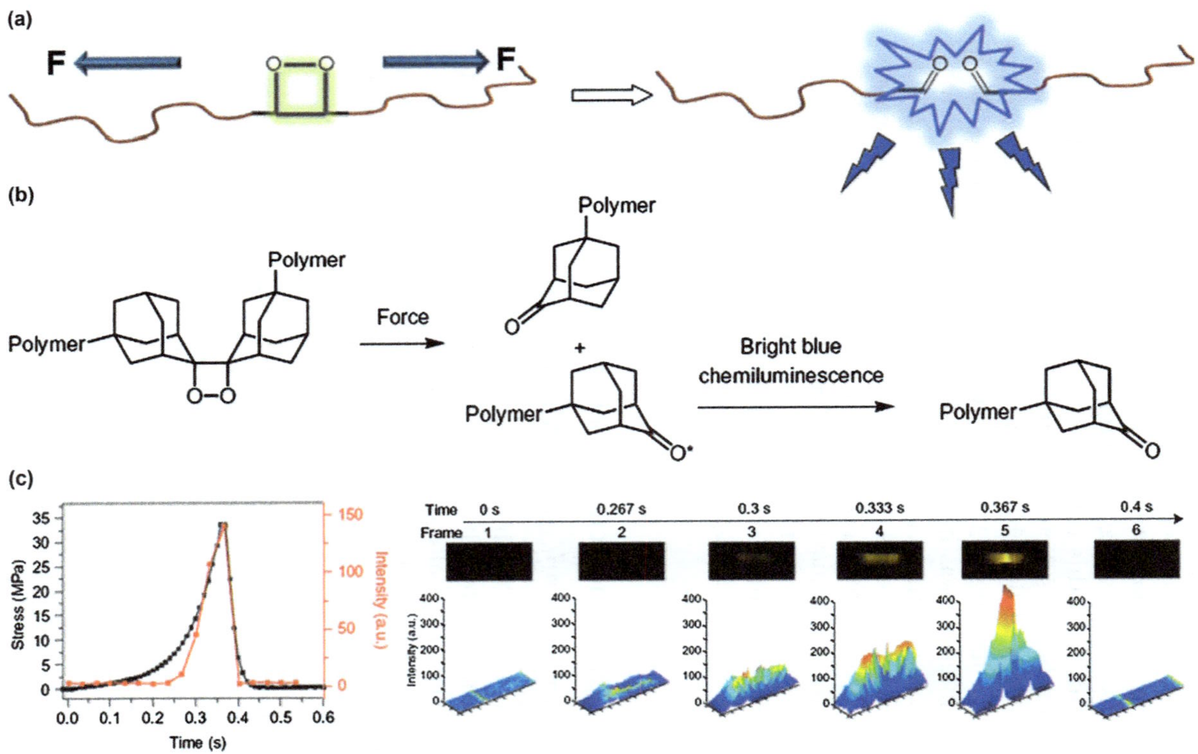

Figure 4.7 (a) Schematic representation of the mechanically induced decomposition of a polymeric bis(adamantyl) dioxetane that results in chemiluminescence when the ketone product relaxes from its excited state to the ground state; (b) molecular basis of the mechanically induced chemiluminescence; (c) stress and light intensity versus time during stretching of bulk film (PMA-bis(adamantyl)-1,2-dioxetane-PMA containing 4,7-di(thiophen-2-yl)benzo[c][1,2,5]thiadiazole). The analysed intensity is based on the same region within the sample (one tick on the x- and y-axes corresponds to 2 mm). The level of the signal at times 0.2 s is non-zero because it derives from a noisy signal of individual camera pixels integrated over the whole film area. Adapted with permission from reference 18. Copyright 2012, NPG.

solutions up to 0.9 GPa. They observed red shifts of the spectral bands in both absorption and photoluminescence, as well as changes in photoluminescence intensity, and they attributed the effects to the conformational planarization changes and consequent changes in effective conjugation length. Due to the steric interaction of the alkyl side chains, the thiophene rings are not all planar, but possess a slight twist.

Hess *et al.* (1993)[38] reported that poly(3-hexylthiophene) (P3HT) shows a reversible decrease in photoluminescence intensity as well as a red shift in the absorption bands with increasing pressure, which is again attributed to the increased backbone planarity because their experimental results ruled out a significant increase in the interchain covalent interaction in P3HT and confirmed the existence of the conjugation length effects. However, Iwasaki *et al.* (1994)[39] reported that, at pressures over 3 GPa, the absorption peak of P3HT started to move back to higher energies. Unlike the polyacetylenes,[40] this blue shift is not permanent and is thought to be caused by a bending of the main chains due to increased interchain sterics.

Kaniowski *et al.* (1998)[41] reported the mechanochromism of regioregular poly(3-octylthiophene) (P3OT)s with 97% head-to-tails as compared to nonregioregular P3OT (Scheme 4.5). The regioregular P3OT experiences less steric interaction of its alkyl side chains and is therefore more planar with an average torsional angle of 25° (For nonregioregular P3OT, the average torsion angle is 40°). This is evidenced by its slightly lower inherent absorption energy. Since the regioregular P3OT is initially more planar, compression causes smaller red shifts in the absorption than that in the nonregioregular sample. Unlike the nonregioregular P3HT, the head-to-tail P3HT shows a linear shift in PL up to 5 GPa with no blue shift.[42] Samuelsen's studies on P3OT suggested that the shifts in this regioregular polymer were due to the increased interchain interaction.[43]

The photoluminescence of head-to-tail P3HT also shows interesting effects. Yamanoto *et al.* (2001)[44] have examined changes in the optical properties of highly regioregular head-to-tail P3HT with a head-to-tail content higher than 98.5% under high pressure by using a diamond-anvil pressure

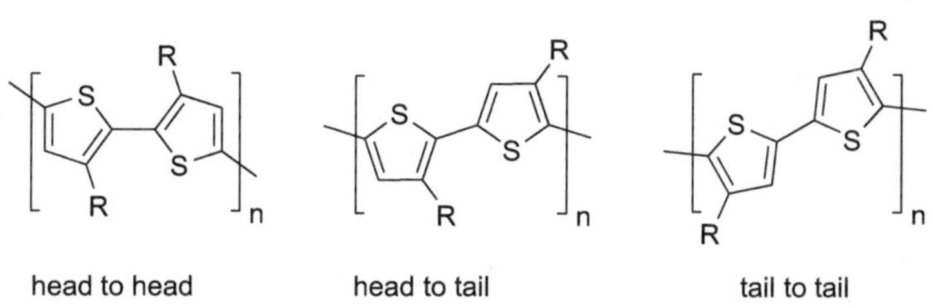

Scheme 4.5 The three possible ring conformations for thiophene polymers with respect to their alkyl side chains. In nonregioregular polythiophenes, significant amounts of head to head and tail to tail conformations are present. The steric interactions of the side chains within a subunit (head to head) or between subunits (tail to tail) cause significant deviations from planarity in the backbone.

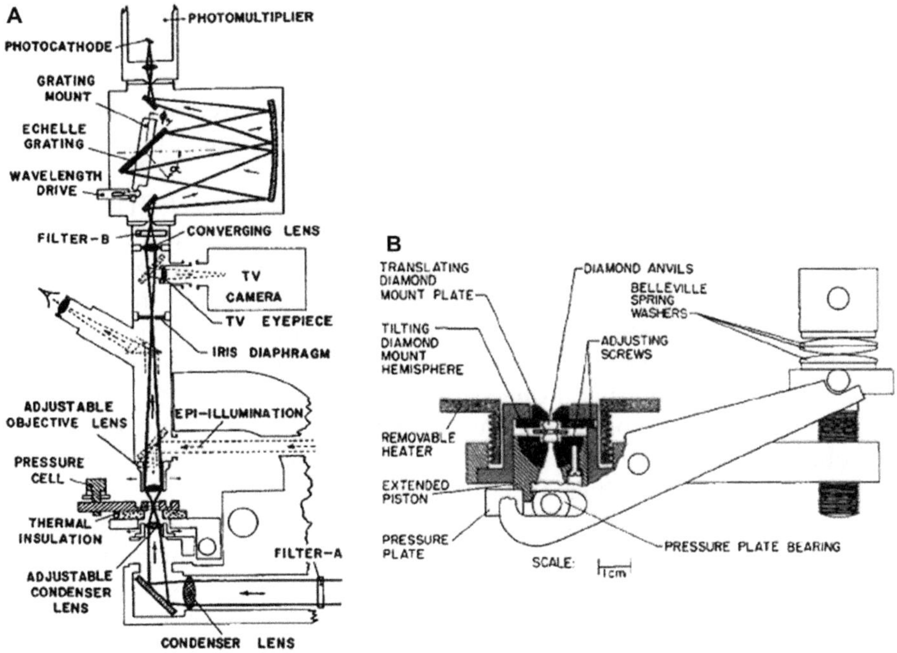

Figure 4.8 (A) Schematic diagram of the optical features of the pressure measuring system. (B) Cut-away cross section drawing of the Waspaloy pressure cell showing improved alignment design, heater assembly, and spring washer configuration.
Reproduced with permission from reference 45. Copyright 1973, AIP publishing.

cell. Solid samples were placed in a diamond-anvil pressure cell and pressed. A small piece of ruby was added to the solid sample and the applied pressure was determined based on the pressure shift in the sharp R-line photoluminescence of ruby. On applying high pressure, the powdery sample became film. PL and UV-vis spectra were measured with an optical system consisted of a standard microscope and a monochromator with an associated photodetection system (Figure 4.8).[45]

Figure 4.9 shows the pressure dependence of the PL peak of the regioregular head-to-tail P3HT. As shown in Figure 4.9b, the energy of the PL peaks of the regioregular head-to-tail P3HT decrease linearly with an increase in pressure up to 5 Gpa, unlike the nonregioregular P3HT. Samuelsen *et al.* presented interesting results on the solid structure of P3OT, with 75% head-to-tail content under high pressures up to about 8 GPa.[43] They found that both the distance between polythiophenen main chains separated by the alkyl side chains and the face-to-face distance between polythiophene planes were shortened at high pressure, *e.g.*, about 80% shortening of the face-to-face distance at about 8 GPa. The shortening of the face-to-face distance will increase the intermolecular electronic interaction between

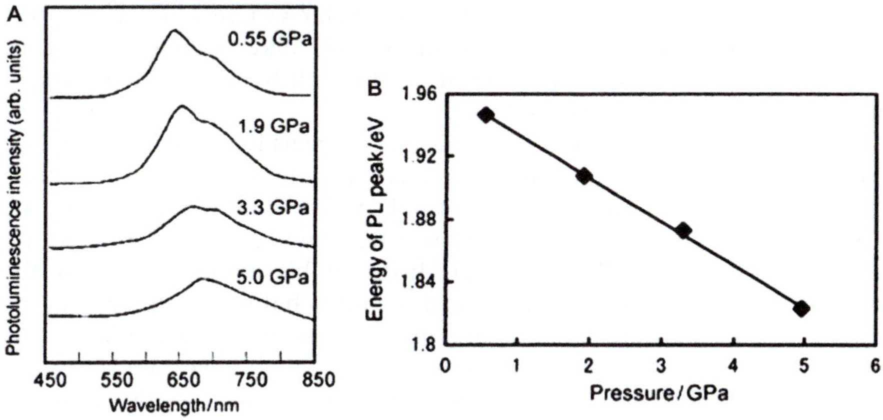

Figure 4.9 Shift of PL peak (A) and its corresponding energy (B) at high pressure. Reproduced with permission from reference 44. Copyright 2001, Elsevier.

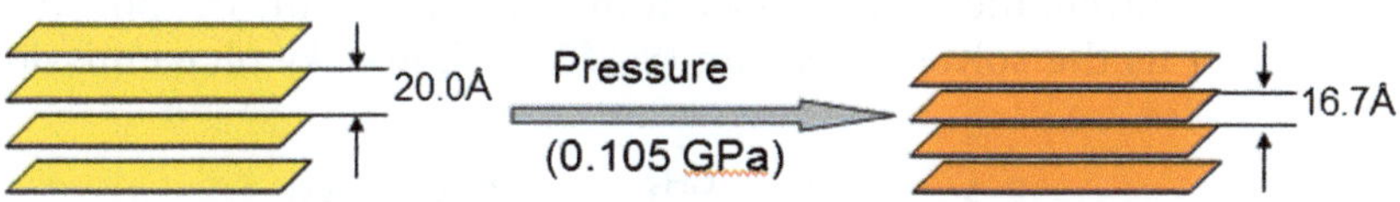

Figure 4.10 Schematic representation of the decrease of the face-to-face distance under pressure.

P3OT molecules. Thus, Yamanoto *et al.* suggest that the increased interchain interaction due to the decrease of the face-to-face distance may be the reason for lowering the band-gap energy for the regioregular head-to-tail P3HT determined based on the PL peak at a high pressure (Figure 4.10).

The head-to-head poly(3-(1-dodecynyl)thiophene-2,5-diyl), HH-P3(C≡CDec)Th, reported by Yamamoto *et al.* (2008)[46] is considered to be planar due to the lack of steric interactions between the alkyne and the thiophene ring. Under a high pressure, HH-P3(C≡CDec)Th exhibited clear piezochromism; the UV–vis absorption peak at about 605 nm observed under atmospheric pressure shifted to 672 nm under 10.71 GPa, suggesting an increase in electronic interaction between the polymer molecules under high pressure. The experimental results indicated that the π–π^* transition energy decreased by 0.028 eV/GPa in the range of 0–5.2 GPa.

4.3.2.1.2 Diphenylacetylene Polymer Derivatives. Among the many types of conjugated polymers, diphenylacetylene polymer derivatives have highly twisted chain conformations due to the steric repulsion between the two bulky phenyl groups, and their backbones are essentially non-coplanar with the side phenyl ring. Thus, in general, diphenylacetylene polymers are worm-like chain molecules with nonplanar geometry, leading to unstacked structures in the solid state. Kwak *et al.* (2011)[47] designed and synthesized a series of extremely high molecular weight and main chain rigidity poly(diphenylacetylene) derivatives with alkyl side chains (polymers 1–3 in Scheme 4.6), which could exhibit well lyotropic liquid crystallinity with an aid of adequate solvent.

The lamellar layer distances of polymers 1–3 are as large as 22.0, 14.9 and 13.0 Å, respectively, and are much greater than the interplanar distances (3.0–4.0 Å) needed for an effective intermolecular π–π interaction. Thus, the highly ordered liquid crystalline structures should be constructed on the basis of a much weaker interchain hydrophobic interaction due to the presence of alkyl side chains rather than the interchain π–π interaction. The ordered structures could be easily disrupted by external mechanical stress because the molecular construction was based on the very weak van der Waals force interaction between the alkyl side chains, resulting in the changes of their emission properties. As expected, the polymers underwent remarkable emission color changes during the grinding process. The emission color could be completely recovered by chain reorganization *via* solvent inducing. As shown in Figure 4.11a, Polymer 1 shows an orange emission in the mechanically processed film (MP-film), while the polymer in the solution or as the solvent cast film (SC-film) shows sky blue emissions.

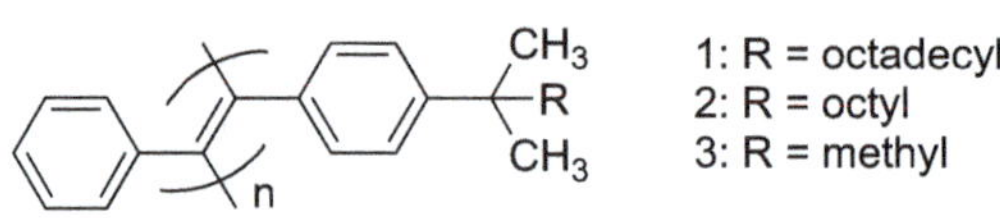

Scheme 4.6 Chemical structures of Polymers 1–3.

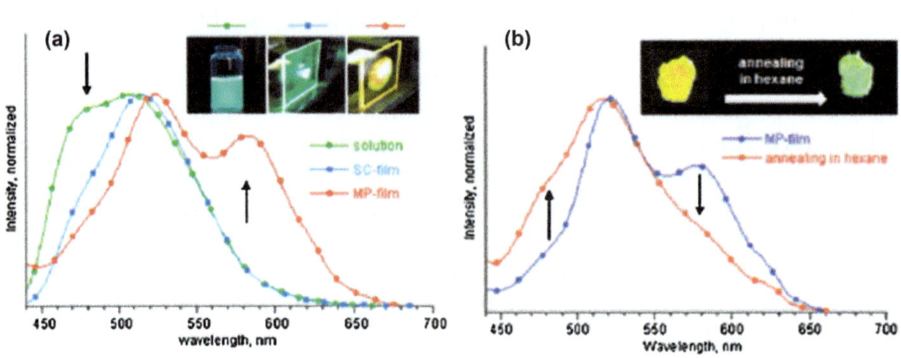

Figure 4.11 (a) Fluorescence emission spectra of 1 in the solution (concentration $\approx 1\times10^{-6}$ mol L^{-1} in toluene), the SC-film (film thickness $\approx$ 30 mm), and the MP-film (film thickness $\approx$ 100 mm). Inset: digital photographs (excited at 420 nm). (b) Variation in fluorescence emission spectra of 1 in the MP-film (film thickness $\approx$ 100 mm). Inset: digital photographs (excited at >365 nm). SC-film: solvent cast film; MP-film: mechanically processed film.
Reproduced with permission from reference 47. Copyright 2011, Royal Society of Chemistry.

The new fractional emission band around 580 nm originates from the intramolecular excimer resulting from a certain phase transition induced by the mechanical stress.

The MP-film of Polymer 1 retained its solubility in organic solvents such as chloroform and toluene. Re-precipitating of its solution to non-solvents regenerated the original bulk solid and fully restored the original sky blue emission. Moreover, the original emission color was also able to be restored simply by annealing the MP-film with hexane liquid and/or its saturated vapor, indicating the typical mechanofluorochromism. Figure 4.11b shows the reversible change in the fluorescence emission spectra of the Polymer 1. The 580 nm shoulder band disappears and the 475 nm monomer emission band regenerates after the solvent annealing. Thus, the original sky blue emission was completely restored after the solvent annealing, as shown in the inset. This chromism was considered to be due to the fact that the polymer chains reorganize during the solvent annealing process in order to reconstruct the ordered structure. However, the reversible change could not be realized by thermal annealing sat around 100 °C for several hours, indicating this efficient restoration in solvent annealing rather than in thermal annealing because of their intrinsic lyotropic liquid crystallinity. The experimental results of wide-angle X-ray diffraction (WAXD) and polarizing optical microscopy (POM) indicated that the SC-film exhibited a smectic phase with a chevron texture and a very sharp diffraction signal at a small angle of 4.0°, which was spontaneously formed *via* the critical concentration point during the slow solvent evaporation in the process of the SC-film preparation, and the MP-film exhibited a nematic texture without highly ordered layer structure. The authors thus considered that the high-order to low-order phase transition occurred during the mechanical processing, which caused the polymer chains further random entanglement and physical crosslinking, providing the chance to form intermolecular π–π stack structure in the meta-stable state of the low-order phases (Figure 4.12). The newly formed intermolecular π–π stack structure could act as an energy trapping site. Thus, intermolecular energy transfer should occur from the excited species with the higher energy (shorter wavelength emission) to the excited species with the lower energy (longer wavelength emission).

4.3.2.1.3 MEH-PPV. Yang *et al.* (1999)[48] investigated the reversible mechanofluorochromism of poly(2-methoxy-5-(2′-ethylhexoxy)-*p*-phenylenevinylene) (MEH-PPV). On the basis of systematic investigations of the

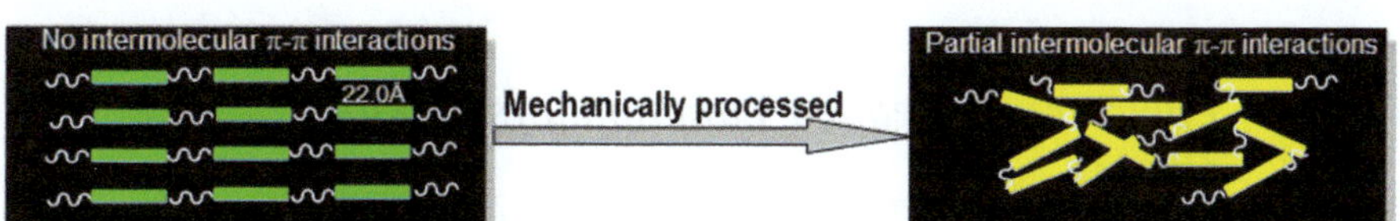

Figure 4.12 Schematic representation of the phase transition, molecular packing and interaction.

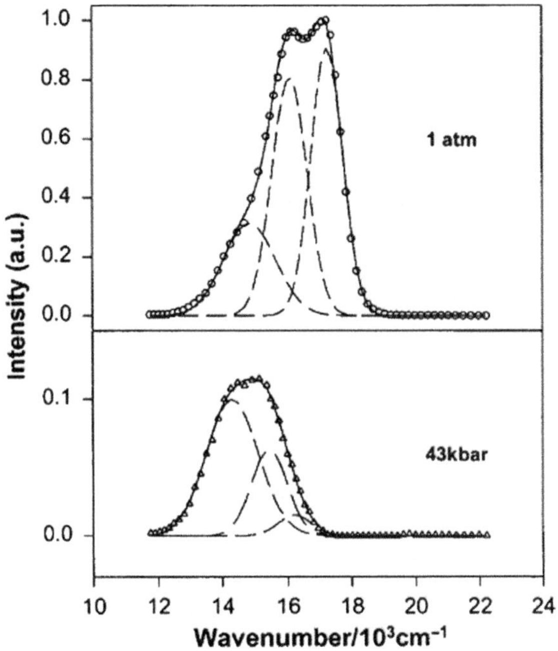

Figure 4.13 Emission spectra for neat MEH-PPV at 1 atm and 43 kbar with fitting bands: (○, △) experimental data; (- - -) fitting band; (-) sum of fitting bands.
Reproduced with permission from reference 48. Copyright 1999, American Chemical Society.

polymer, it was suggested that the emission changes upon compression (Figure 4.13) are due to increased chain interaction, which provides non-radiative pathways for relaxation.

4.3.2.1.4 Anthraquinone Polymers. In conjunction with their work on P3HT, Yamanoto *et al.* (2001)[44] also investigated the mechanochromism of poly(5,8-dihexadecyloxyanthraquinone-1,4-diyl). By applying a high pressure (3.3–11 GPa), the color of the pressed film sample changes from yellow to dark red. A later study by the same group reported that similar anthraquinones with shorter alkyl side chains did not show such significant mechanochromism.[49] Unfortunately, the photoluminescence of the compounds is too weak to observe the mechanofluorochromism.

4.3.2.2 Polymers Embedded Luminophors in Main or Side Chains

4.3.2.2.1 Pyrene-labelled Polydimethylsiloxane. Holder *et al.* (2009)[50] synthesized pyrene-substituted polyhydromethylsiloxanes (PHMS-Pyx) by the hydrosilylation reaction of prop-3-enyloxymethylpyrene with polyhydromethylsiloxane. The ratio of pyrene substituent to Si–H unit was varied to afford a range of pyrene-functionalised polysiloxanes. These copolymers were subsequently incorporated into polydimethylsiloxane (PDMS) elastomers by curing *via* either Pt(0) catalysed hydrosilylation with divinyl-terminated PDMS and tetrakis(dimethylsiloxy)silane, or Sn(II) catalysed condensation with α,ω-dihydroxyPDMS and tetraethoxysilane. An alternative method involving the synthesis and integration of [3-(pyren-1-ylmethoxy)propyl]triethoxysilane (Py-TEOS) into PDMS elastomers was also investigated: a mixture of α,ω-dihydroxyPDMS, tetraethoxysilane and Py-TEOS was cured using an Sn(II) catalyst. The experimental results of differential scanning calorimetry and dynamic mechanical analysis showed that no significant changes were observed in the thermal or mechanical properties of the elastomers containing pyrene when compared to otherwise identical samples not containing pyrene. All of the pyrene-containing elastomers were demonstrated to be fluorescent under suitable excitation in a photoluminescent spectrometer. To monitor the effects of stretching, two elastomer samples were mounted into an extensometer fitted inside a fluorimeter. The set-up of the extensometer and fluorimeter is illustrated schematically in Figure 4.14A. The extensometer was graded in 0.5 mm increments and stretching of the elastomers was performed by increasing and decreasing the extension in 0.5 mm increments up to a maximum of 3 mm; the region of the elastomer between the clamps was 15 mm and hence the maximum extension was 20%. The elastomer samples were placed in the special photoluminescence spectrometer and subjected to cycles of extension and relaxation (strain = 0–16.7%) while changes in the emission spectra were monitored. The resulting spectra of the elastomer containing PHMS-Py50 copolymers were variable and inconsistent. However, the emission peaks of elastomers containing Py-TEOS displayed clear and reproducible changes in fluorescence intensity upon stretching and relaxation. The intensity of the monomer and excimer emission peaks was observed to increase with elongation of the sample and decrease upon relaxation (Figure 4.14B). Furthermore, the ratio of the intensities of the excimer: monomer peak decreased with elongation and increased with relaxation. The above results suggested that fluorescent labeling of elastomers is a valid approach for the non-invasive *in situ* monitoring of stress and strain in such materials.

4.3.2.2.2 Cyano-OPV-labelled Polydimethylsiloxane. Weder *et al.* (2006)[13] investigated thermoplastic polyurethanes (TPUs) that comprise cyano-OPVs as the built-in deformation sensors. Their study involved both physical blends of a commercial TPU and small amounts of various cyano-OPVs, as well as copolymers produced by the covalent incorporation of a cyano-OPV into the TPUs based on poly(tetramethylene glycol), butanediol and 4,4′-methylene-bis(phenylisocyanate) (Scheme 4.3). The latter approach afforded the elastomers which displayed a predominant excimer emission in the non-stretched state (red-orange luminescence), and a pronounced

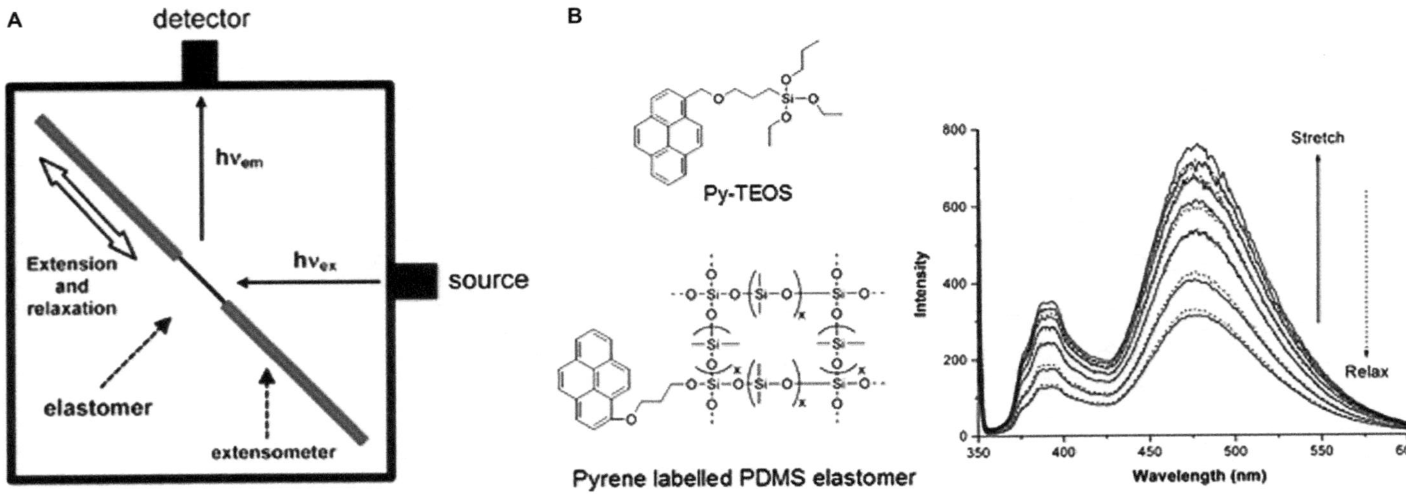

Figure 4.14 (A) Illustration of extensometer set-up inside the fluorescent spectrophotometer. (B) Change in emission spectra of the elastomer obtained by Sn(II) catalysed reaction (0.5 wt% Py-TEOS) with stretching (to 3.0 mm, 16.7% strain) and relaxation at 0.5 mm increments.
Reproduced with permission from reference 50. Copyright 2009, Royal Society of Chemistry.

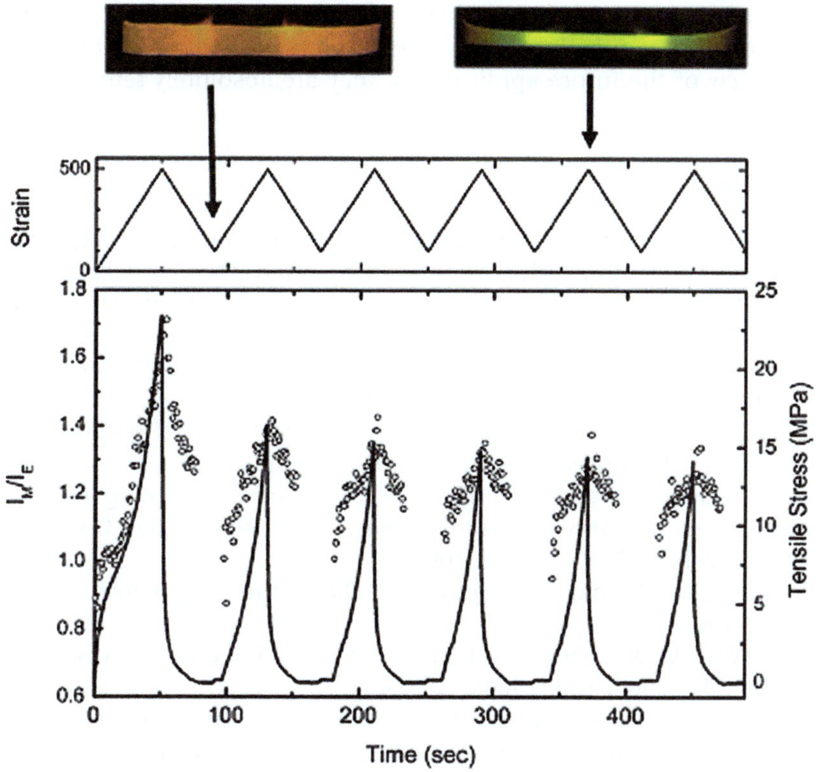

Figure 4.15 Ratio of unassociated to excimer emission, I_M/I_E (circle, measured at 540 and 650 nm), and tensile stress (solid line) for a covalent incorporation of cyano-OPV chromophores into polyurethane under a triangular strain cycle between 100% and 500% at a frequency of 0.0125 Hz and pictures ($\lambda = 365$ nm) of the same film in the unstretched (red) and stretched (green) states.
Adapted with permission from reference 1. Copyright (2013) The Royal Society of Chemistry.

fluorescence color change upon deformation (yellow-green luminescence). The optical change appeared mostly reversible, and reflected the stress-strain behavior of the elastomeric materials (Figure 4.15). The authors believed that the new materials appeared to nicely complement the intriguing, nonfluorescent mechanochromic TPUs reported by Rubner and co-workers[51–53] and Reneker's group[54] which comprise cross-linked diacetylenes or cis-azobenzene moieties, respectively.

4.4 Conclusion

In this chapter, we present an overview of mechanoflurochromism of covalently linked dye-polymers including reactive mechanophore-polymers and packing-changeable polymers. Their basic preparation method and mechanochromic mechanism have been described. The mechanoflurochromic covalently linked polymers are very limited. However, from the point of view of the future applications, they are absolutely tempting.

References

1. F. Ciardelli, G. Ruggeri and A. Pucci, *Chem. Soc. Rev.*, 2013, **42**, 857–870.
2. A. Pucci and G. Ruggeri, *J. Mater. Chem.*, 2011, **21**, 8282–8291.
3. J. N. Brantley, K. M. Wiggins and C. W. Bielawski, *Polym. Int.*, 2013, **62**, 2–12.
4. K. M. Wiggins, J. N. Brantley and C. W. Bielawski, *ACS Macro Lett.*, 2012, **1**, 623–626.
5. J. M. G. Cowie and V. Arrighi, *Polymers: Chemistry and Physics of Modern Materials*, CRC Press, Boca Raton, FL, 3rd ed, 2008.
6. K. Matyjaszewski and A. H. E. Müller, *Controlled and Living Polymerizations: From Mechanisms to Applications*, Wiley–VCH Verlag GmbH & Co. KGaA, Weinheim, Germany, 2009.
7. M. Beija, M.–T. Charreyre and J. M. G. Martinho, *Prog. Polym. Sci.*, 2011, **36**, 568–602.
8. S. L. Potisek, D. A. Davis, N. R. Sottos, S. R. White and J. S. Moore, *J. Am. Chem. Soc.*, 2007, **129**, 13808–13809.
9. D. A. Davis, A. Hamilton, J. Yang, L. D. Cremar, D. Van Gough, S. L. Potisek, M. T. Ong, P. V. Braun, T. J. Martinez, S. R. White, J. S. Moore and N. R. Sottos, *Nature*, 2009, **459**, 68–72.
10. V. Percec, T. Guliashvili, J. S. Ladislaw, A. Wistrand, A. Stjerndahl, M. J. Sienkowska, M. J. Monteiro and S. Sahoo, *J. Am. Chem. Soc.*, 2006, **128**, 14156–14165.
11. Y. Chen, A. J. H. Spiering, S. Karthikeyan, G. W. M. Peters, E. W. Meijer and R. P. Sijbesma, *Nat. Chem.*, 2012, **4**, 559–562.
12. M. Kinami, B. R. Crenshaw and C. Weder, *Chem. Mater.*, 2006, **18**, 946–955.
13. B. R. Crenshaw and C. Weder, *Macromolecules*, 2006, **39**, 9581–9589.
14. B. A. Beiermann, S. L. B. Kramer, J. S. Moore, S. R. White and N. R. Sottos, *ACS Macro Lett.*, 2012, **1**, 163–166.
15. G. O'Bryan, B. M. Wong and J. R. McElhanon, *ACS Appl. Mater. Interfaces*, 2010, **2**, 1594–1600.
16. C. K. Lee, D. A. Davis, S. R. White, J. S. Moore, N. R. Sottos and P. V. Braun, *J. Am. Chem. Soc.*, 2010, **132**, 16107–16111.
17. J. M. Lenhardt, A. L. Black, B. A. Beiermann, B. D. Steinberg, F. Rahman, T. Samborski, J. Elsakr, J. S. Moore, N. R. Sottos and S. L. Craig, *J. Mater. Chem.*, 2011, **21**, 8454–8459.
18. Y. Chen, A. J. H. Spiering, S. Karthikeyan, G. W. M. Peters, E. W. Meijer and R. P. Sijbesma, *Nat. Chem.*, 2012, **4**, 559–562.
19. T. A. Skotheim, R. L. Elsenbaumer and J. R. Reynolds, *Handbook of Conducting Polymers*, Wiley, New York, 2nd edn, 1997.

20. W. R. Salaneck, D. T. Clack and E. J. Samuelsen, *Science and Applications of Conducting Polymers*, Adam Hilger, Bristol, 1991.
21. K. Yoshino, S. Nakajima, H. B. Gu and R. Sugimoto, *Jpn J. Appl. Phys.*, 1987, **26**, L2046.
22. O. Inganäs, W. R. Salaneck, J.-E. Österholm and J. Laskso, *Synth. Met.*, 1988, **22**, 395–406.
23. T. Yamamoto, D. Komarudin, M. Arai, B.-L. Lee, H. Suganuma, N. Asakawa, Y. Inoue, K. Kubota, S. Sasaki, T. Fukuda and H. Matsuda, *J. Am. Chem. Soc.*, 1998, **120**, 2047–2058.
24. H. S. Nalwa, *Handbook of Organic Conductive Molecules and Polymers*, vol. 2, Wiley, Chichester, 1988.
25. O. Inganäs, G. Gustofsson and W. R. Salaneck, *Synth. Met.*, 1989, **28**, C377.
26. W. R. Salaneck, O. Inganäs, B. Jhemans, J.-O. Nilsson, B. Sjögren, J.-E. Österholm, J.-L. Bridas and S. Svensson, *J. Chem. Phys.*, 1998, **89**, 4613–4619.
27. K. Yoshino, K. Nakao and M. Onoda, *Jpn J. Appl. Phys.*, 1989, **28**, L323.
28. B. C. Hess, G. S. Kanner and Z. Vardeny, *Phys. Rev. B*, 1993, **47**, 1407–1411.
29. K.-I. Iwasaki, H. Fujimoto and S. Matsuzaki, *Synth. Met.*, 1994, **63**, 101–108.
30. T. Kaniowski, S. Niziol, J. Sanetra, M. Trznadel and A. Pron, *Synth. Met.*, 1998, **94**, 111–114.
31. Z. Bao, A. Dodabalapur and A. L. Lovinger, *Appl. Phys. Lett.*, 1996, **69**, 4108–4110.
32. H. Sirringhaus, P. J. Brown, R. H. Friend, M. M. Nielsen, K. Bechgaard, B. M. W. Langeveld-Voss, A. J. H. Spiering, R. A. J. Janssen, E. W. Meijer, P. Herwig and D. M. de Leeuw, *Nature*, 1999, **401**, 685–688.
33. I. McCulloch, M. Heeney, C. Bailey, K. Genevicius, I. MacDonald, M. Shkunov, D. Sparrowe, S. Tierney, R. Wagner, W. Zhang, M. L. Chabinyc, R. Klein, M. D. McGehee and M. Toney, *Nature Mater*, 2006, **5**, 328–333.
34. H. Kishida, K. Hirota, T. Wakabayashi, H. Okamoto, H. Kokubo and T. Yamamoto, *Appl. Phys. Lett.*, 2005, **87**, 121902/1–121902/3.
35. K. Yoshino, K. Nakao, M. Onoda and R. Sugimoto, *Solid State Commun.*, 1988, **68**, 513–516.
36. H. Isotalo, M. Ahlskog and H. Stubb, *Synth. Met.*, 1992, **48**, 313–323.
37. K. Yoshino, K. Nakao, M. Onoda and R. Sugimoto, *J. Phys.: Condens. Matter.*, 1989, **1**, 1009–1012.
38. B. C. Hess, G. S. Kanner and Z. Vardeny, *Phys. Rev. B*, 1993, **47**, 1407–1414.
39. K.-I. Iwasaki, H. Fujimoto and S. Matsuzaki, *Synth. Met.*, 1994, **63**, 101–108.
40. D. Moses, A. Feldblum, E. Ehrenfreund, A. J. Heeger, T.-C. Chung and A. G. MacDiarmid, *Phys. Rev. B*, 1982, **26**, 3361–3369.

41. T. Kaniowski, S. Niziol, J. Sanetra, M. Trznadel and A. Pron, *Synth. Met.*, 1998, **94**, 111–114.
42. Y. Muramatsu, T. Yamamoto, M. Hasegawa, T. Yagi and H. Koinuma, *Polymer*, 2001, **42**, 6673–6675.
43. E. J. Samuelsen, J. Mårdalen, O. R. Konestabo, M. Hanfland and M. Lorenzen, *Synth. Met.*, 1999, **101**, 98–99.
44. Y. Muramatsu, T. Yamamoto, M. Hasegawa, T. Yagi and H. Koinuma, *Polymer*, 2001, **42**, 6673–6675.
45. J. D. Barnett, S. Block and G. J. Peirmarini, *Rev. Sci. Instr.*, 1973, **44**, 1–9.
46. T. Sato, T. Yagi, H. Tajima, T. Fukuda and T. Yamamoto, *React. Func. Polym.*, 2008, **68**, 369–375.
47. W. E. Lee, C. L. Lee, T. Sakaguchi, M. Fujiki and G. Kwak, *Chem. Commun.*, 2011, **47**, 3526–3528.
48. G. Yang, Y. Li, J. O. White and H. G. Drickamer, *J. Phys. Chem. B*, 1999, **103**, 5181–5186.
49. T. Yamamoto, Y. Muramatsu, B.-L. Lee, H. Kokubo, S. Sasaki, M. Hasegawa, T. Yagi and K. Kubota, *Chem. Mater.*, 2003, **15**, 4384–4393.
50. N. A. A. Rossi, E. J. Duplock, J. Meegan, D. R. T. Roberts, J. J. Murphy, M. Patel and S. J. Holder, *J. Mater. Chem.*, 2009, **19**, 7674–7686.
51. R. A. Nallicheri and M. F. Rubner, *Macromolecules*, 1991, **24**, 517–525.
52. M. F. Rubner, *Macromolecules*, 1986, **19**, 2114–2128.
53. M. F. Rubner, *Macromolecules*, 1986, **19**, 2129–2138.
54. S. J. Kim and D. H. Reneker, *Polym. Bull.*, 1993, **31**, 367–374.

CHAPTER 5

Mechanofluorochromic Polymer Composites

SIWEI LIU AND ZHENGUO CHI*

PCFM Lab, State Key Laboratory of Optoelectronic Materials and Technologies, School of Chemistry and Chemical Engineering, Sun Yat-sen University, Guangzhou, China
*Email: chizhg@mail.sysu.edu.cn

5.1 Introduction

Polymeric composites, or dye-doped polymers, are a representative family of mechanochromic fluorescent materials. Depending on the dye's self-assembling characteristics and miscibility with polymers, fluorophores are generally considered to be aggregated into supramolecular assemblies above a particular concentration through weak non-covalent interactions among the planar aromatic backbones, such as π-π interactions. During aggregation, the interactions between the excited state of an aromatic molecule and the ground state of the same molecule would result in the formation of an excited dimer (excimer). Upon tensile deformation, the macromolecular chain slippage and reorganization promote the breakdown of non-covalent interactions among fluorophore molecules and their molecular mixing within the polymer matrix, which results in the transformation from an excimer to a monomer (Figure 5.1). Fluorescence from the excimer state generally has a lower energy than the corresponding monomer emission.[1,2] Therefore, the mechanically induced aggregation-to-disaggregation process of the dye assemblies often results in a significant blue shift in emission.

RSC Smart Materials No. 8
Mechanochromic Fluorescent Materials: Phenomena, Materials and Applications
Edited by Jiarui Xu and Zhenguo Chi

Published by the Royal Society of Chemistry, www.rsc.org

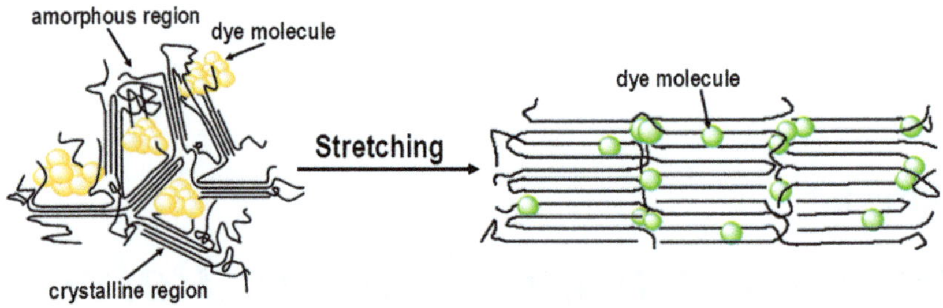

Figure 5.1 Schematic representation of dye molecules incorporated into a polymer matrix in an aggregated fashion (left) and in a molecularly dispersed (right) and photofluorescence color change under stretching. Adapted with permission from reference 2. Copyright 2011, The Royal Society of Chemistry.

In the case of functional dyes covalently tethered to the polymer chains, the mechanical solicitation induces ring-opening reactions causing selective chemical transformations of the mechanophore at the atomic level and the consequent color change. Conversely, for the dye-doped binary systems in which the active dye is physically dispersed in the polymer matrix, the optical responses towards mechanical stimuli are modulated in terms of intensity and selectivity by controlling interphase interactions and supramolecular conformations. The latter approach, even if providing materials with less structural stability than the former, appears to be a more sustainable route in material development as it allows for the use of different kinds of mechanochromic dyes already available on the market.[1]

Not all dye-doped polymer systems show mechanochromic fluorescent properties because of the complexity of molecular interactions, *i.e.*, the coactions between dye molecules and the interactions between dye molecules and the polymer matrix. Some criteria have been recently reviewed to guide the design of dyes for dye-doped polymer systems, whose key features are the:[1]

1. presence of a rigid-rod-shaped central part, which is highly aromatically conjugated, that can promote π-π stacking interactions between dye molecules;
2. existence of electron-withdrawing and/or electron-donating groups conjugated to the aromatic core of the dye, which can increase the dipole moment of the chromophore and modulate its optoelectronic characteristics;
3. absence of bulky functional groups or sterically hindered alky chains that can hamper dye self-assembly. Peripheral linear alkyl chains are sometimes introduced into dye molecules in order to increase their solubility in organic solvents and dispersibility in polymeric matrices. Moreover, these alkyl chains can render dye aggregates more ductile and favor their disruption upon mechanical deformation.

Meanwhile, the requirements for the polymers are as follows.

1. The polymer should be soluble in solvents or processable in melt to be mixed with the dyes, with the latter characteristic preferred in the case of industrial scale-up.
2. The material should be a thermoplastic semicrystalline polymer. Dye molecules are located in the amorphous phase of a polymer or in adsorbed state on the surface of polymer crystallites only. Thus, the presence of a crystalline phase helps in the formation of dye aggregates and is fundamental for mechanochromic mechanism.
3. The semicrystalline polymer should present a substantial amorphous phase below ambient temperature. This characteristic provides enough translational mobility of the dyes within the polymer matrices, as well as the generation of aggregates during solvent evaporation or upon cooling from the melt.

Therefore, the polymer matrix is crucial in dispersion, aggregation and disaggregation of the dyes, functioning much more than a mere passive supporting phase. Due to the strict criteria for the dyes and polymer matrices, mechanochromic fluorescent polymer blends are rather limited.

5.2 The Preparation of the Polymer Composites[3]

The basal principle of mechanochromic polymer blends is founded on color changes associated with the structural modifications of the molecular assemblies of dyes dispersed in thermoplastic polymeric matrix as a consequence of mechanical stimuli.

Dye–polymer dispersions can be realized either in solution or in the molten polymer mass by using compounding apparatuses, depending on the physico-chemical characteristics of the mixture's components. In the case of polymer matrices, such as poly(methyl methacrylate) (PMMA), polyethylene terephthalate (PET) and polystyrene (PS) composed of functional repeating units which can be highly compatible with the chemical structure of dyes, homogeneous dye–polymer solid mixtures can be obtained by film casting. Film casting involves dissolving polymer granules or powder, along with the desirable amount of dye in a suitable solvent. The solution is then poured into a mold and the film is obtained after solvent evaporation at room temperature or by heating (Figure 5.2A).

Problems can, however, arise when the polymer matrix is completely apolar and non-interacting with the dyes, thus strongly limiting their miscibility and compatibility. This is the case for dye–polyolefin mixtures based on polyethylene (PE) or polypropylene (PP) that can show phase-separation during casting and drying.[4] In this case, dye–polymer blends can be realized by means of continuous mechanical mixers, which are able to disperse the dye thoroughly in the bulk polymer.

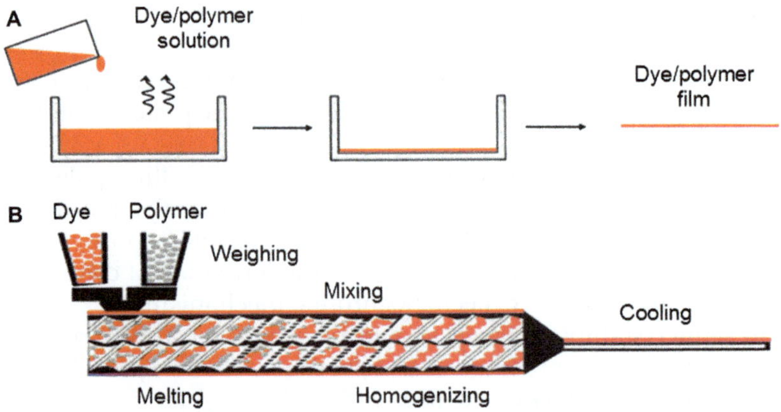

Figure 5.2 (A) Film casting in an open mould. (B) Melt-extrusion process of a dye–polymer material.
Reprinted with permission from reference 3. Copyright 2013, The Royal Society of Chemistry.

In a melt-blending process, the phase morphology is developed as a result of the shearing force overcoming the interfacial tension which tends to resist the deformation and break-up of dye agglomerates (Figure 5.2B).

When the blending process is stopped, the high viscosity of the polymer matrix would prevent the coalescence of dye particles from forming larger phase-separated aggregates.

As for the preparation of the dye-doped polymer blends, the above-mentioned methods only give a preliminary principle, and thus, concrete analysis should be made according to specific circumstances.

Some of the definite examples for the preparation of dye-doped linear low-density polyethylene (LLDPE) film in laboratory are as follows.

1. Guest-diffusion.[4] Films of a thickness of *ca.* 110 μm were produced by melt-pressing the LLDPE (*ca.* 500 mg) between two Mylar foils in a Carver laboratory press at 180 °C with a pressure of 2 tons for 5 min, using a 110 μm spacer. Strips of *ca.* 7 cm×1 cm of the resulting films were immersed for usually 16–18 h in a solution of the dyes in $CHCl_3$ at 60 °C or in toluene at 70 °C. The films were subsequently washed with $CHCl_3$ or toluene and dried in ambient conditions for at least 1 h.
2. Melt-processing.[5] Binary blends of LLDPE and the dyes with a dye concentration of more than 0.18% w/w were prepared by feeding the appropriate amount of dye and 4.0 g of LLDPE into a recycling, co-rotating twin-screw miniextruder (DACA Instruments, Santa Barbara, CA), mixing for 5 min at 180 °C, and subsequent extrusion. Blends comprising less than 0.18% w/w of the dyes were prepared by diluting a concentrated blend (prepared as described above) with neat LLDPE and extruding this material in the same manner. Blends comprising between *ca.* 0.40 and 3.0% w/w dye were prepared by manually mixing

500 mg of LLDPE with 2–15 mg of dye for 10 min in an aluminum dish on a hot stage at *ca.* 180 °C. Films were prepared by compression molding the blends between two aluminum foils and using four 110 μm spacers in a Carver press at 180 °C for approximately 3 min. The samples were immediately quenched after removal from the hot press by immersion in an ice water bath. The resulting blend films had a homogeneous thickness of *ca.* 100 μm.

5.3 Polymer Composite Systems

5.3.1 Cyano-OPV Derivatives

The concept of dye-doped polymers as molecular probes for film deformation was first developed by Weder *et al.* (2002).[4] They prepared two blends by using linear low-density polyethylene (LLDPE) as a matrix and two electron-withdrawing cyano group substituted oligo(*p*-phenylene vinylene) (cyano-OPV) derivative dyes BCMDB and BCMB (Figure 5.3) as the dopants by guest-diffusion or melt-processing (detailed methods are described in the above section). The phase behavior of cyano-OPV/LLDPE blends with 0.01–3 wt% of the dye could be adjusted in terms of the molecular structure of the probe, dye concentration and the processing conditions. Lower concentrations (<0.02 wt%) of the cyano-OPV derivates in the LLDPE films generally displayed monomer fluorescence emission (ranging from 448 nm to 540 nm), and higher concentrations (>0.02 wt%) generally displayed excimer fluorescence emission (ranged from 573 nm to 644 nm). The films containing dye molecules were thermodynamically unstable at ambient conditions, causing slow aggregation of the molecularly dispersed dye molecules, and hence an increase in excimer emission intensity. In order to speed up the aggregation of the dye molecules, the film samples were swelled in solutions of hexane (15 minutes), resulting in instantaneous aggregation. They demonstrated that the tensile elongations of the films displayed deformation-induced color changes along with the increase of elongation ratios (from 0% to 500%) with the monomer to excimer emission ratios I_M/I_E increasing up to a factor of 10. For example, it was found that for the film sample (0.20%wt, BCMB) a decrease in excimer fluorescence with increasing elongation produced a color change from green (excimer emission) to blue (monomer emission). The dye molecules were considered to form aggregates with multi-molecular stacking in the matrix, which results in the emission by excimers as well as dissociation and dispersion of aggregating molecules at the molecular level. The emission by the monomer was attributed to tensile deformation. Upon tensile deformation, the fluorescent color changed from orange to green for the dye BCMDB blend and from green to blue for the dye BCMB blend (Figure 5.3). These studies suggest that the new smart systems have a potential application in preparing sensors to monitor the material deformation.

The versatility of this approach to stress sensing was illustrated by the utilization of the cyano-OPV derivatives in a variety of polymers.

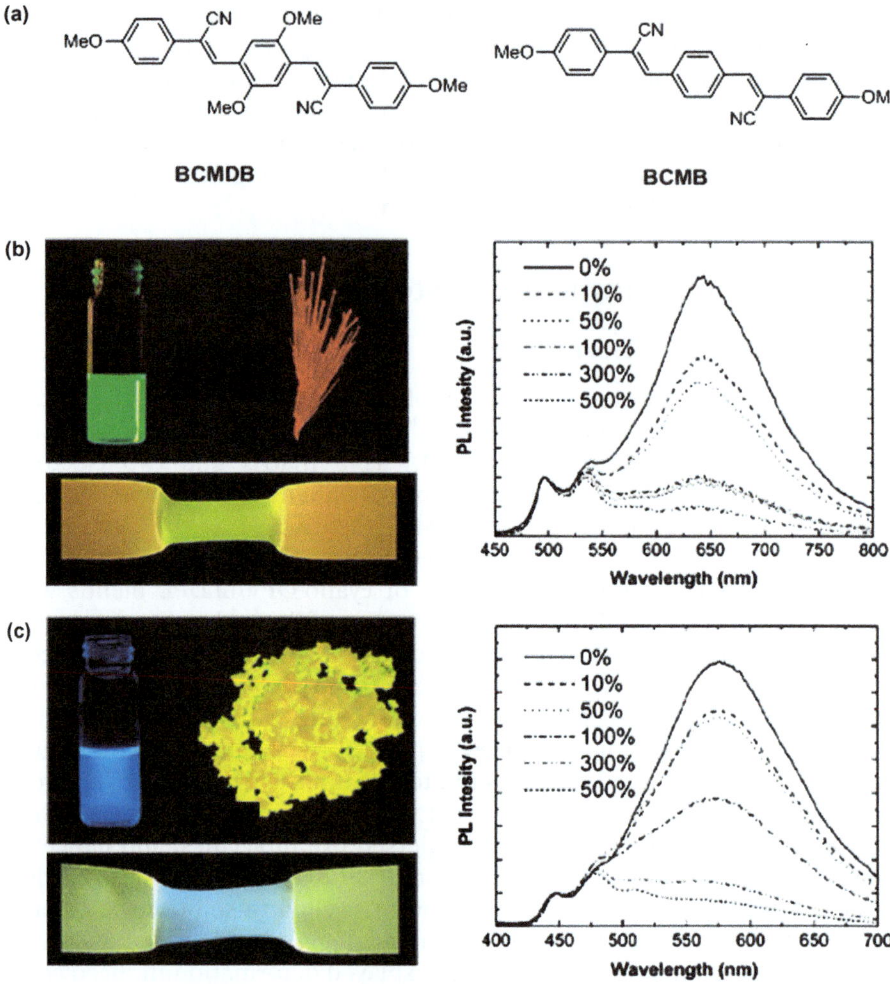

Figure 5.3 Chemical structures of oligo(*p*-phenylenevinylene)-based mechanochromic dyes BCMDB and BCMB (a). Macroscopic fluorescent images recorded under UV 365 nm irradiation for BCMDB (b) and BCMB (c) crystals and chloroform solutions, respectively. Also shown are fluorescence emission spectra recorded for blend films of linear low-density polyethylene (LLDPE) with BCMDB (b) or BCMB (c) upon stretching to varying draw ratios and macroscopic fluorescent images upon stretching to a draw ratio of 500%.
Reproduced with permission from reference 6. Copyright 2003, American Chemical Society.

Weder *et al.*[7–9] used cyano-OPV derivatives incorporated into PMMA or poly(bisphenol A carbonate) (PC) polymers and also into poly(ethylene terephthalate) (PET) or poly(ethylene terephthalate glycol) (PETG) *via* melt-processing techniques, for use as threshold temperature and deformation sensors. When the concept of mechanofluorochromism in dye-doped

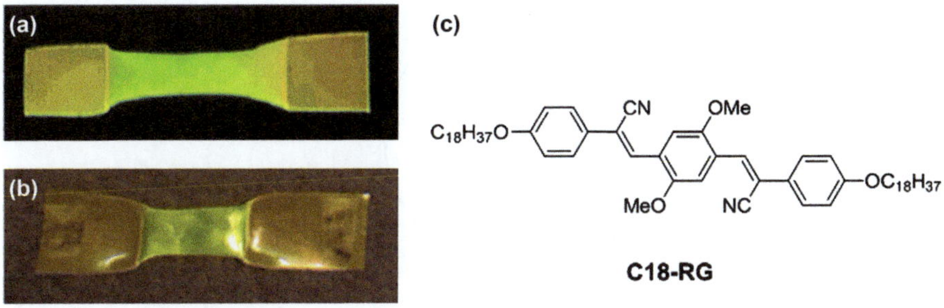

Figure 5.4 The pictures of the stretched blends were taken under 365 nm UV light: (a) PET and 0.9% w/w dye C18-RG; and (b) LLDPE and 1.0% dye **C18-RG**; (c) chemical structure of C18-RG. Imagines (a) reproduced with permission from reference 7.
Copyright 2006, American Chemical Society; (b) reproduced with permission from reference 9. Copyright 2006, Wiley-VCH.

polymers was more recently applied to PET and PETG dye-doped polymer blends, these polymers required the modification of dye BCMDB (Figure 5.4) because the compound was extremely soluble in the PET-based matrix.[7] Therefore, these polymer blends failed to undergo microphase separation and excimer formation. For this reason, the methyl group at the ends of dye BCMDB was replaced with an 18-carbon long alkyl chain. And thus a new dye product was obtained, C18-RG, which was used to improve phase separation and excimer formation. The blend of LLDPE and dye C18-RG also exhibited fluorescent color changes (Figure 5.4).[9] These results further indicate that the matching of the dye and the polymer is very important in the fabrication of a mechanofluorochromic blend system.

5.3.2 Bis(benzoxazolyl)stilbene

Bis(benzoxazolyl)stilbene (dye BBS), a commercially available chemical, has advantages over synthetic cyano-OPV dyes because it has a high resistance to solvent extraction, high thermal stability and low toxicity. Furthermore, the dye BBS is an ideal additive meeting the US Food and Drug Administration's (FDA) regulations for indirect food and consumer packaging materials. A series of further studies by Pucci *et al.* used fluorescent bis(benzoxazolyl)stilbene (BBS) as the dye dispersed into poly(propylene) (PP),[10] LLDPE,[11] poly-(1,4-butylene succinate) (PBS)[12] and other polyolefins[13] to study their mechanochromic, dichroic, anisotropic and thermal stimulation characteristics. BBS was dispersed into various polymer films (thickness from 80 μm to 200 μm) *via* melt processing techniques (*e.g.*, 0.02 wt% to 0.5 wt%). It was typically found that at low concentration (<0.02 wt%) the films displayed blue monomer emission and higher concentrations (>0.02 wt%) displayed green excimer emission. The inset in Figure 5.5 shows that the film samples were uniaxially stretched for PBS/BBS (0.1 wt% BBS), which displays emission color changes from green to blue.[12] This could be

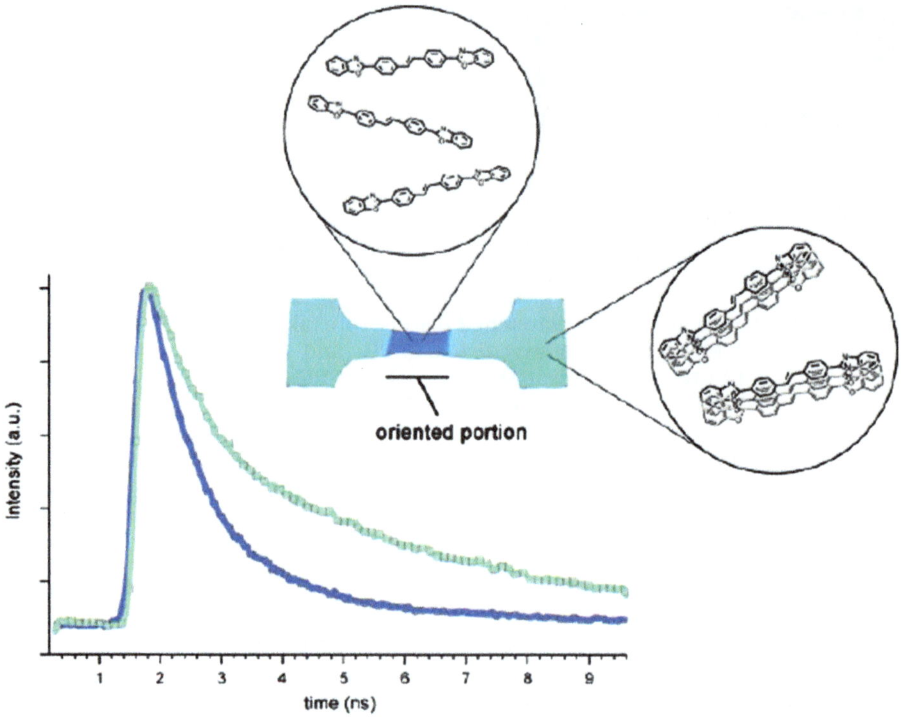

Figure 5.5 The chemical structure of BBS and the photoluminescence decay lifetime experiments performed on a dye BBS/PBS film containing the 0.1 wt% of dye BBS, before (open squares, green line) and after uniaxial deformation (open circles, blue line, 50% elongation). In the inset, picture of the same film taken under excitation with 366 nm UV light. Adapted with permission from reference 12. Copyright 2001, The Royal Society of Chemistry.

attributed to the separation of molecular dye BBS aggregates (excimers) as a consequence of PBS host matrix orientation and tensile film elongation. This result was confirmed by the lifetime experiments. Prior to orientation, the film was characterized by lifetime parameters that were typical of stilbene derivatives in isolated and rigid environments as well as in excimer-type arrangements. After film drawing, the long-lived component greatly reduced its contribution, from 10.4 ns to 0.95 ns, which resulted in a photoluminescent trace that was typical of non-interacting dye BBS mechanophores.

The thermal stability of the BBS supramolecular structure in PP had been qualitatively investigated by Pucci *et al.* (2005)[10] by placing the film containing 0.5 wt% of BBS mechanophores in contact with a hot plate at a temperature of 130 °C for more than 10 min. The films maintained their green excimer emission (Figure 5.6a and b) under the excitation of a long-range UV lamp (366 nm), different to the recently investigated oligo(*p*-phenylene vinylene) derivatives dispersed in linear low-density polyethylene (LLDPE), in which the excimer emission was sensitive to heating.[6]

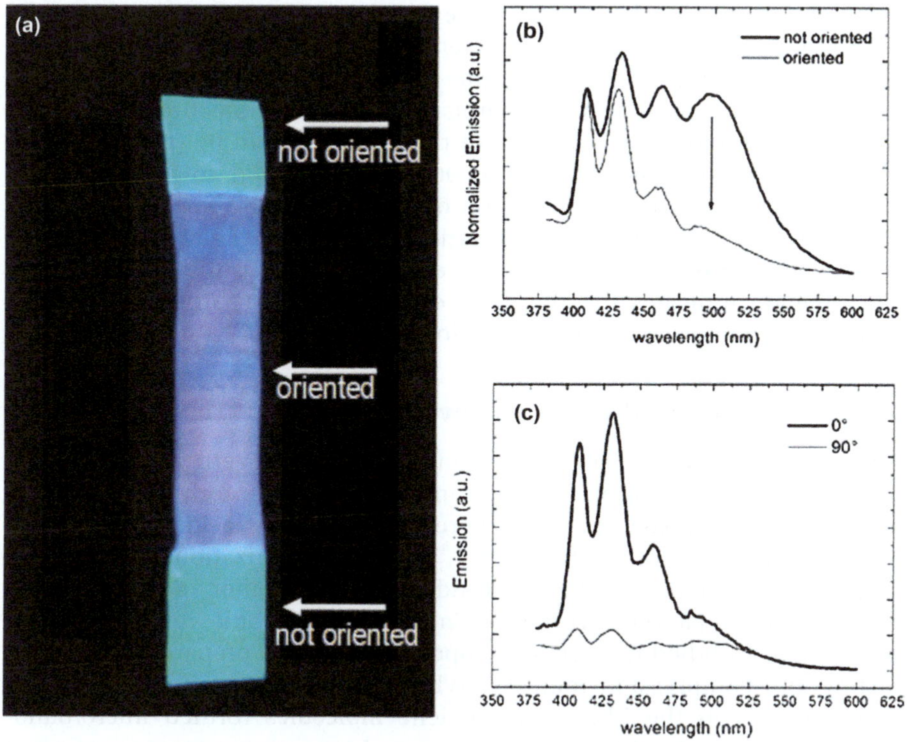

Figure 5.6 (a) Digital image of the uniaxially oriented PP/BBS-0.5 film taken under excitation with a long-range UV lamp ($\lambda = 366$ nm). (b) Fluorescence emission spectra ($\lambda ex = 277$ nm) of a PP/BBS-0.5 film containing 0.5 wt% of BBS molecules, before and after solid-state drawing (Dr = 8). The spectra are normalized to the intensity of the isolated BBS molecules peak (409 nm). (c) Fluorescence emission spectra ($\lambda ex = 277$ nm) of the uniaxially oriented PP/BBS-0.5 film recorded with polarisation parallel (0°) and perpendicular (90°) to the drawing direction, respectively. Reproduced with permission from reference 10. Copyright 2005, Wiley-VCH.

In this case, the different preparation procedure of the PP films (slow cooling to room temperature) with respect to the LLDPE samples (rapid quenching of the films at 0 °C after removal from the hot press)[6] may lead to the formation of more thermodynamically stable aggregates of the dye molecules. The combination of the preparation procedure with the type of dye molecules (low solubility in PP and a melting point of approx. 360 °C) and with the polymer properties (PP instead of LLDPE) renders BBS aggregates more stable to thermal stress even at high temperature (up to 130 °C). The PP/BBS film has also been uniaxially stretched at different draw ratios at the temperature of 130 °C, and the oriented portion of the film changes its emission from green to blue, which is the typical luminescence of the isolated BBS molecules. The effective anisotropic distribution of BBS molecules along the oriented macromolecular PP fibers after solid state drawing is confirmed by fluorescence spectroscopy measurements under the polarized light. These have been performed by exciting the stretched tape with an unpolarized radiation and monitoring the dichroism of the emission through a linear polarizer. By rotating the direction of the polarizer from 0 ° (parallel to the stretching direction of the film) to 90 ° (perpendicular to the stretching direction), the emission contribution of the single BBS molecules strongly decreases with a dichroic emission ratio of about 10 (evaluated as the ratio between the emission at 0 ° and the emission at 90 °) (Figure 5.6c). This phenomenon indicates the very high dichroic distribution of the isolated BBS chromophores after PP deformation.

5.3.3 Perylene and Its Derivatives

Perylene consists of five planar benzene rings fused together, whose structure and electronic conformation strongly promote the formation of luminophor aggregates (such as J-type aggregates) through stacking interactions between the bulky conjugated cores.[14] The self-organization of functional perylenes was reported to be fundamental to tailor defined multichromophoric materials with modulable opto-electronic response. Pucci *et al.* (2009)[15] studied the perylene-doped LLDPE films prepared by solution casting and compression molding. When the perylene content was more than 0.1 wt% in the films, the perylene molecules formed micro/nanostructured aggregates. These aggregates were J-type aggregates, as confirmed by the UV-vis absorption spectra and the emission color changes of the films from bright blue (non-interacting dyes) to low-intensity pale yellow (interacting dyes). After drawing the pale yellow emission film, the stretching part transformed into blue emission. The reorganization of LLDPE macromolecules was considered to break the perylene supramolecular organization (J-type aggregation) during film drawing, which led to the abundant formation of blue emission from the monomeric perylene molecules.

The versatility and reactivity of perylene tetracarboxylic acid bisanhydride allows perylene dyes to be easily functionalized by a multitude of aromatic and/or aliphatic moieties in order to enhance their opto-electronic properties or to strongly increase their solubility in several dispersing media.[16–19] Therefore, perylene derivatives were also used to prepare dye-doped mechanofluorochromic polymer blends.[20] Two different perylene bisimides, namely dyes EE and ES, were doped into LLDPE at specific contents from 0.01 wt% to 0.1 wt%. Both EE and ES dyes were likewise found to generate supramolecular aggregates (J-type aggregates) with a relatively high content, and were promoted by π-π intermolecular interactions between the conjugated planar structures. This phenomenon was confirmed by spectroscopic investigations on dispersions into the LLDPE polymer matrix. The films exhibited yellow-green and red emissions in low and high dye contents, respectively (Figure 5.7a). The red shift was also attributed to the formation of J-type aggregates of the dye molecules. The more planar and less hindered

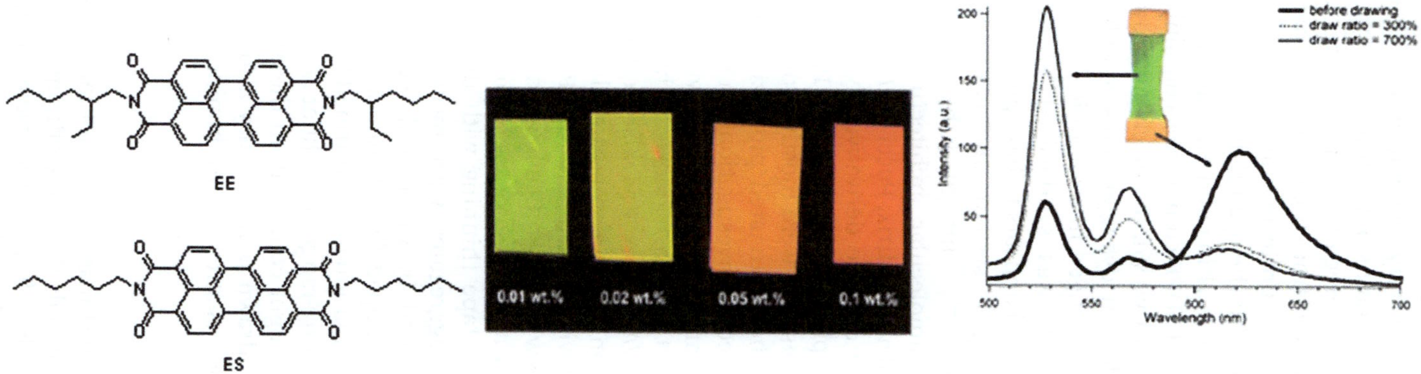

(a) (b) (c)

Figure 5.7 (a) Chemical structures of dyes EE and ES; (b) images of dye EE-LLDPE films at different dye concentrations taken under irradiation at 366 nm; (c) emission spectra (excited at 300 nm) of dye EE/LLDPE films containing the 0.1 wt% of dye before and after orientation at different draw ratios (300% and 700%). In the inset, a picture of the oriented film (300%) under irradiation at 366 nm.
Adapted with permission from reference 20. Copyright 2008, American Chemical Society.

dye ES, which was caused by the linear alkyl chains, induced the formation of chromophoric aggregates even at very low concentrations (0.01 wt%) as compared with the more distorted chains of the branched dye EE derivative, which promoted luminophor aggregation at higher concentrations (>0.05 wt%). However, the aggregates from the dye ES were more resistant to mechanical stretching and had a less uniform orientation along the polymer drawing directions as compared with the dye EE-doped LLDPE films, which showed excellent response upon stretching and emission changes from red to yellow-green (Figure 5.7b). For LLDPE composites containing the highest concentration of dye EE (that is, 0.1 wt %), the mechanical drawing was not able to disrupt all the intermolecular dye aggregates completely. Nevertheless, the latter was shown to be arranged in a dichroic fashion.

Water soluble perylene diimide was synthesized by several research groups since this class of compound appears to have potential for antitumor drugs, fluorescence tags of biological systems and as the elements in the self-assembly of photoactive films.[21–28] Intrigued the possibility of generating multi-chromophoric supramolecular architectures even when dispersed into a semicrystalline polymer matrix, Pucci *et al.* (2009) reported the use of a water soluble perylene diimide derivative (dye PZPER) as an innovative luminescent probe for plastic films. The dye PZPER was found to generate strong H-type aggregates that were mostly promoted by hydrogen bonding interactions between the protonated piperazine side units. This result was confirmed by spectroscopic investigations on the dispersions into the poly(vinyl alcohol) (PVA) and poly(ethylene-co-vinyl alcohol) (EVA) matrices at low loadings (from 0.1 wt% to 1.1 wt%).[29] The stability of PZPER aggregates dispersed into a PVA matrix was also tested by applying a mechanical uniaxial deformation to the composite film at 110 °C. As reported in Figure 5.8a, the UV-vis spectrum of the PVA film containing 0.5 wt% of PZPER did not show any significant variations in the optical features after deformation; only a decrease of the total absorption was detected as a consequence of the tape thinning with drawing. However, a significant rupture of PZPER aggregates with mechanical deformation seemed to occur, as demonstrated by the partial recovery of the luminescence from the oriented PVA-doped film (Figure 5.8b). The increase of the emission intensity after deformation was confirmed by exposing the PVA composite film to excitation by long-range UV light at 366 nm (Figure 5.8c). The more intense orange-yellow light coming from the oriented portion of the film with respect to the undrawn portions may also suggest potential applications of PVA/PZPER tapes as stress–strain luminescent indicators.

5.3.4 AIE Compound

Aggregation induced emission (AIE) materials are an important class of anti aggregation-cause quenching materials that more efficiently emit while in the aggregated state than in the dissolved forms, which were reported for the first time by Tang's group in 2001.[30] Since then, AIE materials have attracted

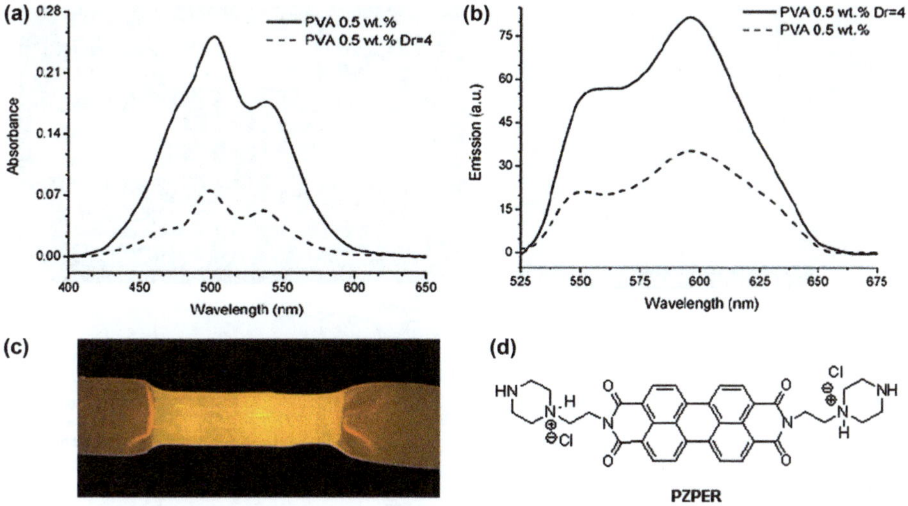

Figure 5.8 (a) UV-Vis absorption and emission (lexc = 450 nm) of a PVA film containing the 0.5 wt% of PZPER before and after orientation (Dr = 4); (c) digital image of the oriented film (Dr = 4) under irradiation at 366 nm; (d) the chemical structure of PZPER.
Reproduced with permission from reference 29. Copyright 2009, The Royal Society of Chemistry.

considerable research attention for their potential application in various fields, such as organic light-emitting devices and chemosensors.[31–40] A number of AIE compounds have also been recently synthesized in Chi and Xu's group, and interestingly, most of them have been found to possess mechanochromic luminescent nature. Hence, AIE has been considered as a key for opening a treasure chest of mechanochromic luminescent materials.[41] In Chapter 6 of this book, AIE mechanochromic luminescent materials are described in detail. In this chapter, an AIE dye-doped PMMA blend is presented as an example.

The AIE compound reported to exhibit highly enhanced emission in the solid state compared with the solution by Park *et al.* (2010) was a cyano-substituted distyrylbenzene derivative (DBDCS, Figure 5.9).[42] In the solid state, it could interchange between two different phases: a metastable green-emitting G-phase and a thermodynamically stable blue-emitting B-phase, upon the application of external stimuli, such as heat, mechanical stress and solvent vapor. As a practical fast-responding demonstration of multi-stimuli luminescence tuning, a green-emitting (G-phase) composite film of DBDCS in PMMA (20 wt%) was prepared by spin coating on glass. The surface morphology of the film indicates a phase separation with small grains of about 10 nm size (Figure 5.9a). When thermal stimulus (125 °C for 10 s with hot letter stamp) was applied, the green fluorescence changed to blue, in accordance with the G- to B-phase transition. The thermally annealed B-phase film showed high pressure-sensitivity, changing the emission color immediately to green, even with a very small shear force, thus allowing for

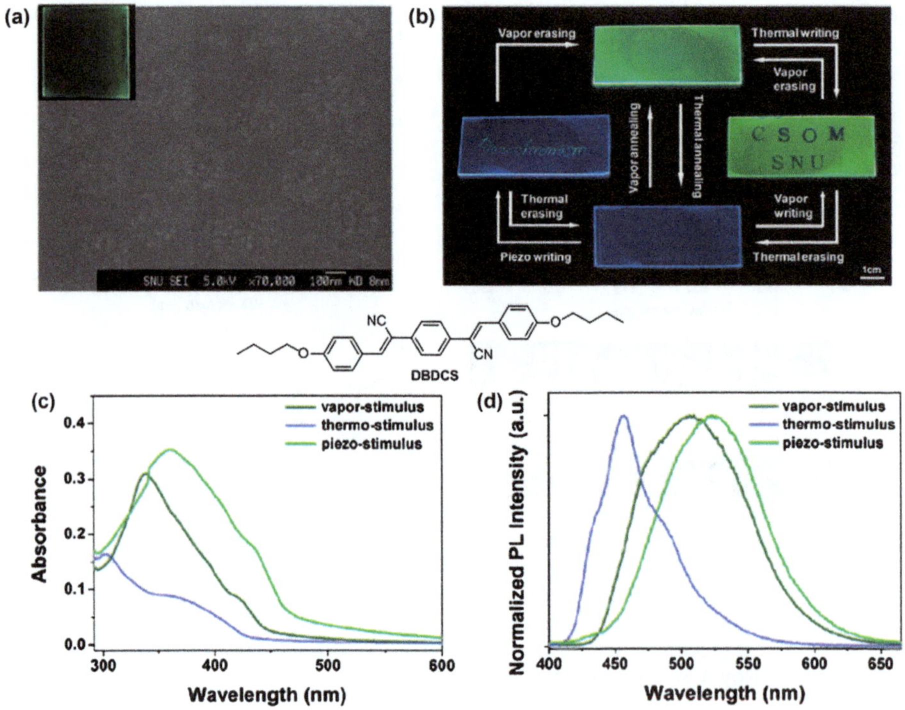

Figure 5.9 (a) The surface morphology of DBDCS (20 wt%) and PMMA composite film which was prepared by spin coating on glass. Inset shows a green fluorescence image of the film under UV light; (b) photos of the luminescence writing/erasing cycle of the film; (c) UV-visible absorption spectra of the film under vapor- (dark green line), thermo- (blue line), piezo- (green line) stimulus; (d) PL spectra of the film under vapor- (dark green line), thermo- (blue line), piezo- (green line) stimulus.
Reproduced with permission from reference 42. Copyright 2010, American Chemical Society.

sensitive piezo-writing, as shown in Figure 5.9. Upon exposing the whole films to the organic vapor (in CH_2Cl_2 for 30 s), it was able to erase this piezo-writing, changing the entire emission color to green (Figure 5.9b). The possibility of vapor adsorption in altering the fluorescence color is excluded by vacuum treatment after vapor exposure. In any case, it should be noted that this G-phase induced by the exposure to vapor is basically the same, but is slightly different from the G-phase induced by shear force, as evident from the comparison of the emission spectra in Figure 5.9d as well as the absorption spectra in Figure 5.9c. This implies that vapor exposure also promotes an alternation of the slip-stacking, although to a somewhat different degree as compared to the case of pressure application. The authors demonstrated that, under all these conditions, a reproducible and fully reversible two-color fluorescence writing/erasing process was achieved and both emission colors of blue and green were found to be persistent over one year of observation under ambient conditions, and they believed that it could be used as a new rewritable fluorescent optical recording media.

5.3.5 Organometallic Complex

Luminescent organometallic complexes in solid state have attracted the attention of material scientists and chemists because of their potential applications in various fields, such as organic light-emitting diodes.[43,44] In the solution state, the luminescence properties depend on the chemical structure of the organometallic complex,[45] while in the solid state, they also rely on the molecular orientation.[46,47] External stimuli, such as mechanical pressure and exposure to volatile molecules would affect the molecular orientation; therefore, they can change the luminescence in solid state. For example, organometallic complexes exhibit dramatic luminescence color changes through mechanochromism and vapochromism.[48–50] Since the chromisms are caused by changes in the intermolecular interactions such as π-π stacking and metal-metal interactions, planar structures of pincer complexes are suitable for demonstrating interesting luminescence color changes.[51] Thus, in this, we will focus on the development of mechanochromic luminescent organometallic complexes.

Although many organometallic complexes with mechanochromic luminescent properties have been reported, mechanochromism of organometallic complex doped polymer has been much less studied. Here, we present the case reported by Rowan's group (2012)[52] studying the mechanochromic phenomenon of a square-planar platinum(II) complex of the 4-dodecyloxy-2,6-bis(N-methyl benzimidazol-20-yl) pyridine ligand ([Pt(1)Cl](PF6)) dispersed in PMMA. It was found that scratching any of the films containing [Pt(1)Cl](PF6) resulted in a change in the emission color from yellow to orange at the site of abrasion. The effect was most apparent when the film was acutely deformed, such as by scratching/cutting with a razor blade, sharply striking with the hemispherical head of a ball-peen hammer, as shown in Figure 5.10c for a 10 wt% [Pt(1)Cl](PF6) in solution cast PMMA film. For each of the pressed films, little-to-no visible color change was observed, however, the emission shifted to longer wavelengths and increased in intensity in a manner somewhat reminiscent of the response to acetonitrile vapor (Figure 5.10b). For example, the emission spectra of a PMMA film with 20 wt% [Pt(1)Cl](PF6) are shown in Figure 5.10d before and after compression. While the emission of the compressed film is more symmetric, less structured and red shifted ($\lambda max = 592$ nm) as compared to the as-cast film ($\lambda max = 575$ nm, Figure 5.10a), it is apparent that the differences are not as dramatic as those observed for similar films exposed to acetonitrile vapor. In an attempt to probe the origin of this mechanochromic behavior, the structural properties of the platinum complexes in the compressed films were characterized using a wide-angle X-ray scattering (WAXS). The observed scattering pattern for a PMMA film with

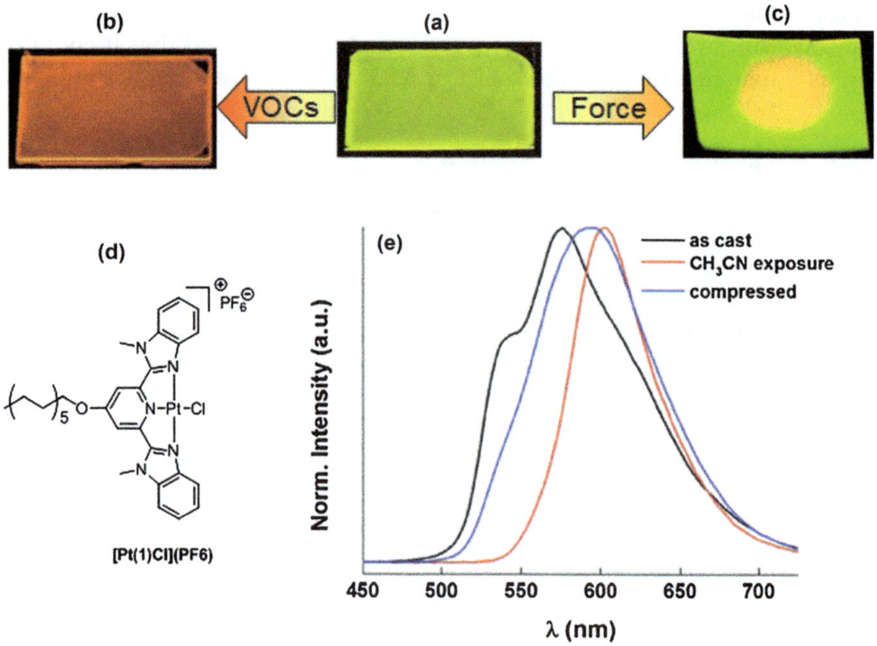

Figure 5.10 Chemical structure of the Pt(II) complex [Pt(1)Cl]PF6, image of the mechanochromic response of a 10 wt% [Pt(1)Cl](PF6) in PMMA which was hit sharply with a ball-peen hammer and normalized emission spectra of a pressed 20 wt% [Pt(1)Cl](PF6) in PMMA film compared to the as-cast and acetonitrile exposed emission.
Adapted with permission from reference 52. Copyright 2009, The Royal Society of Chemistry.

20 wt% [Pt(1)Cl](PF6) shows four distinct peaks occurring at $2\theta = 24.4$, 15.46, 11.02 and 5.52°, which equate to distances of 3.38, 5.73, 8.03 and 16.01 Å, respectively. The two shortest distances match very closely to those of the uncompressed films (3.37 and 5.72 Å) corresponding to alternating the short and long Pt–Pt distances. The larger distances are consistent with longer Pt–Pt distances which may be related to complexes separated by one or a multiple of other complexes. Thus, the mechanism behind the mechanochromism has been shown to be related to a force-induced structural rearrangement, which results in an increased number of shorter intermolecular Pt–Pt interactions.

5.3.6 Polymeric Dyes

Molecular chain relaxation, orientation and residual stress in the final polymer products have a strong impact on their performances. For example, polymer products made by injection, extrusion, compression and blow molding usually have residual stresses that may considerably influence their properties, such as modulus, strength and thermal expansion. Also, the stresses cause fracture, deformation, flow marks, and cracking on exposure to an aggressive environment. The measurement of the residual stresses, therefore, is important for understanding the reliability of the products. The most popular method for monitoring the stresses is the layer-removal procedure, whereas mechanical shears damage the polymer during the layer removal and affect the stresses. Thus far, a variety of nondestructive methods including X-ray diffraction, laser Raman spectroscopy and photoelasticity have been proposed. These methods, however, have some disadvantages. X-ray diffraction and laser Raman methods are utilized only for measuring stresses >1 GPa. The photoelastic method is only suitable for polymeric materials.[53] As a probe of the photoelastic method, polymeric dye has been used widely in the practical analysis.

Here, some representative examples have been chosen to describe the above-mentioned applications of the polymeric dyes.[54]

Polybutadiene tagged with anthracene was synthesized and used as a fluorescent molecular probe by Bur *et al.* (1992)[55] to study shear-induced orientation in a matrix polymer melt. With the tagged polybutadiene doped into a polybutadiene matrix at 0.1 wt% concentration, steady-state fluorescence anisotropy measurements were carried out under zero shear and under finite shear conditions using an optically instrumented cone and plate rheometer. Measurements were made over a shear rate range for which the specimen displayed non-Newtonian behavior, 2.64×10^{-3} to 5.3 s^{-1}. Anisotropy was observed to decrease with increasing applied shear stress. The magnitude of the effect is small and is attributed to the shear-induced orientation of the probe molecule which is engaged in the entanglement network of the host polymer. Diluting the entanglement network using plasticizer produced a smaller effect. For polybutadiene plasticized with 50% cetane, it was observed that anisotropy was independent of the applied shear stress, indicating that the probe molecule was not participating in the orientation of the matrix entanglement network. A relationship between anisotropy, chromophore relaxation time and orientation factors was derived and used to deduce an orientation distribution of fluorescent absorption dipoles and illustrate the difference between shear and extension stress observations. Extension experiments, carried out using a cross-linked polybutadiene specimen, showed that anisotropy increased as a function of applied extensional stress.

The steady-state fluorescence properties of naphthalene-labeled polymers dispersed in the poly(methyl methacrylate) (PMMA) cast films were studied by Yang *et al.*[56,57] under tensile loadings at 80 °C. The labeled polymers were composed of methyl methacrylate (MMA) and 1-naphthylmethyl methacrylate (NMMA). Three of the copolymers contained NMMA 0.59, 22.0 and 56.7 mol%, respectively. The fluorescence spectra of the 0.59 and 22.0 mol% NMMA films were unchanged during elongation. For the 56.7 mol% NMMA film, the excited monomer emission of naphthyl groups at around 337 nm decreased with increasing applied tensile strain. The strain enhanced the emission ascribed to the excimer of the naphthyl groups in the region of 390–420 nm. The ratio of fluorescence intensities at 400 nm and 337 nm, I_{400}/I_{337}, increased with the applied strain, which indicates that the 56.7 mol% NMMA film is a sensitive probe for detecting the structural changes of polymer matrices. The obtained results mean that the excimer-forming sites in the PMMA films during elongation depend both on the applied strain and the concentration of naphthyl groups in the dispersed polymer probes. The authors believed that the conformation of the chains is frozen in polymer films. An excited aromatic ring attached to the chain can only interact with the nearest ring of the ground state to form an excimer site. The distance between the two aromatic rings for excimer formation is supposed to be in the order of 0.4 nm. It is known that upon drawing, the polymer chains will respond to the imposed strain *via* orientation of the chain segments toward the stretching direction. The orientation makes the chains expand, which is expected to disrupt some of the preformed excimer sites in solid film. On the other hand, the orientation leads to highly dense packing of the macromolecules, which facilitates the aromatic rings approaching each other to form new excimer sites. Hence, it can be seen that excimer formation and dissociation in a polymer film during elongation is a competitive process, as shown in Figure 5.11.

The earliest instance found in the literature of utilization of a fluorescent extended π-conjugated system in stress–strain analysis was demonstrated by Ikawa *et al.* (1998)[58] They used poly(3-alkylthiophenes) with butyl, hexyl, octyl and dodecyl chains to disperse into PMMA. The decay time of fluorescence decreased significantly over the elastic deformation range of PMMA. Over the plastic strain range of PMMA, the steady-state fluorescence showed significant changes with a red shift of the fluorescent maximum and a decrease in fluorescence intensity with strain (Figure 5.12a). Decay times were observed to increase with strain over the plastic deformation range, though the change was relatively small (Figure 5.12b).

Ikawa *et al.* (1997)[59] dispersed poly(N-vinylcarbazole) (PVCz) fluorophores into a polystyrene (PS) matrix at a concentration of 0.05 wt%. The cast films were stretched to strain values of 0–0.8% and the fluorescent emission spectrum showed a general decrease in monomer emission (340–360 nm) and an increase in the partial overlap excimer emission (360–430 nm). The

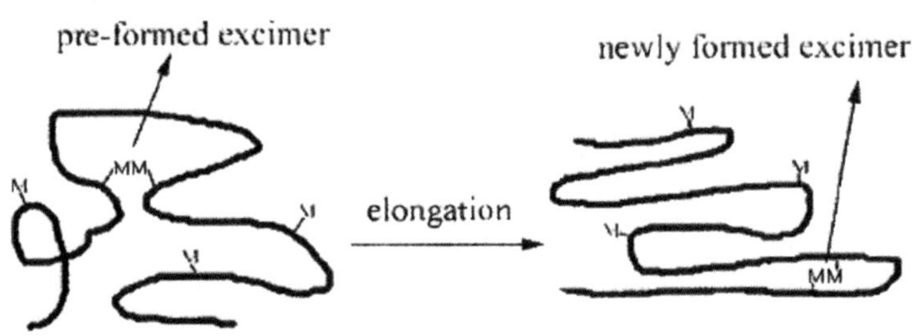

Figure 5.11 Proposed mechanism of excimer formation and dissociation during elongation in a chromophore.
Reproduced with permission from reference 56. Copyright 2001, Wiley-VCH.

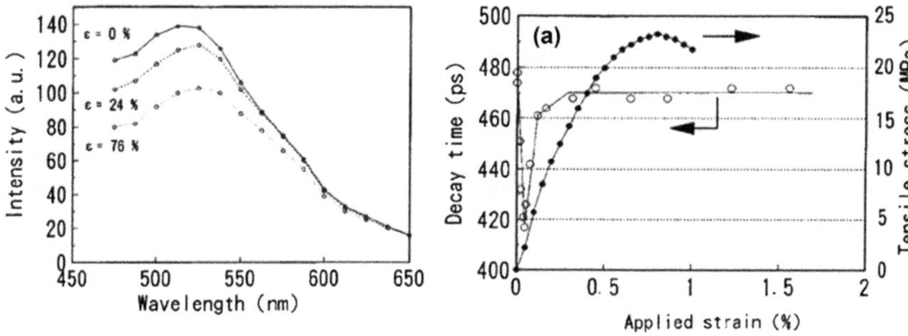

Figure 5.12 Fluorescence spectra and fluorescence decay times of P3HT in PMMA with strain.
Reproduced with permission from reference 58. Copyright 1998, Wiley-VCH

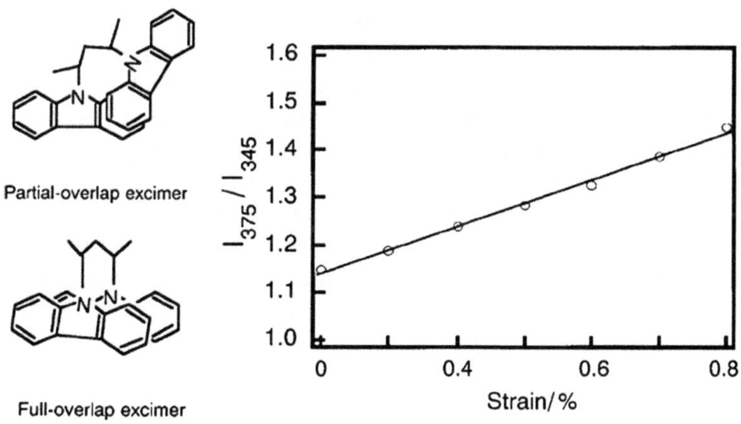

Figure 5.13 Partial- and full-overlap excimers of two carbazolyl chromophores of PVCz (left) and relationship between the ratio of fluorescence intensities at 375 and 345 nm (I_{375}/I_{345}) and applied tensile strain (right).
Reproduced with permission from reference 59. Copyright 1997, Wiley-VCH.

emission due to the full overlap excimer emission (430–500 nm) was found to be unchanged with the increase of strain, and was conceived to be due to the restriction of the molecular motion of the PVCz chains by the rigid PS polymer matrix. The ratio of the fluorescent emission intensities (I_{375}/I_{345}) of the excimer (at 375 nm) to the monomer (at 345 nm) was proportional to the applied strain (Figure 5.13). The film was stretched to a strain value of 0.4% with an increase of the intensity ratio (from 1.15 to 1.23) and held whilst stress–relaxation occurred (from 15 to 10 MPa), during which the intensity ratio remained the same. Upon the removal of the applied strain, the intensity ratio was found to return to its initial value, leading to the assumption that the intensity ratio is only under the influence of strain, not stress.

Spanggaard *et al.* (2003)[60] prepared four different strain probes with one or two carbazole units placed in the middle of a polystyrene–polyisoprene–polystyrene (SIS) type triblock copolymeric elastomer and investigated the use of PVCz-polymers as stress-sensors in commercial SIS type rubbers. The fluorescence spectra of the chromophores in the solution are remarkably similar and all show fluorescence expected for a simple monomeric carbazole type chromophore. This is in contrast to the very similar 2,4-(*N*-carbazolyl)pentanes studied by Vandendriessche *et al.* (1984),[61] where evidence for partial and total overlap between the carbazole units was seen in the excimer type spectra obtained. The strain probes 1–4 were mixed into commercial SIS rubbers at the 0.1% level and made into film strips. When these films were subjected to uniaxial elongation, large changes from 0.7 to 1.1 in the relative fluorescence intensities at two wavelengths (350 and 370 nm) were nonetheless observed (Figure 5.14). The largest change was found for strains around 0–200% and leveled out above 300–400%. Since the results are similar for the mono- and bis(carbazole) probes, changes in molecular overlap giving rise to excimer type fluorescence can be excluded. The observed changes must instead be due to subtle changes in the transition probabilities of the vibronic bands of the carbazole units. This finding casts doubt on the interpretation by Ikawa *et al.* (1997)[59] on the fluorescence spectra for the stretched PVCz films. It was assumed that the changes observed were related to the different ratios of overlap geometries in response to the induced strain. While this may still be the case, it cannot be excluded that the variations observed was instead related to fluorescence changes in the monomer carbazole units themselves. The preparation of single chromophoric units imbedded in the polymer matrix proved to be crucial in separating the factors responsible for the fluorescence changes.

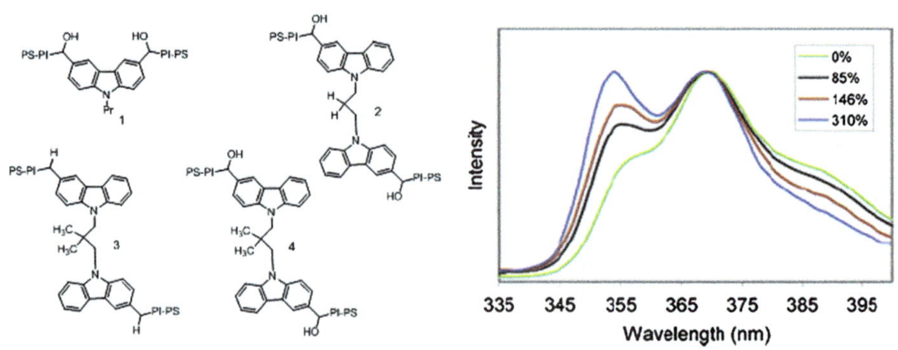

Figure 5.14 Chemical structures of the polymeric dyes and representative normalized PL spectra at varying strain of 2 dissolved in a commercial SIS elastomer.
Adapted with permission from reference 60. Copyright 2003, American Chemical Society.

A cyano-containing poly[(*m*-phenylene ethynylene)-alt-(*p*-phenylene ethynylene)] derivative (CN-PPE) has been prepared and homogeneously dispersed with micro/nano-sized domains in a linear low density polyethylene (LLDPE) by melt processing to form LLDPE films with CN-PPE content from 0.02 wt% to 0.3 wt%. Their dichroic behaviors, caused by anisotropic orientation under tensile elongation were studied by Pucci *et al.* (2005).[62] The films with low concentrations of CN-PPE (<0.1 wt%) displayed a blue monomer emission (~450 nm) and with higher concentrations (>0.1 wt%) a green excimer emission (~480 nm). This was due to the intermolecular or interchain fluorophore interactions caused by the planar conformation of the (CN-PPE) polymer chains in the solid state. The film samples were then elongated to draw ratios (of 10) at temperatures of 100 °C with their fluorescence emission recorded using polarizers. The elongated CN-PPE/LLDPE film (0.2 wt% CN-PPE) displayed dichroic behavior with green (excimer emission) to blue (monomer emission) visible dependent on the polarizer viewing angle (0° parallel and to 90° perpendicular) (Figure 5.15). This was thought to be due to the orientation of the CN-PPE fluorophore molecules caused by the orientation of the host (LLDPE) polymer matrix, in which the long-range fluorophore interactions are very anisotropic. However, no significant change in the emission properties due to the tensile elongation was observed. This was postulated to be due to the resistance of the CN-PPE aggregates to the polymer host deformation, contrasting with the results obtained by Weder *et al.* (2003)[6] for the small molecule cyano-OPV derivatives in LLDPE films.

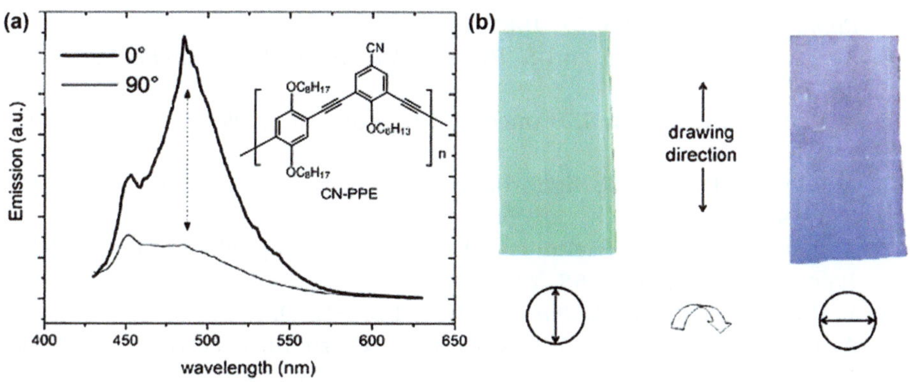

Figure 5.15 (a) Fluorescence emission spectra of an oriented CN-PPE/LLPPE film (Dr = 10) recorded with polarization respectively parallel (0°) and perpendicular (90°) to the drawing direction. (b) Digital images of the same film taken under excitation at 366 nm with the camera objective set with a commercial linear polarizer. The arrows denote the orientation of the polarizer transmission axis with respect to the orientation of the film. The inset in (a) is the chemical structure of CN-PPE.
Adapted with permission from reference 62. Copyright 2005, Elsevier.

5.4 Conclusion

In this chapter, we present an overview of mechanoflurochromism of polymer composite systems, including basal principle, preparation method and mechanochromic mechanism. Because the mechanoflurochromic covalently tethered dye-polymers are very limited, the development of the dye-doped polymer composite systems seems to be of vital importance. Not all dye-doped polymer systems show mechanochromic fluorescent properties because of the complexity of molecular interactions, not only between the dye molecules but also between the dye molecules and matrix polymer molecules. In consideration of the strict criteria for the dyes and polymer matrices, mechanochromic fluorescent polymer blends are still rather limited.

References

1. A. Pucci and G. Ruggeri, *J. Mater. Chem.*, 2011, **21**, 8282–8291.
2. A. Pucci, R. Bizzarri and G. Ruggeri, *Soft Matter*, 2011, **7**, 3689–3700.
3. F. Ciardelli, G. Ruggeri and A. Pucci, *Chem. Soc. Rev.*, 2013, **42**, 857–870.
4. A. Pucci, N. Tirelli, G. Ruggeri and F. Ciardelli, *Macromol. Chem. Phys.*, 2005, **206**, 102–111.
5. C. Lowe and C. Weder, *Adv. Mater.*, 2002, **14**, 1625–1629.
6. C. Weder and B. R. Crenshaw, *Chem. Mater.*, 2003, **15**, 4717–4724.
7. M. Kinami, B. R. Crenshaw and C. Weder, *Chem. Mater.*, 2006, **18**, 946–955.
8. B. R. Crenshaw and C. Weder, *Adv. Mater.*, 2005, **17**, 1471–1476.
9. J. Kunzelman, B. R. Crenshaw, M. Kinami and C. Weder, *Macromol. Rapid Commun.*, 2006, **27**, 1981–1987.
10. A. Pucci, M. Bertoldo and S. Bronco, *Macromol. Rapid Commun.*, 2005, **26**, 1043–1048.
11. A. Pucci, C. Cappelli, S. Bronco and G. Ruggeri, *J. Phys. Chem. B*, 2006, **110**, 3127–3134.
12. A. Pucci, F. Di Cuia, F. Signori and G. Ruggeri, *J. Mater. Chem.*, 2007, **17**, 783–790.
13. A. Pucci, G. Ruggeri, S. Bronco, M. Bertoldo, C. Cappelli and F. Ciardelli, *Prog. Org. Coat.*, 2007, **58**, 105–116.
14. F. Wuerthner, *Chem. Commun.*, 2004, 1564–1579.
15. A. Pucci, F. Donati, G. Ruggeri and F. Ciardelli, *e-Polymers*, 2009, **9**, 716–726.
16. H. Langhals, *Heterocycles*, 1995, **40**, 477–500.
17. F. Wuerthner, *Chem. Commun.*, 2004, 1564–1579.
18. H. Langhals and S. Saulich, *Chem.-Eur. J.*, 2002, **8**, 5630–5643.
19. H. Langhals and S. Kirner, *Eur. J. Org. Chem.*, 2000, 365–380.
20. F. Donati, A. Pucci, C. Cappelli, B. Mennucci and G. Ruggeri, *J. Phys. Chem. B*, 2008, **112**, 3668–3679.
21. K. Peneva, G. Mihov, A. Herrmann, N. Zarrabi, M. Boersch, T. M. Duncan and K. Muellen, *J. Am. Chem. Soc.*, 2008, **130**, 5398–5399.

22. M. Franceschin, C. M. Lombardo, E. Pascucci, D. D'Ambrosio, E. Micheli, A. Bianco, G. Ortaggi and M. Savino, *Bioorg. Med. Chem.*, 2008, **16**, 2292–2304.
23. T. Tang, K. Peneva, K. Muellen and S. E. Webber, *J. Phys. Chem. A*, 2007, **111**, 10609–10614.
24. T. Tang, A. Herrmann, K. Peneva, K. Muellen and S. E. Webber, *Langmuir*, 2007, **23**, 4623–4628.
25. M. Franceschin, E. Pascucci, A. Alvino, D. D'Ambrosio, A. Bianco, G. Ortaggi and M. Savino, *Bioorg. Med. Chem. Lett.*, 2007, **17**, 2515–2522.
26. A. Alvino, M. Franceschin, C. Cefaro, S. Borioni, G. Ortaggi and A. Bianco, *Tetrahedron*, 2007, **63**, 7858–7865.
27. T. Tang, J. Qu, K. Muellen and S. E. Webber, *Langmuir*, 2006, **22**, 26–28.
28. T. Tang, J. Qu, K. Muellen and S. E. Webber, *Langmuir*, 2006, **22**, 7610–7616.
29. F. Donati, A. Pucci and G. Ruggeri, *Phys. Chem. Chem. Phys.*, 2009, **11**, 6276–6282.
30. J. Luo, Z. Xie, J. W. Y. Lam, L. Cheng, H. Chen, C. Qiu, H. S. Kwok, X. Zhan, Y. Liu, D. Zhu and B. Z. Tang, *Chem. Commun.*, 2001, 1740–1741.
31. Y. Hong, J. W. Y. Lam and B. Z. Tang, *Chem. Commun.*, 2009, 4332–4353.
32. Y. Hong, J. W. Y. Lam and B. Z. Tang, *Chem. Soc. Rev.*, 2011, **40**, 5361–5388.
33. X. F. Li, Z. G. Chi, B. J. Xu, H. Y. Li, X. Q. Zhang, W. Zhou, Y. Zhang, S. W. Liu and J. R. Xu, *J. Fluoresc.*, 2011, **21**, 1969–1977.
34. X. Q. Zhang, Z. Y. Yang, Z. G. Chi, M. N. Chen, B. J. Xu, C. C. Wang, S. W. Liu, Y. Zhang and J. R. Xu, *J. Mater. Chem.*, 2010, **20**, 292–298.
35. B. K. An, D. S. Lee, J. S. Lee, Y. S. Park, H. S. Song and S. Y. Park, *J. Am. Chem. Soc.*, 2004, **126**, 10232–10233.
36. H. Y. Li, Z. G. Chi, B. J. Xu, X. Q. Zhang, Z. Y. Yang, X. F. Li, S. W. Liu, Y. Zhang and J. R. Xu, *J. Mater. Chem.*, 2010, **20**, 6103–6110.
37. X. Q. Zhang, Z. G. Chi, H. Y. Li, B. J. Xu, X. F. Li, S. W. Liu, Y. Zhang and J. R. Xu, *J. Mater. Chem.*, 2011, **21**, 1788–1796.
38. Z. Yang, Z. G. Chi, B. J. Xu, H. Y. Li, X. Q. Zhang, X. F. Li, S. W. Liu, Y. Zhang and J. R. Xu, *J. Mater. Chem.*, 2010, **20**, 7352–7359.
39. B. J. Xu, Z. G. Chi, Z. Yang, J. B. Chen, S. Z. Deng, H. Y. Li, X. F. Li, Y. Zhang, N. S. Xu and J. R. Xu, *J. Mater. Chem.*, 2010, **20**, 4135–4141.
40. H. Y. Li, Z. G. Chi, X. Q. Zhang, B. J. Xu, S. W. Liu, Y. Zhang and J. R. Xu, *Chem. Commun.*, 2011, **47**, 11273–11275.
41. Z. G. Chi, X. Q. Zhang, B. J. Xu, X. Zhou, C. P. Ma, Y. Zhang, S. W. Liu and J. R. Xu, *Chem. Soc. Rev.*, 2012, **41**, 3878–3896.
42. S. J. Yoon, J. W. Chung, J. Gierschner, K. S. Kim, M. G. Choi, D. Kim and S. Y. Park, *J. Am. Chem. Soc.*, 2010, **132**, 13675–13683.
43. M. A. Baldo, D. F. O'Brien, Y. You, A. Shoustikov, S. Sibley, M. E. Thompson and S. R. Forrest, *Nature*, 1998, **395**, 151–154.
44. Y. Chi and P.-T. Chou, *Chem. Soc. Rev.*, 2010, **39**, 638–655.

45. Z. Wang, E. Turner, V. Mahoney, S. Madakuni, T. Groy and J. Li, *Inorg. Chem.*, 2010, **49**, 11276–11286.
46. Y. Sagara and T. Kato, *Nat. Chem.*, 2009, **1**, 605–610.
47. M. Kato, *Bull. Chem. Soc. Jpn.*, 2007, **80**, 287–294.
48. X. Q. Zhang, Z. G. Chi, Y. Zhang, S. W. Liu and J. R. Xu, *J. Mater. Chem. C*, 2013, **1**, 3376–3390.
49. H. Ito, T. Saito, N. Oshima, N. Kitamura, S. Ishizaka, Y. Hinatsu, M. Wakeshima, M. Kato, K. Tsuge and M. Sawamura, *J. Am. Chem. Soc.*, 2008, **130**, 10044–10045.
50. T. Abe, T. Itakura, N. Ikeda and K. Shinozaki, *Dalton Trans.*, 2009, 711–715.
51. S. J. Choi, J. Kuwabara, Y. Nishimura, T. Arai and T. Kanbara, *Chem. Lett.*, 2012, **41**, 65–67.
52. J. R. Kumpfer, S. D. Taylor, W. B. Connick and S. J. Rowan, *J. Mater. Chem.*, 2012, **22**, 14196–14204.
53. T. Ikawa, T. Shiga and A. Okada, *J. Appl. Polym. Sci.*, 2002, **83**, 2600–2603.
54. D. R. T. Roberts and S. J. Holder, *J. Mater. Chem.*, 2011, **21**, 8256–8268.
55. A. J. Bur, R. E. Lowry, S. C. Roth, C. L. Thomas and F. W. Wang, *Macromolecules*, 1992, **25**, 3503–3510.
56. J. Yang, H. X. F. Li, G. H. Wang and B. L. He, *J. Appl. Polym. Sci.*, 2001, **82**, 2347–2351.
57. J. Yang, G. C. Wang, H. X. F. Li, L. S. Chen and B. L. He, *J. Appl. Polym. Sci.*, 2000, **78**, 1869–1872.
58. T. Shiga, T. Ikawa and A. Okada, *J. Appl. Polym. Sci.*, 1998, **67**, 259–266.
59. T. Ikawa, T. Shiga and A. Okada, *J. Appl. Polym. Sci.*, 1997, **66**, 1569–1573.
60. H. Spanggaard, M. Jorgensen and K. Almdal, *Macromolecules*, 2003, **36**, 1701–1705.
61. J. Vandendriessche, P. Palmans, S. Toppet, N. Boens, F. C. De Schryver and H. Masuhara, *J. Am. Chem. Soc.*, 1984, **106**, 8057–8064.
62. A. Pucci, T. Biver, G. Ruggeri, L. I. Meza and Y. Pang, *Polymer*, 2005, **46**, 11198–11205.

CHAPTER 6

Aggregation-induced Emission Mechanofluorochromic Materials

ZHENGUO CHI AND JIARUI XU*

PCFM Lab, State Key Laboratory of Optoelectronic Materials and Technologies, School of Chemistry and Chemical Engineering, Sun Yat-sen University, Guangzhou, China
*Email: xjr@mail.sysu.edu.cn

6.1 Introduction

In the preceding chapters, the mechanofluorochromism of polymers, dye-doped polymers, metal complexes and some organic compounds has been discussed. It is already known that the fluorescent properties of molecules in the solid state depend mainly on the molecular arrangement, conformational flexibility and intermolecular interactions, such as monomer-excimer transformation, dipole-dipole interaction or conformational planarization, in which mutual inter-conversions, *e.g.*, crystal-crystal, crystal-amorphous or crystal-liquid crystal may be involved. Any modification in the molecular packing and conformation of the fluorophore would affect the HOMO-LUMO energy levels and alter the fluorescent properties. Thus, controlling the mode of molecular packing (*i.e.*, the aggregation states) to achieve the dynamic control of highly efficient and reversible solid-state fluorescence is more attractive for both fundamental research and practical applications because of its advantages of low pressure demand and good

RSC Smart Materials No. 8
Mechanochromic Fluorescent Materials: Phenomena, Materials and Applications
Edited by Jiarui Xu and Zhenguo Chi

Published by the Royal Society of Chemistry, www.rsc.org

reversibility. However, as described in the previous chapters, mechanochromic fluorescent materials that are dependent on changes in physical molecular packing modes are still rare. This rarity may be attributed to two major issues.[1] First, predicting and designing materials that exhibit polymorphism is difficult. Each identified compound reported seems to be an isolated event, which makes the identification of a general characteristic difficult. Secondly, the fluorescence efficiency of organic fluorescent materials often becomes very weak when they are in the solid state because of the aggregation-caused quenching (ACQ) effect. Consequently, observing the mechanochromic fluorescent phenomenon becomes quite difficult.

In 2001, Tang *et al.*[2] reported on some "aggregation-induced emission" (AIE) materials, which is an important class of anti-ACQ materials that emit more efficiently when they are in the aggregated state than in the dissolved form. Since then, AIE materials have attracted considerable research attention for their potential application in various fields, such as organic light-emitting devices and chemosensors.[3] And recently, a number of AIE compounds with different AIE moieties have been found to possess mechanochromic luminescence properties. Hence, utilizing the AIE moieties to construct new compounds is now an important strategy in the synthesis of various mechanochromic luminescent materials.

This chapter will review published studies on the recent advances in organic mechanochromic luminescent materials, which include mechanochromic mechanism, structure-property relations and most reported mechanochromic AIE compounds.

6.2 Mechanochromic AIE Concept and Typical Mechanism

6.2.1 PAIE Concept

As a class of stimuli-responsive materials, mechanochromic compounds containing AIE moieties exhibit an off/on switching property of emission intensity upon external stimuli, such as organic solvent vapor, pressure, heat and so on.[4–7] Tang *et al.* reported that several AIE compounds possess bright-dark switching properties between crystalline and amorphous states, such as dyes **1**[8] and **2**.[9] For example, (4-biphenylyl)phenyldibenzofulvene (**2**) is weakly luminescent in the amorphous phase but becomes highly emissive upon crystallization. This unusual crystallization-induced emission enhancement effect allows its emission to be repeatedly switched between dark and bright states by fuming-heating or heating-cooling processes (Figure 6.1). In 2010, Park *et al.* reported the mechanofluorochromism of a cyano-distyrylbenzene derivative (**3**) that is an AIE compound.[10] However, at that time, it was not well recognized that there exists a relationship in the molecular structures between the AIE compound and the mechanofluorochromic nature. Almost within the same period, Chi and Xu's group

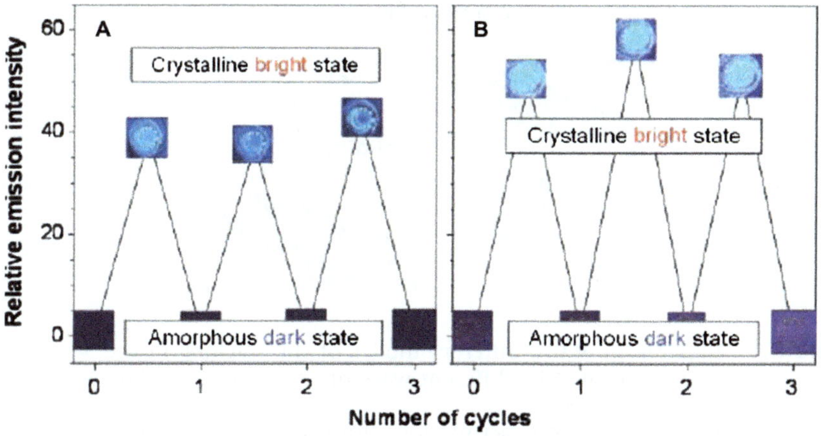

Figure 6.1 Repeated switching between dark and bright states of the emission of the thin solid films of **2** coated on quartz plates by (A) fuming-heating and (B) heating-cooling cycles. The photos were taken under illumination of a UV lamp.
Reproduced with permission from reference 9. Copyright 2007, The Royal Society of Chemistry.

synthesized and reported[11] (in 2011) a number of new mechanofluorochromic compounds with AIE nature and pointed out that mechanochromism should be the common property for most of AIE compounds. These compounds were thus named by the authors "the piezofluorochromic aggregation-induced emission (PAIE) compounds" at that time, because they possess both the piezofluorochromic (mechanofluorochromic) and aggregation-induced emissive properties.

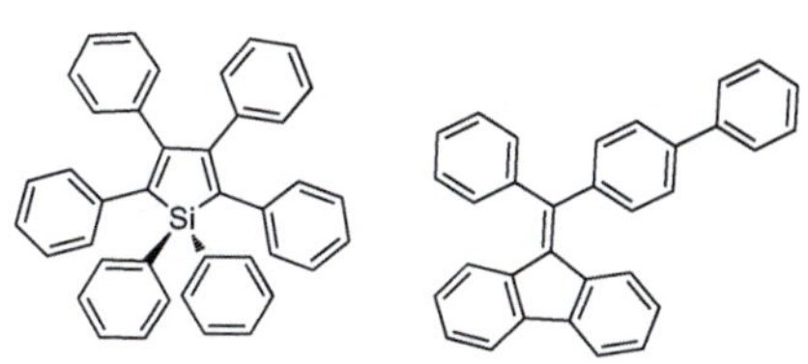

1 **2**

6.2.2 Mechanochromic Mechanism Based on Molecular Sheets[10]

Based on the cyano-distyrylbenzene derivatives, Park *et al.* (2010) concluded that in the solid state, the dye formed highly fluorescent "molecular sheets", which were assisted by multiple C–H···N and C–H···O hydrogen bonds with stacking and shear-sliding capabilities *via* external stimuli. The "molecular sheets" exhibited two-color fluorescence luminescence that switches in response to pressure, temperature

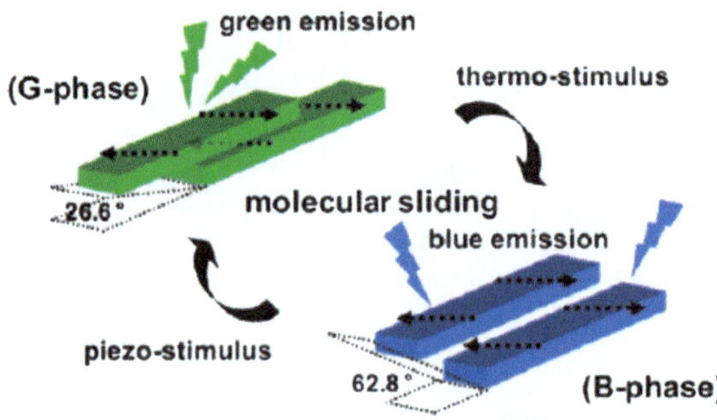

Figure 6.2 Illustration of two different modes of slip-stacking in dye **3** molecular sheets, dictated by different ways of antiparallel/head-to-tail coupling of local dipoles.
Reproduced with permission from reference 10. Copyright 2010, American Chemical Society.

and solvent vapor. Based on the structural, optical, photophysical and computational studies, they identified two different phases, *i.e.*, the metastable green-emitting G-phase and the thermodynamically stable blue-emitting B-phase. They elucidated the phase transition pathways as well as their spectroscopic implications. In the G-phase, the antiparallel coupling of the local dipoles kinetically stabilized the structure with a rather moderate excitonic coupling, but with an efficient excimer formation. Upon annealing, a smooth slip of the molecular sheets with a low activation barrier formed the B-phase with a head-to-tail arrangement of the local dipoles (Figure 6.2). In the study, the excimer formation was diminished, whereas the excitonic interaction substantially increased.

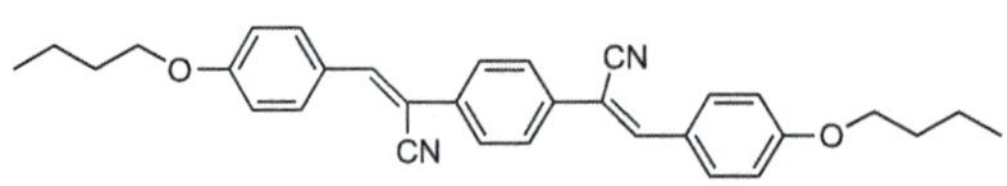

3

6.2.3 Mechanochromic Mechanism Based on Planarization of the Molecular Conformation[11]

A possible mechanism based on the planarization of the molecular conformation (Figure 6.3) was also proposed by Chi and Xu's group to explain the mechanochromic phenomenon, which differed from that of Park *et al.* It has been realized that numerous reported AIE compounds, such as typical triphenylethylene, tetraphenyl-ethylene, silole and cyano distyrylbenzene derivatives,[12–17] were found to possess one common structural feature, that is, the multiple phenyl peripheries are linked to an olefinic core *via* rotatable carbon-carbon single bonds to form an AIE moiety.

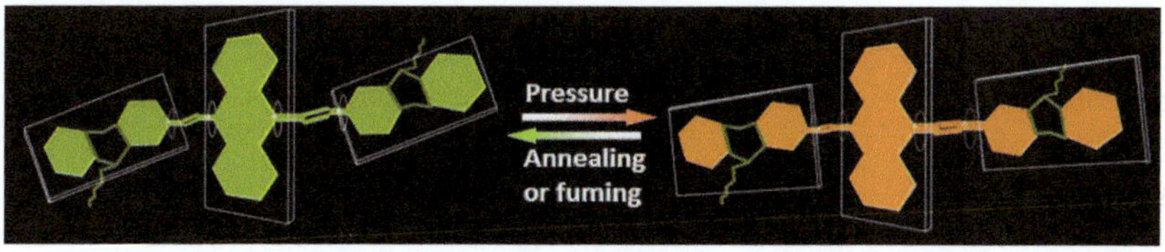

Figure 6.3 Mechanochromic mechanism based on planarization of molecular conformation (the example molecule is compound **39**).

This steric effect between the phenyl rings forces the AIE moieties or the molecules to take a twisted conformation. The twisted conformation and the weak π–π interactions make the molecular packing relatively loose and generates a number of defects (cavities), resulting in a low lattice energy. Clearly, the cavities are the feeblest parts of the crystalline structures. These two structural features, low lattice energy and formation of cavities, render the crystals easily destroyed by the planarization of the molecular conformation or slip deformation under external pressure. The planarization of molecular conformation after the release of twist stress would result in increased molecular conjugation, thereby facilitating a red shift of the photoluminescence (PL) spectrum.[64] According to this hypothesis, they have synthesized a number of AIE compounds and found that most of them have machanofluorochromism. The common structure-property relationships to be set up will be helpful in identifying and synthesizing more novel mechanofluorochromic materials.

6.2.4 Mechanochromic Mechanism Based on Changes in the Molecular Aggregation States[18]

Tian *et al.* (2012) reported an effective mechanism for mechanochromic luminescence on the basis of the molecular aggregation state of 9,10-bis((E)-2-(pyrid-2-yl)vinyl) anthracene (**4**). The compound exhibited spectacular luminescence characteristics: grinding and the exertion of external pressure on the powder led to a change in its photoluminescence color from green to red. Three crystal polymorphs with different emission colors were obtained. The mechanochromic luminescence observed indicates that when pressure is applied, the compound powder undergoes a transformation between the molecular aggregation states of the three crystals (Figure 6.4). In this process, external pressure impels the molecular aggregation state of the compound powder to transform from J-type aggregation to H-type aggregation, and further to aggregated dimers stacked in a more tightly bound face-to-face arrangement. Meanwhile, the intermolecular π–π interaction strengthens gradually and thus induces the PL spectrum of the powder to change from a green emission (no π–π interaction) to an orange emission (weak π–π interaction) and then to a red emission (strong π–π interaction).

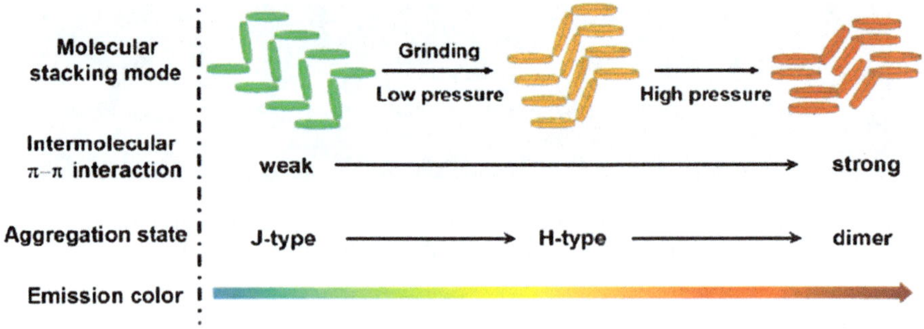

Figure 6.4 Stacking modes and corresponding emission colors for the various molecular aggregation states in dye **4** powder.
Reproduced with permission from reference 18. Copyright 2012, Wiley-VCH.

Therefore, the ability of the molecular aggregation state to change upon grinding or under pressure leads to the changeable fluorescence color and the mechanochromic effect in the compound powder.

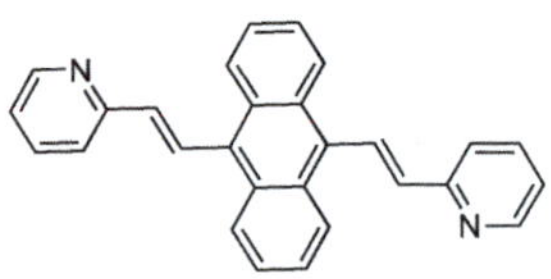

4

6.3 Relationship Between Structure and Properties of Mechanofluorochromic Compounds

A full understanding of structure–property relationships is necessary, in other words, what structural features effectively induce mechanofluorochromic properties in organic materials. The importance of donor–acceptor materials in the design of efficient mechanofluorochromic materials has been previously reported.[19,20] A recent design strategy that demonstrates the importance of both twisted conformations and intermolecular interactions from donor–acceptor molecules has provided some aids to better understand the structure–property relationships.[21] However, there is still ambiguity in the literature, such as the interesting, yet curious, cases of 9,10-bis[(fluorene)vinyl]-anthracenes[22] and 9,10-bis[(carbazole)vinyl]anthracenes.[23] In the former case, increasing the alkyl chain length at the fluorine 9-position was observed to increase the grinding induced fluorescence shifts, while for the latter case increasing the alkyl chain length at the carbazole N-position was observed to decrease the grinding induced fluorescence shifts. This shows that it is still an important challenge to better understand the structure–property relationships in the organic mechanofluorochromic materials.

6.3.1 Relationship Between Crystallinity and Mechanochromism

According to the above-mentioned mechanisms, the crystallinity of the AIE compounds will directly affect their mechanochromic properties since the mechanochromism of AIE compounds is due to the changes of their aggregation state structures.

The butterfly-shaped AIE dyes **5** and **6** derived from tetraphenylethylene and carbazole were synthesized by Chi and Xu *et al.* (2012).[24] Two different aggregates were obtained from the solutions of dye **5** in different solvent systems *via* rotary evaporation. The sample obtained from the dichloromethane/*n*-hexane mixed solvent (1 : 3, v/v) solution is a white crystalline aggregate with a strong blue emission (451 nm). The sample obtained from the dichloromethane solution is a light-green amorphous aggregate with a strong green emission (479 nm) (Figure 6.5A). The results indicate that dye **5** has a better polymorph-forming ability. However, dye **6** only produces blue-emission crystals in the same concentration conditions. When the crystalline samples of dye **5** are briefly pressed in an infrared pellet at 1500 psi for 1 min, ground using a mortar and pestle, or when their melt is quenched in liquid nitrogen, all samples are converted to their amorphous form (Figure 6.5B). The emission of the pressed (ground) sample is at 479 nm, whereas that of the quenched sample is at 493 nm. Thus, the quenched sample has a longer PL wavelength than that of the pressed sample. The results indicate that dye **5** possesses solid morphology-alterable emission and mechanofluorochromic properties. However, dye **6** had no such properties because it had excellent crystallization capability. When dye **6** is evaporated from either dichloromethane or dichloromethane/*n*-hexane (1 : 3, v/v) solutions, it always forms the crystalline state not in the amorphous state (Figure 6.5C). In other words, if an AIE compound has a strong intention to crystallize to form stable crystals, no changes from the crystalline to amorphous phase will occur, and thus it would not exhibit mechanofluorochromism.

For dye **5**, the time to cause the rapid decrease in the PL peak intensity of the ground sample changes from 30 s to 120 s when fumed with dichloromethane. The intensity gradually increases then with prolonged fuming time after 120 s (Figure 6.5D). During the fuming process, the permeation of good solvent could weaken the interaction of the packing molecules because of solvation, which results in increased intramolecular rotational and vibrational motions, increased non-emissive decay of excited-state energy, and decreased PL intensity. Simultaneously, the molecules undergo a solvent-induced crystallization process. As the degree of crystallization increases, the intramolecular vibrations and rotations are gradually restricted. Moreover, the non-emissive decay of the excited-state energy gradually weakens, which results in a substantial increase in PL intensity. Thus, the two opposite effects result in a V-shaped curve depending on which effect plays the dominant role in the entire PL behavior.[24] This finding indicated

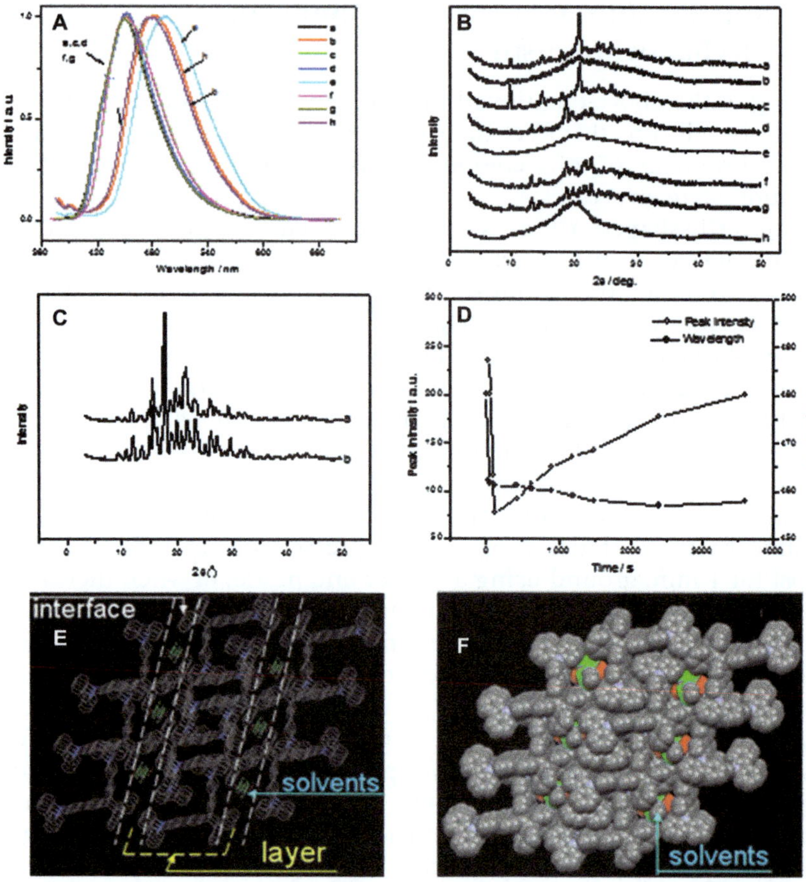

Figure 6.5 (A) PL spectra of the dye **5** samples: (a) obtained from dichloromethane/*n*-hexane (1 : 3); (b) crystalline sample was pressed at 1500 psi; (c) pressed sample was fumed with dichloromethane for 12 h; (d) pressed sample was annealed at 260 °C for 5 min; (e) melt quenched to room temperature; (f) quenched sample was fumed with dichloromethane for 12 h; (g) quenched sample was annealed at 260 °C for 5 min; and (h) obtained from dichloromethane. (B) XRD curves of the dye **5** samples: (a) obtained from dichloromethane/*n*-hexane (1 : 3); (b) crystalline sample was pressed at 1500 psi; (c) pressed sample was fumed with dichloromethane for 12 h; (d) pressed sample was annealed at 260 °C for 5 min; (e) melt quenched to room temperature; (f) quenched sample was fumed with dichloromethane for 12 h; (g) quenched sample was annealed at 260 °C for 5 min; and (h) obtained from dichloromethane. (C) XRD curves of the dye **6** samples obtained from (a) dichloromethane/*n*-hexane (1 : 3) and (b) dichloromethane. (D) PL peak intensity and wavelength of dye **5** *vs.* fuming time with dichloromethane. Molecules packing of dye **5** in single crystal: (E) capped sticks style and (F) spacefill style, showing inclusion of dichloromethane/*n*-hexane between the layers (the hydrogen atoms have been omitted for clarity).
Adapted with permission from reference 24. Copyright 2012, The Royal Society of Chemistry.

that: (1) the crystallization of the ground amorphous sample could be induced by solvent vapor to achieve a reversible change from the amorphous to the crystalline state, and (2) the amorphous phase is a meta-stable phase and could be immediately converted to a more stable crystalline phase *via* solvent-induced crystallization.

The single-crystal analysis results indicate that the molecules are packed *via* the synergetic effect of weak π–π and C–H···π interactions that form lamellar layer structures. The layers are connected *via* the antenna parts of the butterfly-shaped molecules with weak π–π interactions (partially π-overlapping). The interfaces between the layers are relatively loose, and a number of defects (cavities) are formed where the solvent molecules are filled (Figure 6.5E and F). As mentioned earlier, the structural features of the dye cause the crystal to be easily destroyed by the planarization of the molecular conformation or slip deformation when an external pressure is applied, which results in mechanofluorochromism.

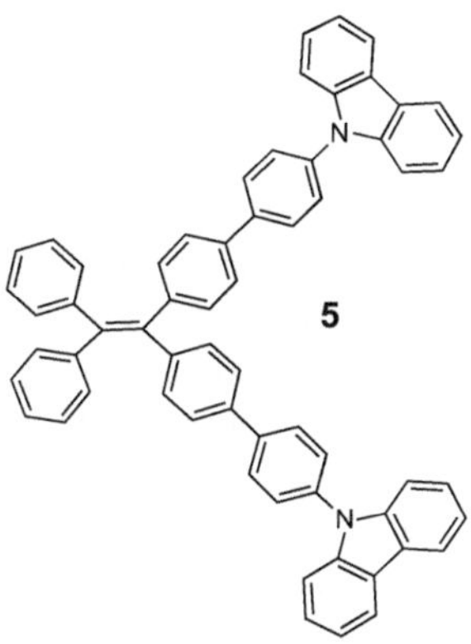

5

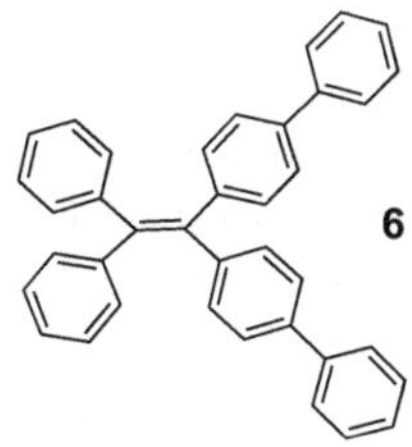

6

As mentioned above, the structure of dye **5** crystals has a relatively loose molecular packing and contains a number of defects. This finding was supported by a study on dye 7 (Figure 6.6), which was designed and synthesized by Tang *et al.* (2012).[25] Two single crystals (O, orange, and R, red) were successfully obtained in the appropriate conditions. The crystallographic data indicate that crystals O and R have reasonable hollow structures inside, in which different solvent molecules are selectively encapsulated.

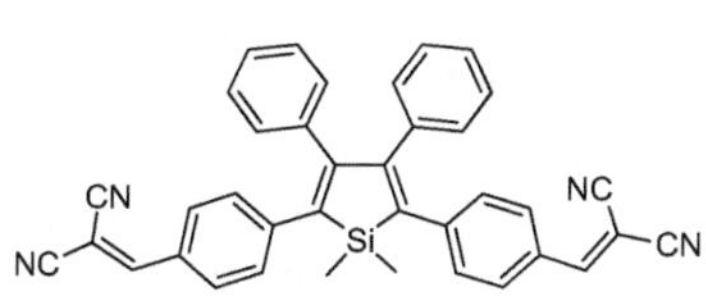

7

Chi and Xu *et al.* (2012) further confirmed that AIE compounds, namely the carbazolyl substituted distyrylanthracene derivatives, with strong

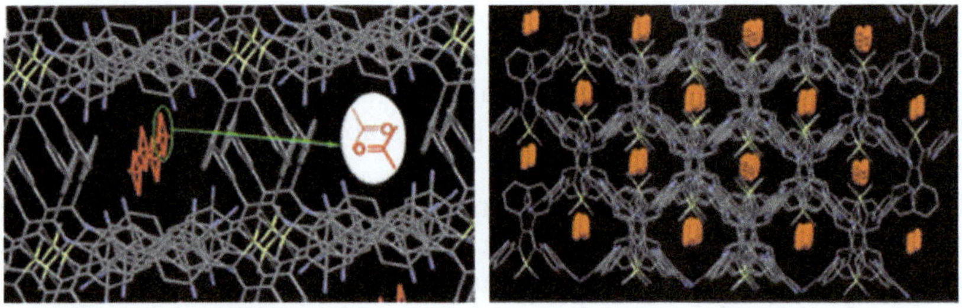

Figure 6.6 X-Ray crystal structures of dye **7** for (a) crystal R at 173 K, perspective view of the framework and the 1-dimensional channel in the *c* axis. The guest acetone molecules inhabit in the 1-dimensional channels highlighted in the inset; (b) crystal O at 133 K, a perspective view in the azimuth angle of $-45°$ to the *a* axis to show the guest hexane molecules in the 1-dimensional channels.
Reproduced with permission from reference 25. Copyright 2012, The Royal Society of Chemistry.

crystallizability, display no mechanofluorochromism.[26] The distyrylanthracene derivatives with monocarbazolyl group (**8**), tricarbazolyl group and without carbazolyl group were synthesized. The results show that the mechanofluorochromic properties of these three AIE compounds are significantly influenced by the variety of carbazolyl groups in the molecular structures. The compound images taken at room temperature under UV light subject to various conditions are shown in Figure 6.7. Compound **8** without carbazolyl group did not exhibit a distinct mechanofluorochromic property. On the other hand, the strong green-emitting solid of **9** (with momocarbazolyl group) can be converted into a solid with strong yellow luminescence when the compound is ground, whereas for the solid of **10** (with tricarbazolyl groups), the yellowish green luminescence is changed to orange. After the ground samples of **9** and **10** are annealed or fumed with CH_2Cl_2 vapor, colors close to their originals are achieved, thus demonstrating good reversibility in the emission spectra.

The PL spectra of the samples obtained under different conditions (original, grinding, and annealing or fuming of the ground samples) are shown in Figure 6.7. The maximum emission wavelengths of **8** under different conditions remain almost unchanged with a wavelength change of only 1 nm. Thus, compound **8** has no mechanofluorochromic property. However, significant red shifts up to 24 and 32 nm were observed in compounds **9** and **10**, respectively. After the ground samples of these two compounds were annealed or fumed with CH_2Cl_2 vapor, the PL wavelengths were blue-shifted by 14 nm, and were unable to return to the wavelength of the unground samples, probably due to the steric effect generated by the carbazolyl group(s). However, the reversible color change was observed in

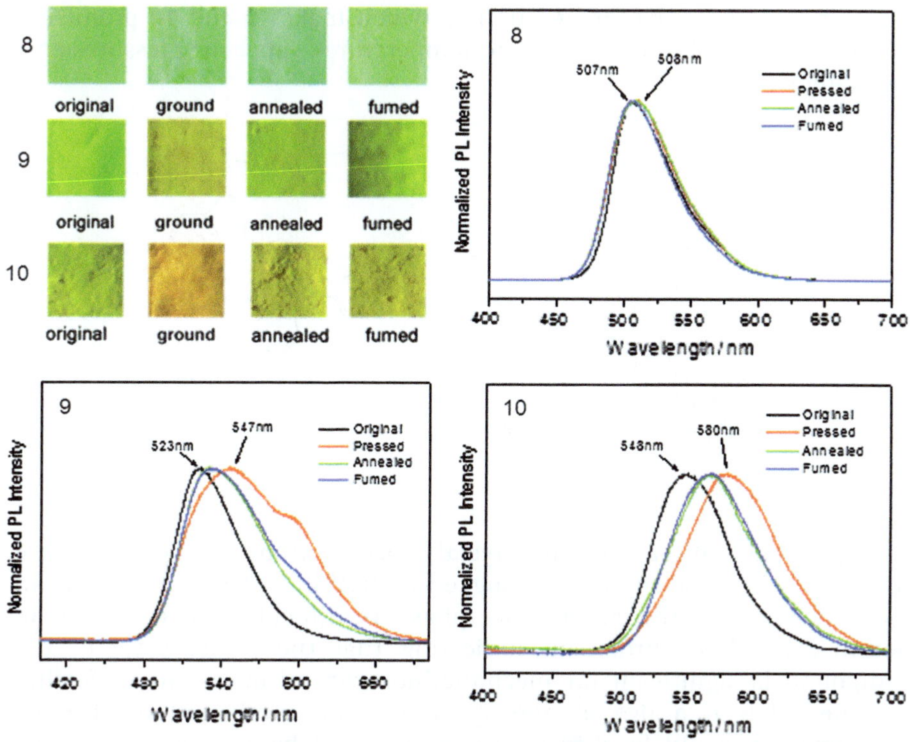

Figure 6.7 The images taken under 365 nm UV light and normalized PL curves of **8**, **9** and **10** samples obtained under different conditions. Original: original sample; ground: ground sample; annealed: after annealing the ground sample at 200 °C for 5 min; fumed: after fuming the ground sample in CH_2Cl_2 vapor for 5 min.
Reproduced with permission from reference 26. Copyright 2012, American Chemical Society.

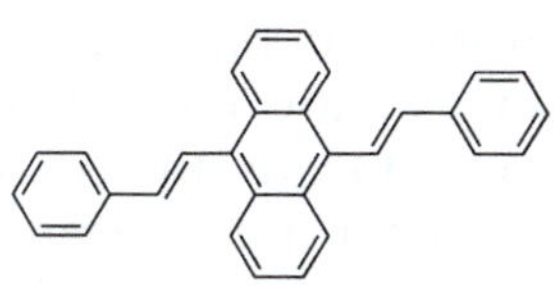

8

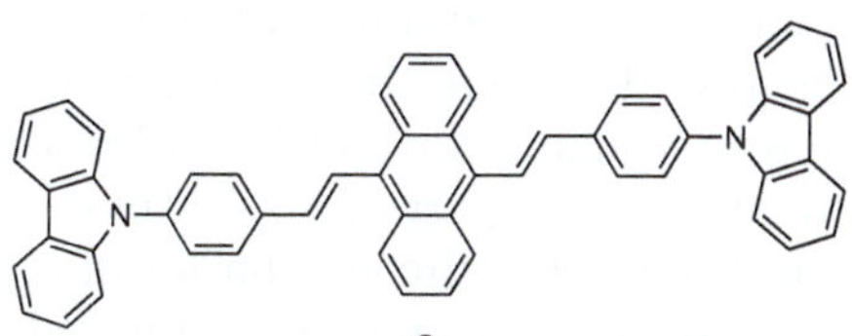

9

annealing-grinding or fuming-grinding, which makes **9** and **10** promising candidates for optical recording and temperature-, vapor- or pressure-sensing systems.

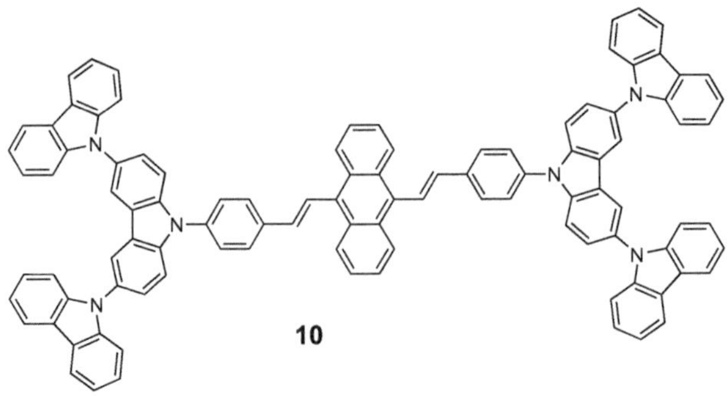

10

As previously discussed, the crystals formed from dyes **6** and **8** are very stable because of their strong crystallizability. Thus, change from the crystalline to the amorphous phase is difficult, which results in non-mechanofluorochromism, indicating that the initial state of the compound is significant for mechanofluorochromism. It can be further extended that if a dye is non-crystalline, the mechanofluorochromic phenomenon would be difficult to be observed because no crystalline-amorphous transition in the solid phase will exist. This ratiocination was confirmed by the experimental results on the series of dyes **11** to **14**, which were designed and synthesized by Chi and Xu *et al.* (2011).[27] All of these compounds are AIE-active because they contain triphenylethylene or distyrylanthracene AIE moiety. The distyrylanthracene derivatives **11** and **13** exhibited obvious mechanofluorochromic phenomena (Figure 6.8). Before and after grinding, the emission wavelength increased from 534 nm to 572 nm and from 566 nm to 580 nm for dyes **11** and **13**, respectively. However, the diphenylanthracene derivatives **12** and **14** were non-mechanofluorochromic. It has been verified by XRD that the initial states of dyes **11** and **13** are crystalline to a particular extent, whereas those of dyes **12** and **14** are amorphous. No phase transition would occur upon pressure as the initial states of dyes **12** and **14** are amorphous, and the mechanoflurochromic behavior has been known to depend on the packing change from the crystalline to the amorphous state. Consequently, the two compounds are not mechanoflurochromic. From this study, it is revealed that the existence of a relatively stable crystalline state is a prerequisite for mechanoflurochromism. In other words, AIE compounds with strong crystallizability or non-crystallizability would not be good candidates for the mechanofluorochromic materials. Therefore, designing a strategy for mechanofluorochromic compounds remains a crucial issue.

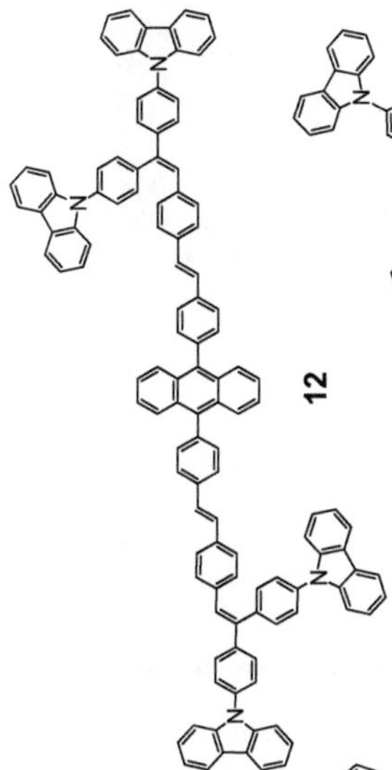

12

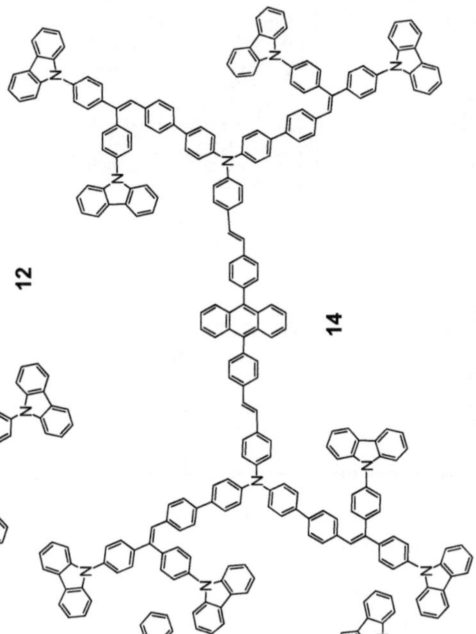

14

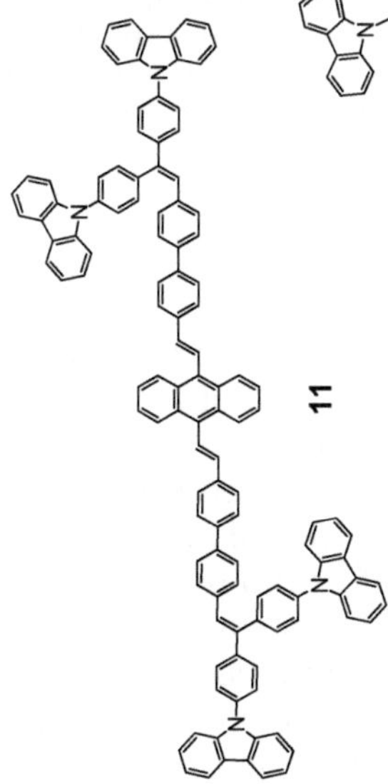

11

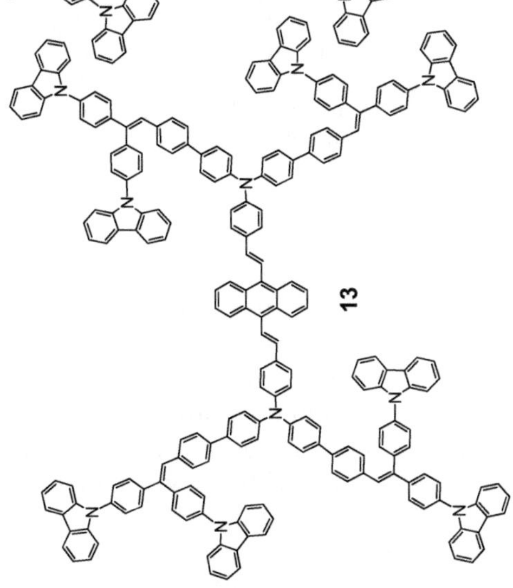

13

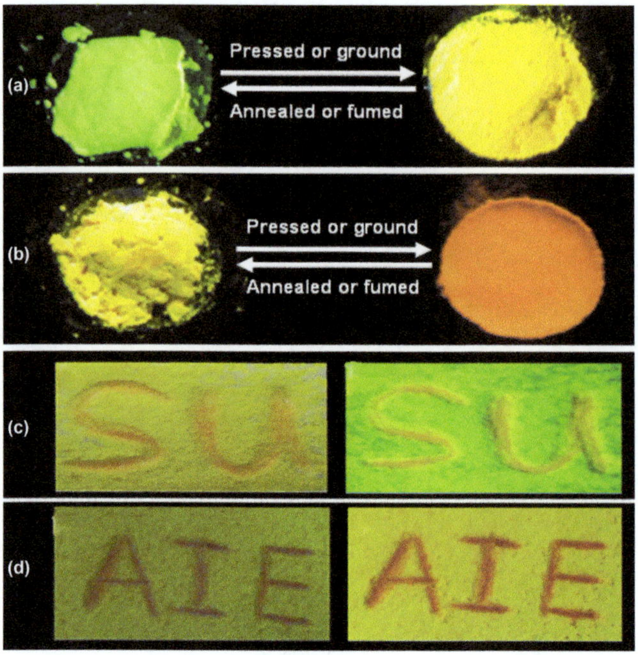

Figure 6.8 The images of (a) dye **11** and (b) dye **13** taken at room temperature under 365 nm UV light: (left) as-synthesized samples or annealed samples (at 300 °C, for 5 min); (right) pressed (1500 psi for 5 min) or ground sample. Dye **11** (c) and dye **13** (d) were cast on filter paper and "SU" and "AIE" were written with a metal spatula at room temperature under ambient light (left) and UV light (right).
Reproduced with permission from reference 27. Copyright 2011, American Chemical Society.

Another well-demonstrated example has been reported by Chi and Xu *et al.* (2011).[28] Dyes **15, 16** and **17,** which contain triphenylamine-anthrylenevinylene and tetraphenylethene moieties with AIE natures, were synthesized and characterized. When the as-synthesized sample of **16** was briefly ground using a mortar and pestle, the emission spectrum of the ground sample exhibited significant red shift of 22 nm from 561 to 583 nm, indicating that the compound has a mechanofluorochromic property. The as-synthesized sample of **16** shows strong yellow emission under 365 nm UV light, and after it is ground it shows strong orange-red emission. When the sample was pressed by streaking a metal spatula across the filter paper containing it, a color path was observed, and the color change occurred only in the streaked area. Under ambient or UV light, the marked "PAIE" can be clearly seen on the streaked **16** sample on the "paper". According to the WAXD measurements, the compound exhibits different structures of molecular aggregation before and after grinding treatment. The diffraction curve of the as-synthesized sample displays sharp and intense reflections which are indicative of crystalline ordered structure. The diffraction curve of the sample after grinding shows a weak, broad and diffuse peak, indicating that the ordered structure has been destroyed and the aggregation structure has changed to amorphous after grinding. This again suggests that the mechanofluorochromic nature is generated through the molecular packing mode change under external pressure, *i.e.,* phase transition from the crystalline state to the amorphous state. Although the three compounds have some similarity in their molecular structures, only **16** exhibits obvious mechanofluorochromic behavior, whereas the emission spectra of **15** and **17** show little change before and after grinding. The diffraction curves of the as-synthesized samples of **15** and **17** likewise show a diffused peak, indicating the aggregation structures of the two as-synthesized samples are amorphous. Since their initial states are amorphous, no ordered structures can be destroyed by external pressure, and consequently, **15** and **17** do not exhibit a mechanofluorochromic property. These results strengthen the previous proposition that the initial phase state is a significant prerequisite for mechanofluorochromism. In other words, the mechanofluorochromic behaviors cannot be observed for all the AIE compounds unless they have appropriate crystallizability.

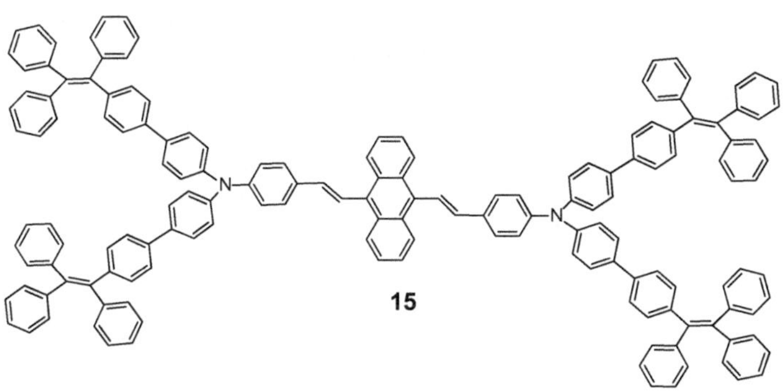

15

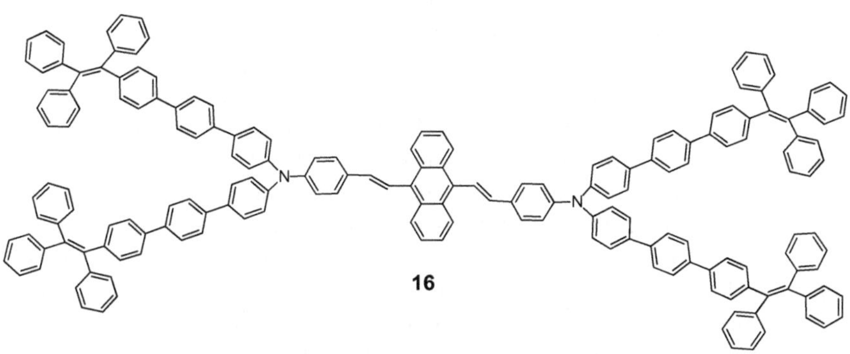

16

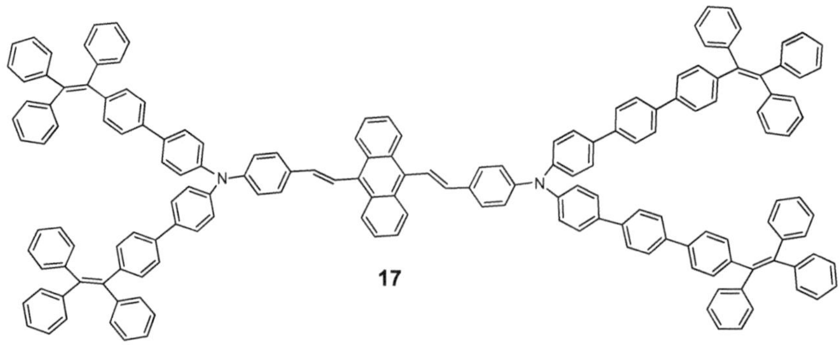

17

For AIE dye **3**, the two-color fluorescence switching behavior was explained by the interchange between the metastable green-emitting G-phase and the thermodynamically stable blue-emitting B-phase with different modes of local dipole coupling (antiparallel and head-to-tail arrangements, respectively). The interchange was facilitated by the two-directional shear-sliding capability of molecular sheets formed *via* intermolecular multiple C–H···N and C–H···O hydrogen bonds. The structures of dyes **18–21**, however, had no such hydrogen bonds. If the dyes are mechanofluorochromism-active, the mechanism proposed by Park *et al.*[10] seems not applicative in explaining the mechanofluorochromism phenomenon. Nevertheless, the mechanofluorochromic phenomenon of this class of AIE compounds could be well explained by the proposed mechanism based on the planarization of the AIE molecular conformation.

Chi and Xu *et al.* (2012)[29] designed and successfully synthesized two distyrylanthracene derivatives capped with tetraphenylethylene groups (dyes **18** and **21**), and the other two derivatives capped with triphenylethylene groups (dyes **19** and **20**). All compounds exhibit strong AIE activities, however, only the derivatives capped with tetraphenylethylene groups (dyes **18** and **21**) show significant mechanofluorochromic properties. Compounds **19** and **20** with triphenylethylene end-group do not exhibit distinct piezochromic or mechanofluorochromic properties, whereas compounds **18** and **21** with tetraphenylethylene end-groups do. For example, a strong green-emitting light-yellow solid of **18** is converted into an orange solid with strong yellow luminescence when the compound is pressed. After the pressed samples are annealed, the color returns to its original appearance, thus demonstrating good reversibility in both the absorption and emission spectra. Similar phenomena are observed for **21**. The PL spectra of the samples obtained under different conditions (pressing and annealing) are shown in Figure 6.9. The maximum emission wavelengths of **19** and **20** under different conditions remain almost unchanged. The wavelengths only change by 1 nm and 3 nm, respectively. Thus, **19** and **20** have no mechanofluorochromic properties. However, significant red shifts that reached up to 68 and 44 nm are observed for **18** and **21**, respectively. After the pressed samples are annealed, the initial PL wavelengths (similar to the

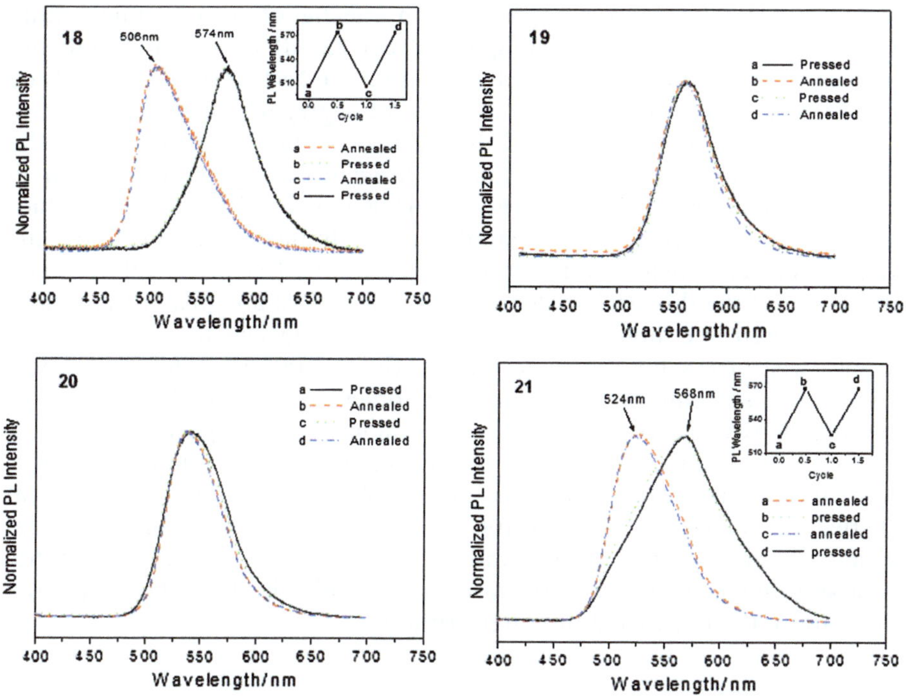

Figure 6.9 The normalized PL curves of dyes **18–21** obtained from different conditions.
Reproduced with permission from reference 29. Copyright 2012, The Royal Society of Chemistry.

unpressed samples) are recovered, thereby demonstrating high reversibility in the emission spectra.

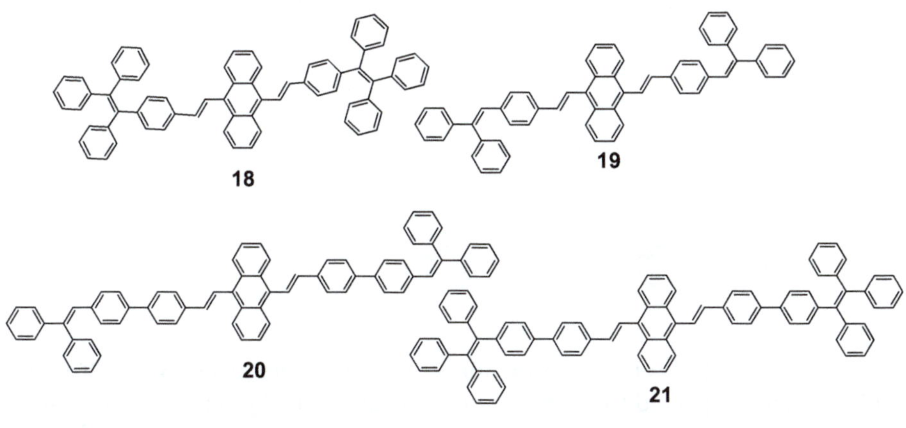

18 **19** **20** **21**

The AIE dyes **22–24** studied by Chi and Xu *et al.* (2011)[30] also had neither heteroatom nor C–H···N and C–H···O hydrogen bond interaction in their molecular structures, which are similar to the last series dyes

18–21. Nevertheless, dye **22** possesses a mechanofluorochromic nature (Figure 6.10A). However, the analogs of AIE dye **22**, AIE dyes **23** and **24** do not possess mechanofluorochromic phenomena. The XRD results demonstrate the discrepancies resulted from the different aggregated states of the dyes. The analogs are amorphous, and dye **22** is crystalline (Figure 6.10B). The data of molecular conformation and packing in the crystal were obtained through single crystal analysis (Figure 6.10C). The dihedral angle values between the aryl rings in the single crystal indicate that the molecule possesses a very twisted conformation in either the free or crystalline state. The values of the dihedral angles in the single crystal are greater than those in the isolated free molecule, and the difference suggests the existence of a strong twist stress for the molecules in the crystalline state. The typical cofacial π–π stacking of the molecules is practically impossible because of the highly twisted conformation. The molecules are packed *via* weak C–H···π interactions in the crystal cell, which leads to the relative looseness of the molecular packing. Such looseness results in the formation of several cavities, as shown in Figures 6.10C (c and d). This feature of the crystal structure enables the compound to exhibit pronounced mechanofluorochromism. The PL spectrum of the as-synthesized dye **22** sample red shifts from 454 nm to 482 nm after grinding.

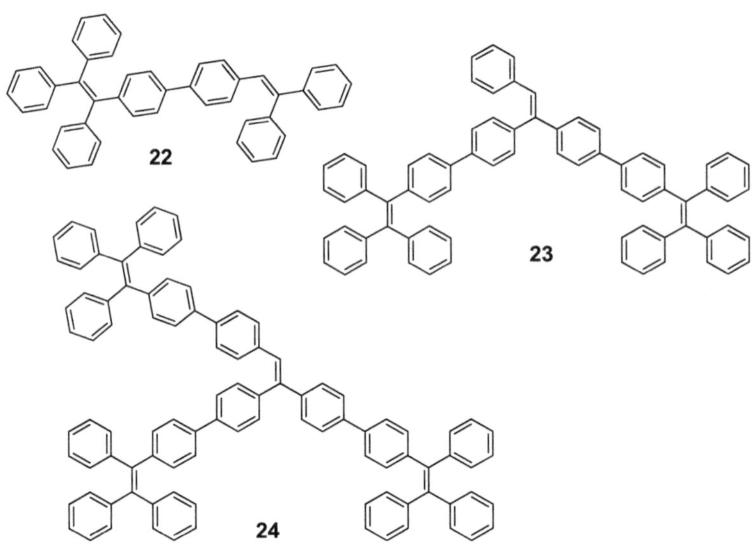

22 **23** **24**

Sun *et al.* (2011)[31] reported on E and Z stereoisomers (dyes **25** and **26**). The molecules are expected to be mechanofluorochromically active because of its propeller tetraphenylethylene moiety properties. Moreover, the molecules exhibited multiple chromic effects, which include mechano-, piezo-, thermo-, vapo- and chronochromisms. The as-synthesized dye **25** is an off-white solid with a blue emission (447 nm). This dye changes to a pale-yellow powder

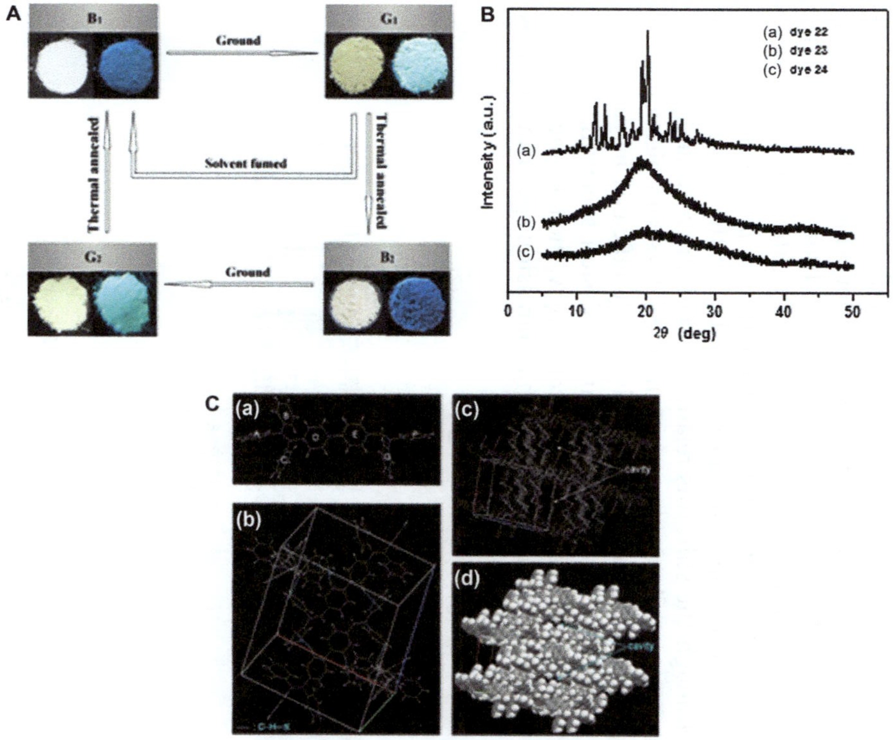

Figure 6.10 (A) Dye **22** taken at room temperature under ambient light (left) and UV light (right). Samples: (B_1) as-synthesized sample; (G_1) ground sample; (B_2) annealed sample; (G_2) re-ground sample. (B) XRD curves of the as-synthesized sample of dyes **22-24**. (C) The molecular conformation of dye **22** in single crystal with defined planes (a and b), and the molecules packing in a crystal cell (c and d).
Reproduced with permission from reference 30. Copyright 2011, Wiley-VCH.

with a bluish-green emission (477 nm) after grinding, which shows a mechanofluorochromic effect. The ground sample is transformed back to the off-white solid with a blue emission after treating at 120 °C for 1 min. The as-synthesized dye **26** is a pale-yellow solid with a bluish-green emission (460 nm). Mechanical grinding of the solid caused little change in the physical appearance or emission color of dye **26**. The grinding process shifts its emission peak from 460 nm to 470 nm (Figure 6.11). Such an increase indicates that grinding a mainly amorphous solid might not result in significant changes in the aggregated structure and in the fluorescence spectrum. The XRD patterns of the as-prepared solid of the dyes demonstrate that the crystallization capability of the Z isomer is lower than that of E isomer, which accounts for their obviously different chromic behaviors. The mechano- and thermochromisms are associated with the aggregated state transformations between the crystalline and amorphous phases. Aside from grinding, pressurization could also cause a red shift in the PL spectrum of the E isomer, although the shift was relatively small (approximately 8 nm). The significant difference in the extent of the mechano- and piezochromic effects indicates that shearing (grinding) is highly efficient in causing a larger change in the aggregated structure and emission spectrum as compared with compression (pressurization). Interestingly, the E isomer shows a novel chronochromic phenomenon, in which its emission spectrum changes

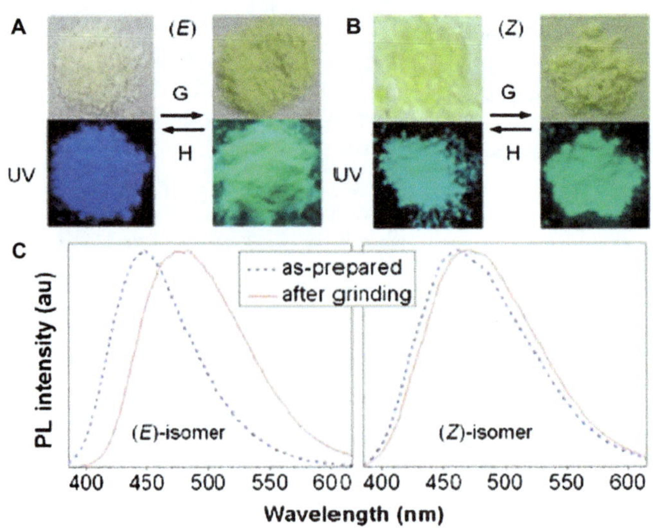

Figure 6.11 (A) Photographs of as-prepared (left) and ground (right) samples of dye **25** taken under room lighting (upper) and UV illumination (lower); G = grinding; H = heating (at 120 °C for 1 min). (B) Mechano- and thermochromic processes of dye **26**, with the photographs arranged in the same order as in panel A. (C) PL spectra of (left) E-conformer **25** and (right) Z-conformer **26** before and after grinding (λex = 332 nm). Reproduced with permission from reference 31. Copyright 2012, American Chemical Society.

with time. The chronochromism indicates that the ground sample is in a metastable state, which slowly transforms back to the thermodynamically stable crystalline state at room temperature. Dye **25** also exhibits a solvent-dependent vapochromic effect. The ground sample of this dye is sensitive to volatile polar solvents, such as chloroform, dichloromethane and tetrahydrofuran. After exposure to chloroform vapor for 1 min, the bluish-green emission of the ground sample is quickly reverted to the blue emission of its crystals caused by the solvent-induced crystallization.

25

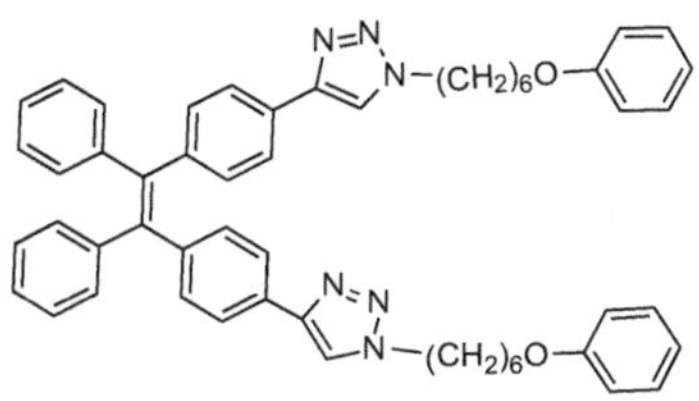

26

Tang *et al.* (2012)[32] locked the phenyl rings of a popular AIE luminogen, tetraphenylethene (**27**), with the "O" bridge step by step, and the Φ_F value of the molecules increased with the number of locked phenyl rings. The emission spectrum and Φ_F of **29** solution fitted well with those of its crystals due to the fully locked phenyl rings and twisted conformation. Emission of **28** crystals changed from bright blue to yellow-greenish upon grinding in a mortar (Figure 6.12). And the emission reverted to blue upon heating. Thus the emission of **28** could be tuned reversibly between blue and yellow-greenish through repeated heating and grinding. The XRD diffraction curves of both the original blue crystals and the annealed samples of **28** display many sharp and intense reflection peaks that coincide with each other, indicating their same crystalline orders and that the ground solid can revert to the original crystalline state upon heating. Diffraction of the ground powder also exhibits some reflections that agree with those of the original crystal and annealed ground sample, but does not have as many or as sharp peaks, hinting a not absolutely amorphous phase, which may be caused by the spontaneous recovery during grinding. The DSC thermogram of ground **28** shows a broad exothermic peak at around 86 °C before melting, which is absent in the DSC curves of both crystalline and annealed samples, indicating that the ground sample is in a metastable amorphous state and can crystallize promptly in the solid state upon heating. After crystallization,

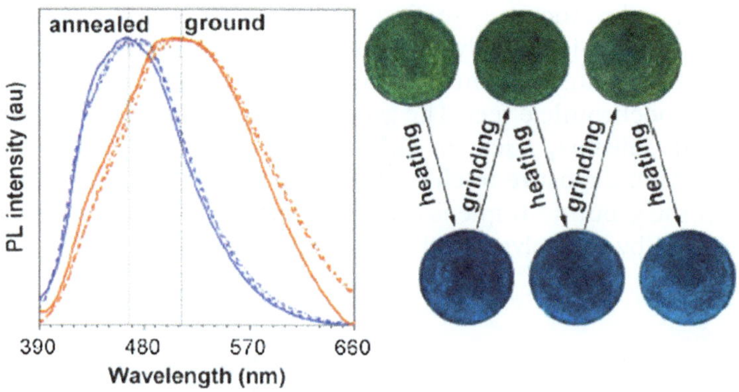

Figure 6.12 Normalized PL spectra (left panel) and photos (right panel) of **28** in three repeating cycles of (red lines) grinding and (blue lines) annealing (170 °C, 25 min).
Reproduced with permission from reference 32. Copyright 2012, The Royal Society of Chemistry.

both the ground and annealed samples melt at a similar temperature to that of the pristine sample, suggesting that the ground sample can revert back to the original crystalline state. Thus the mechanochromic fluorescence of **28** is ascribed to the transformation from a crystalline to an amorphous phase upon grinding. However, the emissions of ground solids of **27** and **29** are the same as their crystals, that is, both **27** and **29** exhibit no response to grinding. **27** crystallizes very quickly, and an amorphous solid of **27** cannot be obtained through grinding or quenching of its melt, thus **27** is unresponsive to the mechanical stimulus. As all the phenyl rings of **29** are locked and the conformation of **29** molecules are kept unchanged in the different aggregation states, **29** does not exhibit mechanochromic fluorescence either.

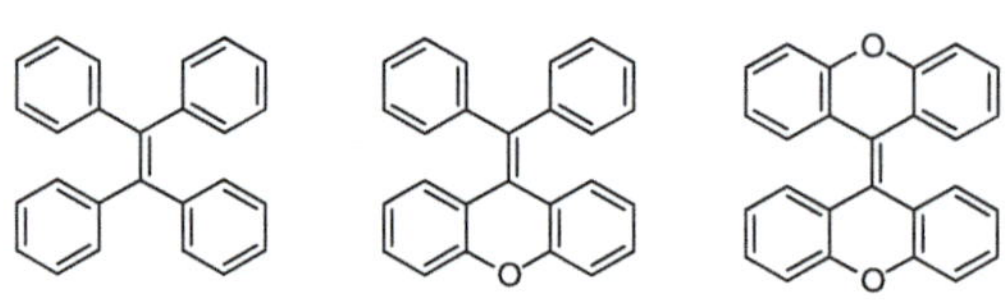

27 **28** **29**

Two crystallization-induced emission enhancement (also a typical AIE phenomenon) dyes **30** and **31** were reported by Dong *et al.*(2011).[33] Dye **30** forms two types of crystals that emit green and yellow light with quantum yields of 82.1% and 56.2%, respectively. By contrast, the amorphous state of the crystal emitted a rather weak orange light with an efficiency lower than 1%. The fluorescence of the crystals of both **30** and **31** could be turned into "dark" and "bright" upon grinding and annealing (Figure 6.13). The yellow-emission crystal could be converted into a green-emission crystal by heating. The emission of dye **30** could be repeatedly switched between "dark" and

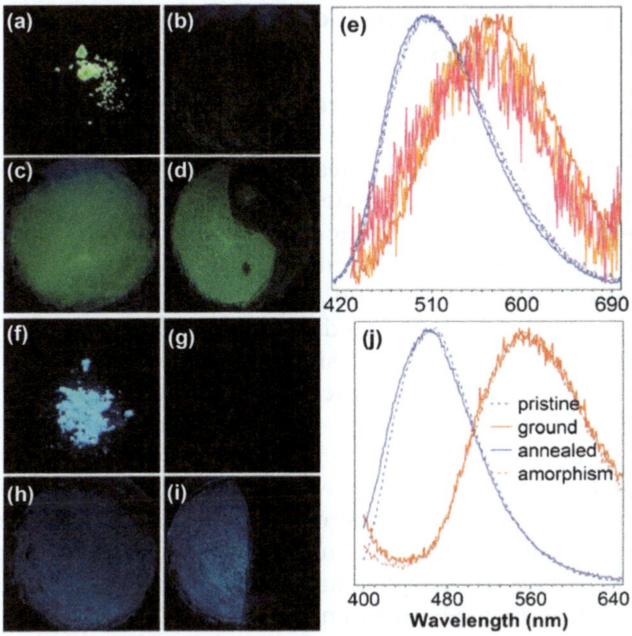

Figure 6.13 Photographs of dye **30** (a–d) and **31** (f–i) before (a, f) and after grinding (b, g). Annealed dye **30** (c) and **31** (h) and regrinding of selected areas (d) and (i), respectively). Photographs are taken under UV illumination. (e) Normalized PL spectra of **30** before grinding (blue solid line), after grinding (red lines), and annealing (blue dashed and dash dot lines) in the three repeating cycles. Normalized PL spectra of dye **31** are shown in panel (j). Annealing detail: 120 and 160 ° C for 1 min for dye **30** and **31**, respectively; excitation wavelength: 370 nm. Reproduced with permission from reference 33. Copyright 2011, Wiley-VCH.

"bright" by controlling the transition between the amorphous state and the green-emission crystal *via* heating and cooling. The emission of the ground sample of dye **30** could spontaneously return to green emission at room temperature, whereas ground dye **31** remained dark for over 24 h in the same condition. This phenomenon was considered a result of the different molecular flexibilities that governed the capability for molecular motion in the solid state.

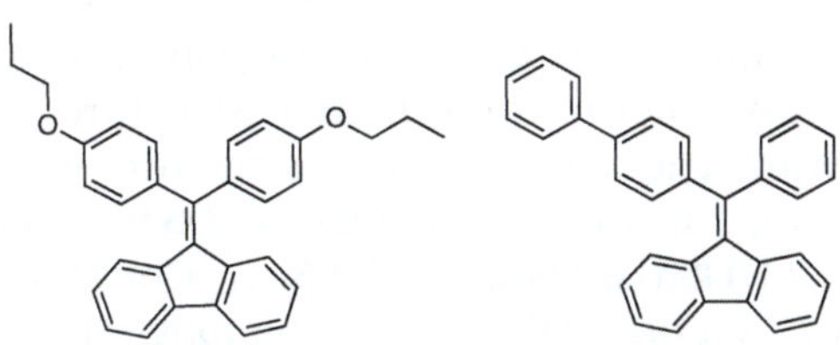

30 **31**

6.3.2 Relationship Between Alky (Alkoxy) Length and Mechanochromism

The changes of the molecular structure and regulation of aggregation state structure could affect and/or alter the chemical and physical properties of materials, which is an important means to develop novel high-efficiency organic photoelectric functional materials. Usually, the introduced outer edge alkyl chain of conjugated organic molecule was helpful to improve the solubility of materials, but some recent studies have demonstrated that the length of alkyl chain had effects on the state aggregation behavior and photoelectric properties of conjugated organic small molecules, which indicated that the alkyl chain played a significant rule to regulate the photoelectric properties. In this section, the effects of alkyl chain length on the aggregation state and morphology of the molecules and mechanofluorochromic properties based on the conjugated organic small molecules with AIE properties will be discussed with several typical examples. The influence of alkyl chain will be considered as an important factor in the synthesis and controllable preparation of mechanofluorochromic materials.[34,35]

Chi and Xu *et al.* (2012)[36] designed and synthesized a series of remarkable multifunctional 9, 10-distyrylanthracene derivatives (**32**, **DSA*n***, $n = 7$–12). All the derivatives possess typical aggregation-induced emission (AIE) property. **DSA$_{10}$**, **DSA$_{11}$** and **DSA$_{12}$** showed significant changes in emission when they were ground, *i.e.*, red shifting 45, 45 and 52 nm, respectively. However, **DSA$_7$**, **DSA$_8$** and **DSA$_9$** exhibited insignificant mechanofluorochromism. After grinding, their emissions only red shifted 4, 10 and 12 nm, respectively (Figure 6.14A). The relationships between the molecular structures of the samples and their mechanofluorochromic properties were further investigated by single crystal structural analysis. The X-ray structural analysis of these single crystals reveals that the compounds crystallize in the triclinic space group $P\bar{1}$ and demonstrate that they have a nonplanar conformation in the crystals. In their single crystals, the conformation structures of **DSA$_{10}$**, **DSA$_{11}$** and **DSA$_{12}$** are symmetrical; thus, the $\theta_{A\text{-}B}$ values are equal to $\theta_{B\text{-}C}$ and the values were 76°, 70° and 74°, respectively. These values are greater than 60°, considering that the larger dihedral angles of the molecules are responsible for the shorter PL wavelengths (from 498 nm to 500 nm) in their original states. However, the **DSA$_7$** and **DSA$_8$** crystals have two distinct conformation structures, namely, symmetric and asymmetric. In the **DSA$_7$** crystals, $\theta_{A\text{-}B}$ and $\theta_{B\text{-}C}$ are 58° in the symmetric conformation structure, and 28° and 66° in the asymmetric conformation structure. In the **DSA$_8$** crystals, $\theta_{A\text{-}B}$ and $\theta_{B\text{-}C}$ are 56° in the symmetric conformation structure, and 28° and 74° in the asymmetric conformation structure. The **DSA$_9$** crystals only have an asymmetric conformation structure with $\theta_{A\text{-}B} = 40°$ and $\theta_{B\text{-}C} = 50°$. In general, the dihedral angles of **DSA$_7$**, **DSA$_8$** and **DSA$_9$** are smaller than those of **DSA$_{10}$**, **DSA$_{11}$** and **DSA$_{12}$**. Consequently, the former three compounds in their original states exhibit longer PL wavelengths (from 515 nm to 531 nm). The analysis of these crystal structures shows that the

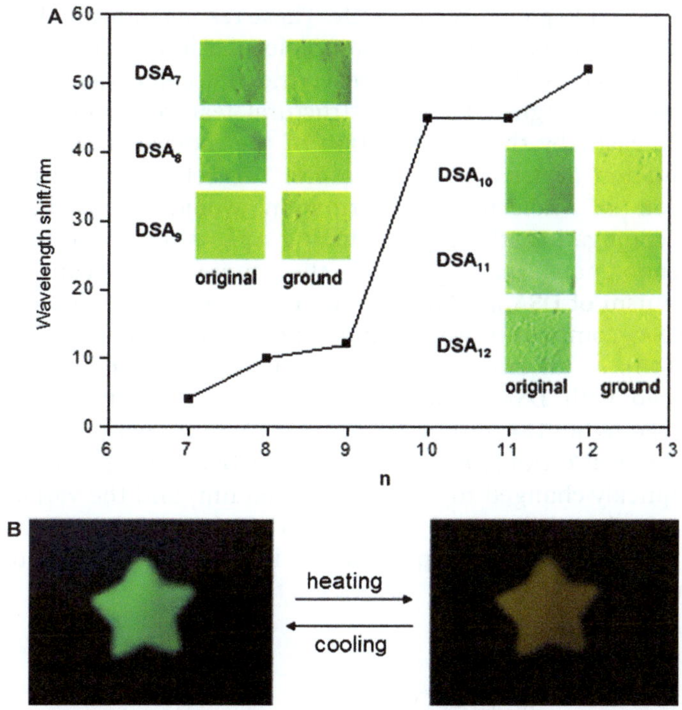

Figure 6.14 (A) Wavelength change of compound **32** series versus *n* after grinding. (B) Reversible color changes of $\mathbf{DSA_{11}}$ under heating and cooling. Reproduced with permission from reference 36. Copyright 2012, The Royal Society of Chemistry.

above-mentioned difference among the molecular conformation structures of the crystals is caused by the supramolecular interactions including Ph→O, Ph→Ph, Ph→An, Oct→Ph, An→An, An→Ph and Vin→An. Among these supramolecular interactions, CH/π interactions are the most important. The backbone of the molecules largely deviates from a plane, and typical cofacial π–π stacking becomes impossible because of the highly twisted conformation, as well as steric hindrance of the bulky aromatic rings in the molecules. Thus, only the $\mathbf{DSA_8}$ and $\mathbf{DSA_9}$ crystals exhibit weak face-to-face π-stacked anthracene rings with small π-overlap areas and contact distances of about 3.64 Å ($\mathbf{DSA_8}$) and 3.56 Å ($\mathbf{DSA_9}$). The $\mathbf{DSA_7}$, $\mathbf{DSA_8}$ and $\mathbf{DSA_9}$ crystals show four, five and four interactions; whereas the $\mathbf{DSA_{10}}$, $\mathbf{DSA_{11}}$ and $\mathbf{DSA_{12}}$ crystals show only one, two and two supramolecular interactions, respectively. This result indicates that the supramolecular interactions in the former three crystals are stronger than those in the latter ones. The strong supramolecular interactions not only make the molecules more planar and stable in the lattice, but also induce tight intermolecular packing. The density *d* of the $\mathbf{DSA_7}$, $\mathbf{DSA_8}$ and $\mathbf{DSA_9}$ crystals ranges from 1.167 g/cm^3 to 1.185 g/cm^3, while for the $\mathbf{DSA_{10}}$, $\mathbf{DSA_{11}}$, and $\mathbf{DSA_{12}}$ crystals, *d* ranges from

1.127 g/cm^3 to 1.146 g/cm^3, respectively. These results indicate that the more twisted conformation and weaker supramolecular interactions in the **DSA$_{10}$**, **DSA$_{11}$** and **DSA$_{12}$** crystals relatively loosen the molecular packings and result in lower lattice energies. The two structural features render the crystals destructible easily by the planarization of the molecular conformation or slip deformation under external pressure. The release of twist stress and planarization of molecular conformation would result in increased molecular conjugation, thereby facilitating a red shift of the PL spectrum. This phenomenon may explain the more significant mechanofluorochromism of **DSA$_{10}$**, **DSA$_{11}$** and **DSA$_{12}$**, as compared to the others.

All the **DSA$_n$** compounds were found to possess thermochromic properties (Figure 6.14B). As an example, a star-like pattern was stamped on a filter paper soaked with **DSA$_{11}$** solution in CH_2Cl_2. After drying, the pattern emitted green light (514 nm) under 365 nm UV illumination. As the pattern was heated to the T_i transition temperature of **DSA$_{11}$** using a hot air gun, its emission quickly changed to dark yellow (566 nm) and the variation in the emission wavelength reached 52 nm. As soon as the hot air gun was removed, the green emission pattern was immediately recovered, showing the excellent reversibility.

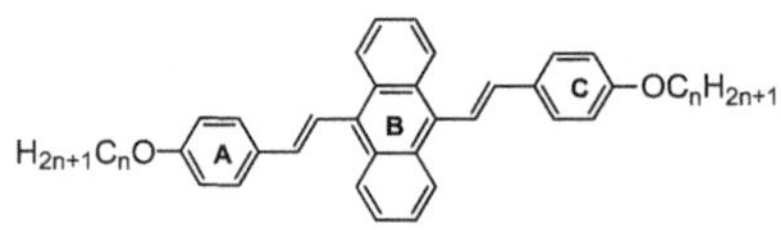

32 DSA$_n$ (n = 7–12)

Yang *et al.* (2013)[37] also systematically demonstrated the chain length-dependent mechanofluorochromic behavior of 9,10-bis(*p*-alkoxystyryl)-anthracenes (**DSA$_n$**, n = 1–10, 12, 14, 16, 18). It is shown that the long alkyl-containing **DSA$_n$** (n ≥ 10) display more remarkable spectral shifts (46–53 nm) than those with shorter alkyl chains (n ≤ 9, 14–20 nm). To obtain remarkable mechanofluorochromic behavior from 9,10-bis(p-alkoxystyryl) anthracene homologs, it was suggested that long peripheral aliphatic tails are needed to balance the intermolecular π–π and aliphatic interactions. The diffuse reflectance absorption and IR spectra evidence that out-of-plane hydrogen bending modes of the phenyl rings have disappeared in the ground long alkoxy-containing **DSA$_n$**. This observation implies a less twisted backbone conformation in the ground state than in the unground state, which is responsible for the dichotomous mechanofluorochromic behavior.

Yang *et al.* (2013)[22] also designed and synthesized a series of new alkyl length-dependent solid-state fluorescence and mechanochromic luminescence AIE-active 9,10-bis[(9,9-dialkylfluorene-2-yl)vinyl]anthracenes (**33**, FLA-Cn) with propyl, pentyl and dodecyl side chains. Pristine **FLAC12**, **FLA-C5** and **FLA-C3** solids emitted green, yellowish-green and yellow fluorescence, respectively. When they were pressed, **FLA-C12** and **FLA-C5** were changed into yellow-emitting solids and **FLA-C3** became yellowish-orange emitting. When the pressed samples were annealed before the isotropic melt transition temperature or exposed to solvent vapor (fuming above dichloromethane), the fluorescence colors were recovered to the original ones. When the fumed or annealed samples were re-pressed, the fluorescence colors were changed again as the first pressing, indicating reversible mechanofluorochromic behavior. It was observed that the longer the alkyl chain of **FLA-Cn**, the more remarkable the fluorescence color change found upon pressing. The pressing-induced spectral shifts of **FLA-C12**, **FLA-C5** and **FLA-C3** solids were 40, 26 and 18 nm, respectively (Figure 6.15A), implying that the **FLA-Cn** solids are also chain length-dependent mechanofluorochromic materials like series **32**, and alkyl chains have played a functional role in tuning the mechanofluorochromic behavior of **FLA-Cn** although such alkyl elements are normally inactive and undesirable for optoelectronic properties. In addition, a just-ground **FLA-C12** solid emitted yellow fluorescence, however, when it was left standing at

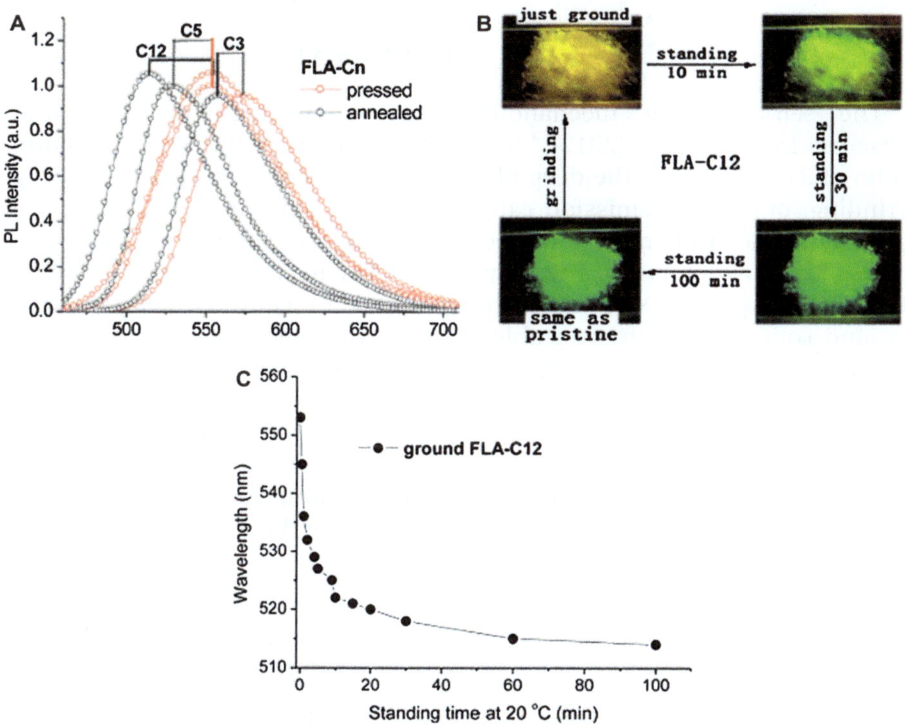

Figure 6.15 (A) Emission spectra of **FLA-Cn** solids mixed with KBr upon pressing and annealing. (B) Fluorescence images of ground pure **FLA-C12** solid at different standing times at room temperature. (C) The peak emission wavelengths of ground pure **FLA-C12** solid at different standing times at room temperature.
Reproduced with permission from reference 22. Copyright 2013, The Royal Society of Chemistry.

room temperature, the fluorescence colour changed with the increase of standing time and gradually recovered to the original green emission (Figure 6.15B). Figure 6.15C shows the curve of the peak emission wavelength change for the ground **FLA-C12** solid versus the standing time. It is seen that the rapid recovering in fluorescence colour and emission happens within the initial 10 min (form 553 to 522 nm), and standing for 60 min, the fluorescence emission (515 nm) is almost identical with the pristine solid (514 nm) and almost unchanged further. However, unlike **FLA-C12**, the fluorescence colors and emission spectra of the ground **FLA-C3** and **FLA-C5** solids remain orange and yellow over 24 h at room temperature. It is then believed that the stable or self-recovering mechanofluorochromic behavior of the 9,10-diarylvinyl-anthracene dyes can be controlled by tuning the alkyl length.

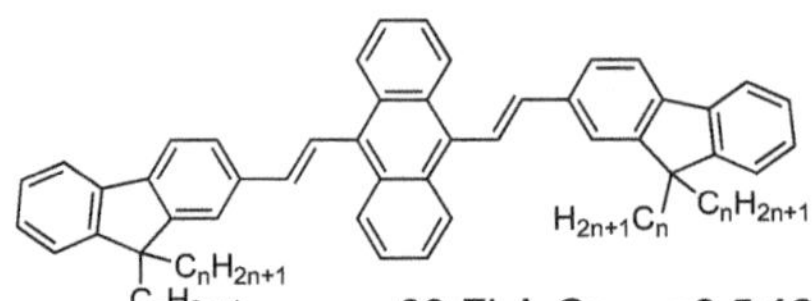

33 FLA-Cn **n=3,5,12**

The self-recovering mechanofluorochromic phenomenon was also observed by Dong *et al.* (2012)[38] for the compounds **34** and **35**. The ground solid of **35** transfers to the deep-blue emissive crystals very quickly. During grinding, only green emission can be observed. Once grinding stops, the ground powder can transfer to deep blue emission state in 30 s at room temperature (about 30 °C). The change is too fast to check its information of PL, DSC and powder XRD about the ground solid. On the other hand, the ground solid of **34** is more stable than **35**, with the emission remaining unchanged in 10 min at 30 °C, but spontaneous transfer occurred to deep-blue emissive crystal for longer periods (Figure 6.16). The authors believed that the longer alkyl groups in luminogen **35** afforded a looser packing, thus enabling the quicker transformation from the ground solid to crystals at room temperature. The comparison of these two compounds also informs us that the self-recovery speed can be tuned by varying the substitute groups on the phenyl rings.

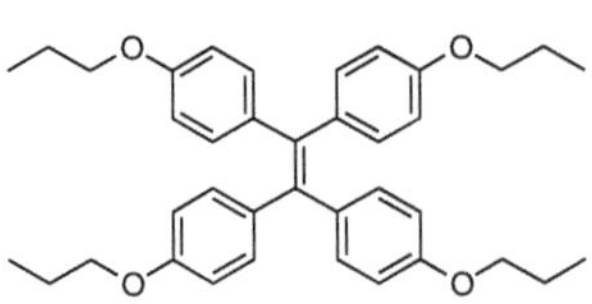

34

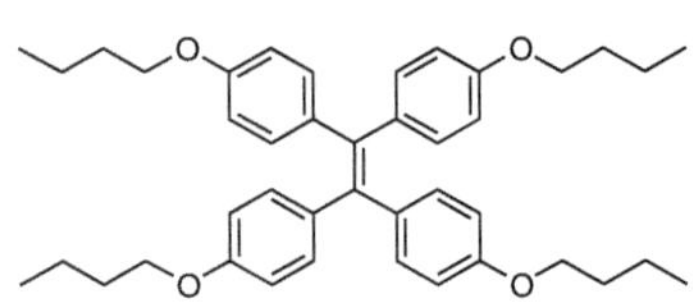

35

Yang *et al.* (2013)[39] reported the mechanofluorochromism of a series of 9,10-bis(N-alkylcarbazol-2-yl-vinyl-2)anthracenes (**36**, **A2Cn**) with changed

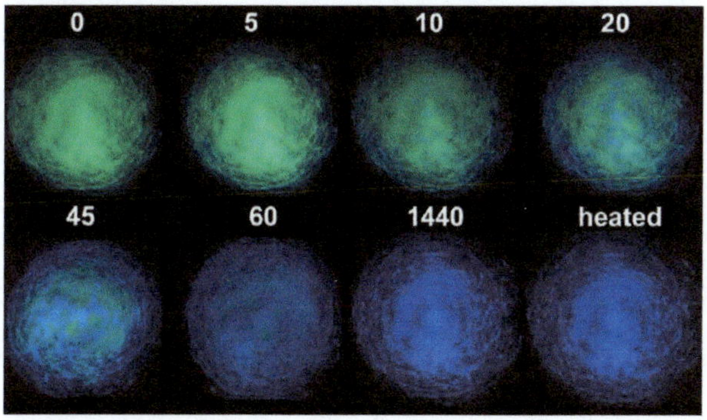

Figure 6.16 Spontaneously recovering of luminogen **34** at room temperature (30 °C) after ground in a mortar. Photos are taken under UV illumination at different time (min).
Reproduced with permission from reference 38 Copyright 2012, American Chemical Society.

Table 6.1 Peak emission wavelengths (λ, in nm) of **A2Cn** samples upon external stimuli.

A2Cn	$\lambda_{pristine}$	$\lambda_{pressed}$	$\lambda_{anealed}$	$\lambda_{repressed}$	λ_{fumed}	$\Delta\lambda^{a}$
A2C1	548	572	536	572	541	36
A2C3	543	579	537	577	540	42
A2C5	535	578	533	575	533	45
A2C12	530	571	527	572	530	44

$^{a}\Delta\lambda = \lambda_{\text{pressed}} - \lambda_{\text{anealed}}$. Reproduced with permission from reference 39. Copyright 2013, Elsevier.

length of N-alkyl chains (methyl, propyl, pentyl and dodecyl). It was observed that all the **A2Cn** samples exhibited fluorescence color change between green and yellow upon external stimuli under UV lamp. The pressed states were yellow emission, which can be changed reversibly into green colors by thermal-annealing (at 120 °C) or solvent-fuming (exposed to CH_2Cl_2 vapor at room temperature) and regenerated by repressing. All the **A2Cn** samples showed remarkable pressing-induced spectral shifts ($\Delta\lambda = \lambda_{\text{pressed}} - \lambda_{\text{anealed}}$) of 36–45 nm (Table 6.1). The peak emission wavelengths of the pristine (as-prepared) solids ($\lambda_{\text{pristine}}$) are moderately blue-shifted (from 548 to 530 nm) with increasing the alkyl lengths. This emission change tendency is consistent with that observed in 9,10-bis[(9,9-dialkyl-fluorene-2-yl)vinyl-2]anthracenes (**FLA-Cn**) described above. However, the alkyl lengths do not affect significantly the peak emission-wavelengths of **A2Cn** samples in the pressed (λ_{pressed}) and annealed states (λ_{anealed}), rendering **A2Cn** samples only slightly increased $\Delta\lambda$ with alkyl lengths. That is, the mechanofluorochromic behavior of **A2Cn** is weakly alkyl length-dependent, which is obviously different from those presented in

FLA-Cn system whose $\Delta\lambda$ are strongly alkyl length-dependent and doubled change.

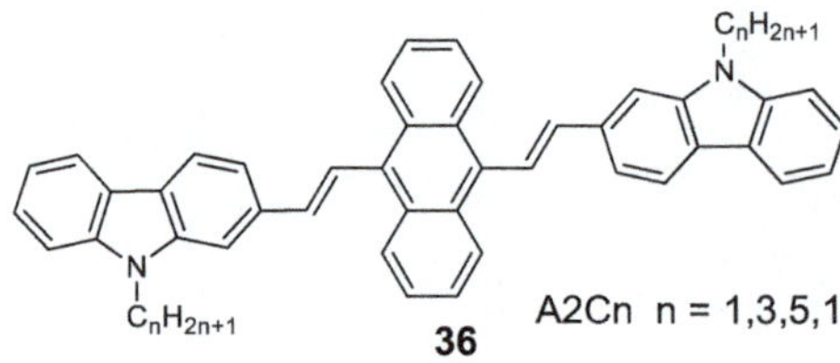

36 A2Cn n = 1,3,5,12)

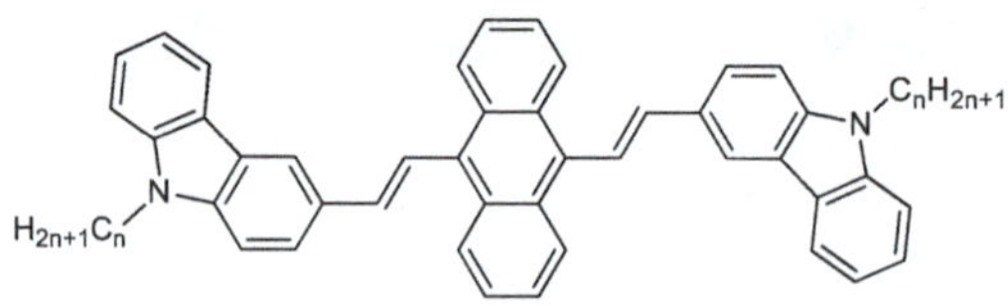

37 ACZn n = 2,3,7,8,12

As the linking isomer of **A2Cn, ACZn** (37) was also synthesized by Yang *et al.* (2013).[23] The unusual chain length-dependent solid-state fluorescence properties under various external stimuli were demonstrated. The shorter the N-alkyl chain, the more remarkable the mechanofluorochromic behavior was found (Figure 6.17) to be, which is opposite to that shown in **A2Cn.** The authors had considered that the long alkyl chains have obstructed not

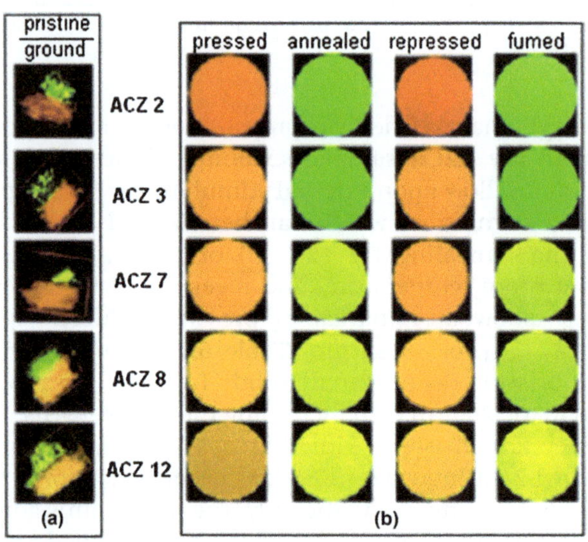

Figure 6.17 The images of series compounds **37** taken at different conditions (under UV light).
Reproduced with permission from reference 23. Copyright 2013, The Royal Society of Chemistry.

only the close stacking of molecules in pristine crystals and aqueous suspensions, but also the sufficient amorphization upon grinding, affording the unusual chain length-dependent solid-state fluorescence properties and mechanofluorochromic behaviors. This work has also demonstrated that the subtle manipulation of the end groups in the 9,10-divinylanthracene derivatives could endow them with unique and tunable solid-state optical properties.

As stated above, the mechanofluorochromic behavior of some alkyl-containing 9,10-bis(aryl-vinyl)anthracenes is highly alkyl length-dependent. For example, 9,10-bis(N-alkylcarbazol-3-yl-vinyl-2)anthracenes with shorter alkyl chains (**37**) and 9,10-bis(9,9-dialkylfluorene-2-ylvinyl-2)anthracenes with longer alkyl chains (**36**) exhibit more remarkable mechanofluorochromic behaviors, and 9,10-bis(p-alkoxy-styryl)-anthracenes only with sufficient long alkyl chains (over decyl) (**35**) can be effective mechanofluorochromic materials. The cause for these dichotomous alkyl length-dependencies can be ascribed to the influence of alkyl length on the molecular backbone conformation and stacking structure. It is quite possible that the linking positions of alkoxy chains at the peripheral aryl rings should significantly affect the molecular backbone conformation and intermolecular stacking structure of the AIE molecules to change the optical properties. Based on such consideration, Yang *et al.* (2013)[40] designed and synthesized a series of 9,10-bis(alkoxystyryl)-anthracenes (**38, OCn**) with different linking positions (*ortho*, *meta* and *para*) and lengths of alkoxy chains (propoxy, heptyloxy and hexadecyloxy) (denoted ***o*OCn**, ***m*OCn** and ***p*OCn**).

Figure 6.18A shows the fluorescence images of ***o*OCn**, ***m*OCn** and ***p*OCn** samples upon a cycle of pressing, annealing, repressing and solvent-fuming, and the corresponding spectroscopic data are summarized in Table 6.2. All the **OCn** solids exhibit fluorescence color change and a spectral shift under external stimuli, but they are obviously both alkoxy position- and length-dependent. For the ***p*OCn** series, the longer alkoxy-containing compound exhibits larger $\Delta\lambda$. However, when changing the linking positions of alkoxy chains at the phenyl rings, the resulting new isomers (***o*OCn** and ***m*OCn**) exhibit different alkoxy-related mechanofluorochromic behavior from that of ***p*OCn**. Instead, short alkoxy-containing **OCn** (***o*OC3** and ***m*OC3**) become highly mechanofluorochromic active. The X-ray single crystal analysis of ***p*OC3**, ***m*OC3** and ***o*OC3** revealed that there are two classes of CH···π hydrogen bonds formed between two **OC3** molecules (Figure 6.18B). One is the hydrogen in the benzene moiety along the long axis of one molecule as the H-donor with the corresponding phenyl ring of the anthrylene moiety of the adjacent molecule as the H-acceptor (interaction type I); another is the OCH moiety of one molecule as the H-donor with the phenyl ring of the adjacent molecule as the H-acceptor (interaction type II). In ***p*OC3**, the interaction is of type I with a short interaction distance of 2.79 Å, and in ***o*OC3**, the interaction is of type II with a short interaction distance of 2.74 Å. There are also two interaction types (I and II) existing among ***m*OC3** molecules, but the interaction distances are relatively long (2.93 and 2.84 Å,

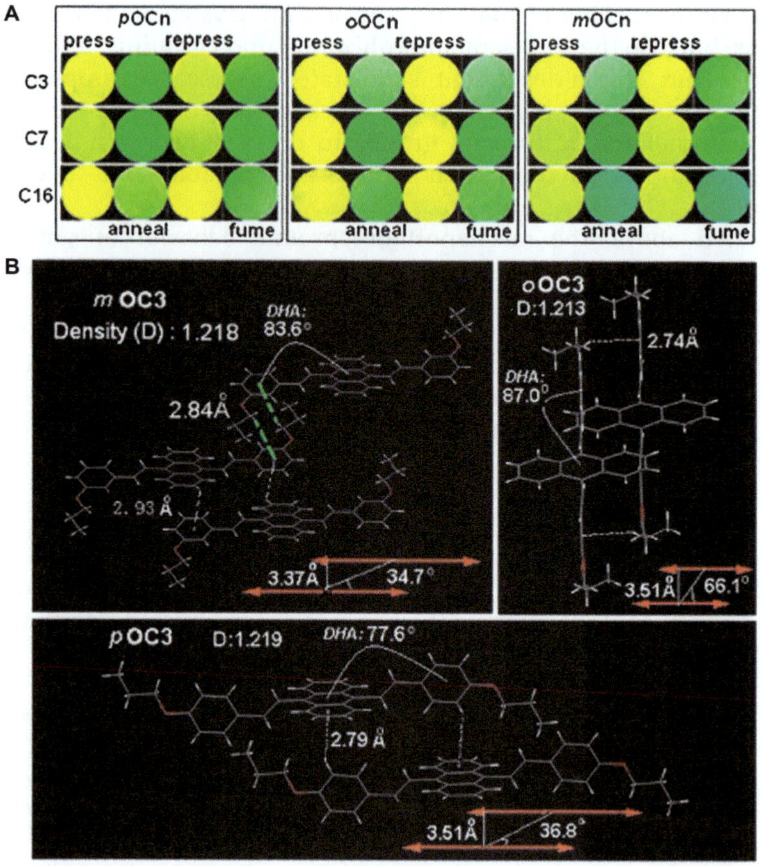

Figure 6.18 (A) Fluorescence images of **OCn** mixed with KBr under a 365 nm UV lamp upon brief pressing, thermal-annealing, re-pressing and solvent-fuming. (B) Schematic molecular conformations and interactions of ***p*OC3**, ***m*OC3** and ***o*OC3** in the crystal.

respectively). Since ***o*OC3** and ***m*OC3** exhibit easy mechanofluorochromism and large $\Delta\lambda$, it seems that the OCH$\cdots\pi$ interaction is weaker and more destructible. This is probably ascribed to the inherent flexibility of the alkoxy chains. It is observed that one molecule slides to its neighbor along the long axis of the molecule with the angles of 36.81, 34.71 and 66.11 and the vertical distance of 3.37, 3.51 and 3.51 Å for ***p*OC3**, ***m*OC3** and ***o*OC3**, respectively. This indicates that the ***p*OC3** and ***m*OC3** molecules are in J-aggregation mode and the *o*OC3 molecules adopt H-aggregation mode in the crystals. On the other hand, ***p*OC3**, ***m*OC3** and ***o*OC3** have similar crystal densities (1.219, 1.218 and 1.213 mg m^{-3}, respectively), but they have different selected dihedral angles (phenyl ring *vs.* anthryl ring), which are in the

Table 6.2 Peak emission wavelengths (λ, in nm) of **OCn** derivatives under various external stimuli.

Compound	$\lambda_{pressed}$	$\lambda_{anealed}$	$\lambda_{repressed}$	λ_{fumed}	$\Delta\lambda^a$
***o*OC3**	550	495	546	495	55
***o*OC7**	529	503	529	503	26
***o*OC16**	530	497	527	495	33
***m*OC3**	538	493	535	492	45
***m*OC7**	514	499	511	498	22
***m*OC16**	523	484	523	484	39
***p*OC3**	525	511	522	511	14
***p*OC7**	525	508	525	509	17
***p*OC16**	549	499	549	501	50

[a] $\Delta\lambda = \lambda_{pressed} - \lambda_{anealed}$. Reproduced with permission from reference 40. Copyright 2013, The Royal Society of Chemistry.

order ***o*OC3** (87.0°)>***m*OC3** (83.6°)>***p*OC3** (77.6°). More twisted backbones will decrease the molecular conjugation degree and endow fluorophores with more blue-shifted emissions. Thus, it was concluded that changing the linking-positions of alkoxy chains at the phenyl rings of **OCn** derivatives can significantly affect the molecular conformations and packing structures, and the strongly twisted conjugated backbone and weaker intermolecular interaction can endow the crystals with stronger inner stress, easily-destructible structures and large conformation change under external stimuli.

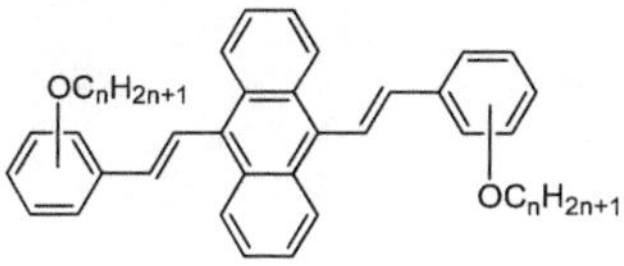

para-linkage (***p*OCn**)
ortho-linkage (***o*OCn**)
meta-linkage (***m*OCn**)

38 n = 3, 7, 16

6.4 Mechanofluorochromic Compounds

6.4.1 Diarylvinylanthracene Derivatives

The diarylvinylanthracene unit is an important AIE structural unit, whose AIE nature was first reported by Tian's group (2009).[41] The machanofluorochromic properties of the diarylvinylanthracene derivatives have been reviewed by Chi and Xu *et al.*[1] In this section, two other examples not included in that review are described.

The spectroscopic properties and morphological structures of dye **39** upon pressing, grinding, annealing or fuming (Figure 6.19A) were reported by Chi and Xu *et al.* (2011).[42] The mechanofluorochromic nature was generated through phase transformation, which was confirmed by XRD and DSC. The results of crystalline structure analysis showed that the dye molecules were

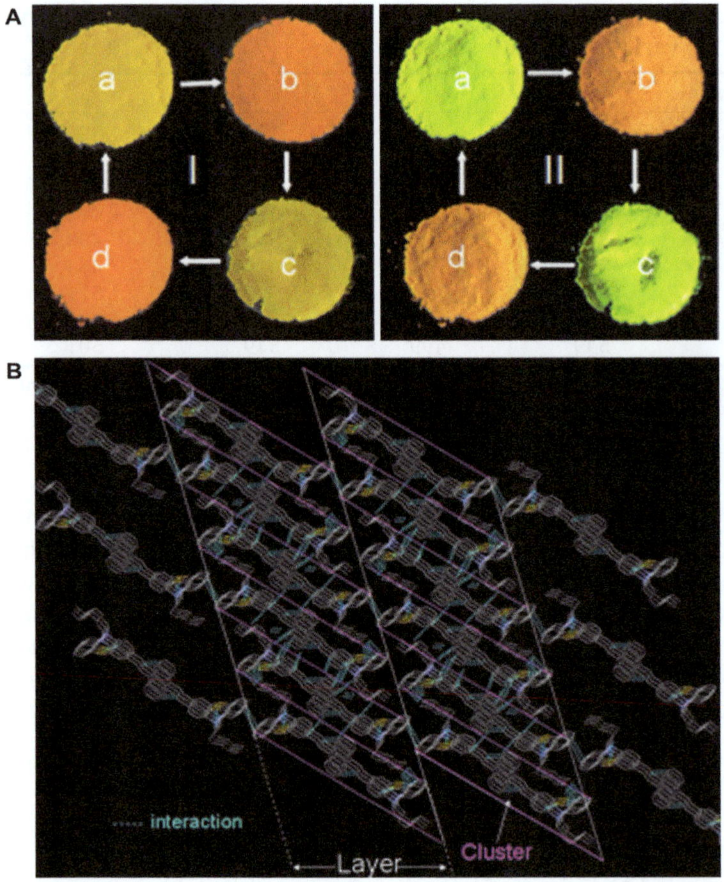

Figure 6.19 (A) Dye **39** taken at room temperature under (I) natural light and (II) UV light. Samples obtained: (a) as-synthesized or fumed sample; (b) pressed or ground sample; (c) annealed sample; (d) pressed or ground sample. (B) Molecules packing of dye **39** in single crystal structure.
Reproduced with permission from reference 42. Copyright 2011, American Chemical Society.

packed in a head-to-head manner (Figure 6.19B). The backbone of the molecule largely deviated from the plane. Typical cofacial π–π stacking became impossible because of the highly twisted conformation and the steric hindrance of the bulky phenothiazinyl groups in the molecule. The molecules were packed in clusters *via* weak C–H···S, S···π and C–H···π interactions. The former also acted to bind the clusters to form lamellar layers. These layers were connected *via* the weak and sparse π–π interactions and partial π-overlaps from the phenyl rings in the phenothiazinyl groups. Thus, the layer-layer and cluster-cluster interfaces were readily destroyed *via* slip deformation by an external force or stimulus, which facilitated mechanofluorochromism.

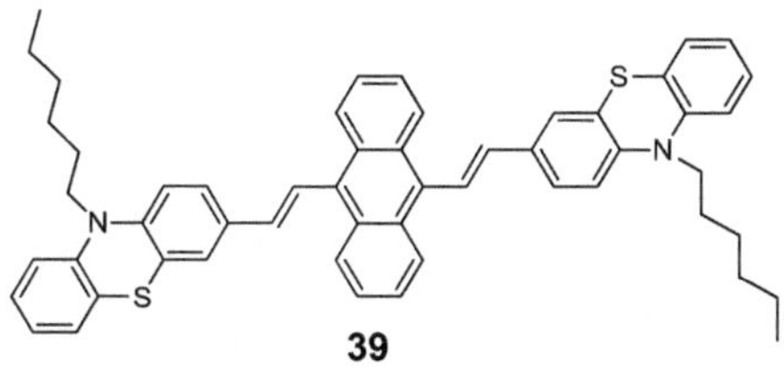

39

In section 6.1.4, the mechanoflurochromic mechanism of compound **4** has been described. Here, a detailed account of its mechano-fluorochromic phenomena is given.[18] After being ground, the **4** powder showed a strong red shift with a yellow emission (λmax = 561 nm), and after being heated above 160 °C, the ground powder recovered to its initial green emission (λmax = 528 nm). The interconversion of the two states with their distinct emission colors is completely reversible through grinding and heating (Figure 6.20a and b). The experimental results of the influence of applied pressure on the luminescence of **4** indicated that, as the applied pressure increased, the fluorescence emission of the **4** powder clearly showed a gradual red shift (Figure 6.20c). The increasing applied pressure from 0 to 8 GPa caused a more noticeable color change in the luminescence of the **4** powder than that observed upon grinding. The observed color change of 124 nm from green (528 nm) to red (652 nm) is the largest mechanofluorochromic shift reported at that time. The PL spectrum with λmax = 561 nm under an applied pressure of 2.43 GPa was quite similar to that of the powder ground with a pestle. An increased applied pressure of 7.92 GPa caused the luminescence to change to a redder emission with λmax = 652 nm, proving apparently that the grinding method is not powerful enough to cause a stronger mechano-fluorochromic effect in the case of **4**. The authors believed that the dramatic change in the color of luminescence of **4** powder under applied pressure was the result of changes in the molecular aggregation state under external pressure.

6.4.2 Tetraphenylethylene Derivatives

Tetraphenylethylene (TPE) derivatives have been hotly investigated in recent years for their aggregation-induced emission (AIE) properties. Recent studies show that some TPE derivatives possess solid state mechanofluorochromic performances, which means that the materials have different fluorescent emissions under mechanical force.

In section 6.2.2, the morphology-dependent multi-color emissions of ethoxy- and butoxy-substituted TPE (**34** and **35**) by different external stimuli in solid state have been described. Other two methoxy substituted TPE

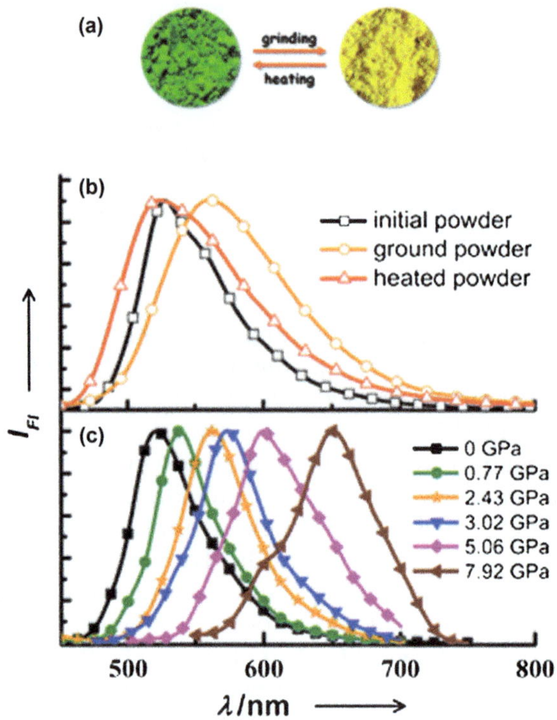

Figure 6.20 (a) Photographs of the ground and heated powders of dye **4** under UV light (365 nm). (b) PL spectra of the initial, ground, and heated powders of dye **4**. (d) PL spectrum of dye **4** powder under external pressure. I_{FL} = fluorescence intensity.
Reproduced with permission from reference 18. Copyright 2012, Wiley-VCH.

derivatives were synthesized by Zhang *et al.* (2013),[43] which are tetra(4-methoxyphenyl)ethylene (**40**) and tetra(3,4-dimethoxyphenyl)ethylene (**41**). The crystalline to amorphous phase transition of **40** and **41** by grinding resulted in good mechanofluorochromic performance: the strong fluorescence of **40** pristine crystals can change from blue (λem: 420 nm) to cyan (λem: 480 nm) by grinding; similarly, the emission of **41** pristine crystals at 460 nm red shifts to 480 nm after grinding (Figure 6.21A). The fluorescence can almost go back to the original again upon thermal treatment. The XRD measurements show that the mechanofluorochromic properties are also directly caused by crystalline to amorphous phase transition. In this case, the results of fluorescence decay experiments show that the lifetimes of all the crystals and their amorphous counterparts are all at a level of several nanoseconds and very small changes in lifetime are found before and after grinding, indicating that the molecular environments do not have significant change, and this also excludes the possibility of intermolecular π–π stacking after grinding. In addition, in this case, the samples in the crystalline and amorphous states show no significant change in quantum yields, and all have comparable high quantum yields. It is known that TPE (**28**) does not have mechanofluorochromic properties, because it cannot undergo a crystalline to amorphous phase transition. The authors thought that although only small methoxy groups are substituted on TPE, more flexible intermolecular interactions of C–H···O and C–H···π are introduced, which make the sliding and deformation of the crystal structure easier, facilitating the transition from crystalline to amorphous phase. In contrast, TPE crystals with only rigid C–H···π interactions between the benzyl groups exhibit fragile character under external pressure or stimuli. Thus, Dong *et al.* believed that the increment of relative "soft interaction" between the molecules which plays an important role to stabilize the metastable state of more planar conformation under pressure is the essential factor for mechanofluorochromic TPE derivatives.[38]

To try out the material in the application as an anti-counterfeiting label in paper money, compound **40** was selected to prepare the "anticounterfeiting ink" due to its more obvious fluorescence change than **41**. Characters "5 yuan (in Chinese)" were printed onto a banknote to demonstrate the application (Figure 6.21B). Selective grinding the character "5" can cause an obvious color contrast with the unground character "元". The authors pointed out that the repeated annealing and grinding can be used to induce the responsive fluorescent color change and increase the complexity of the anti-counterfeiting performance compared with the common fluorescent dyes in use, and this kind of material is a good candidate for practical applications in high-security anti-counterfeiting inks.

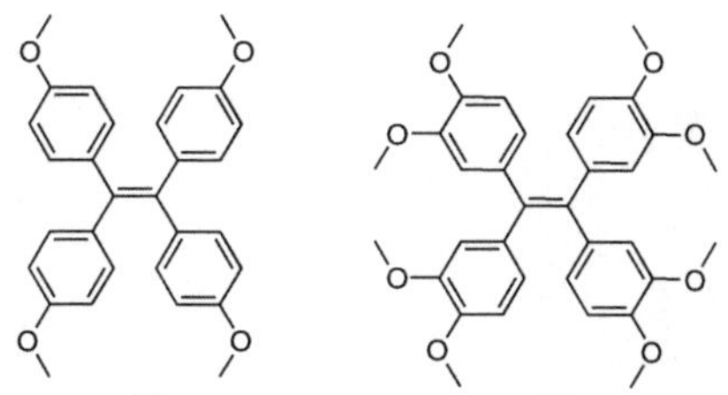

40 **41**

Dong *et al.* (2013)[44] have studied the mechanical response of luminescence of **42** and **43**. The emission of both luminogens could be modulated reversibly between blue and green through repeated grinding and heating processes (Figure 6.22). The powder XRD curves of both original blue crystals and annealed samples display many sharp and intense reflection peaks and coincide with each other, indicating the same crystalline orders. The diffraction curve of ground powders also exhibits some reflections, which agree with but not as many and sharp as those of their original crystals, hinting its partly amorphous essence. Hence, the authors believed that the mechanochromic fluorescences of **42** and **43** are caused by the amorphization upon grinding.

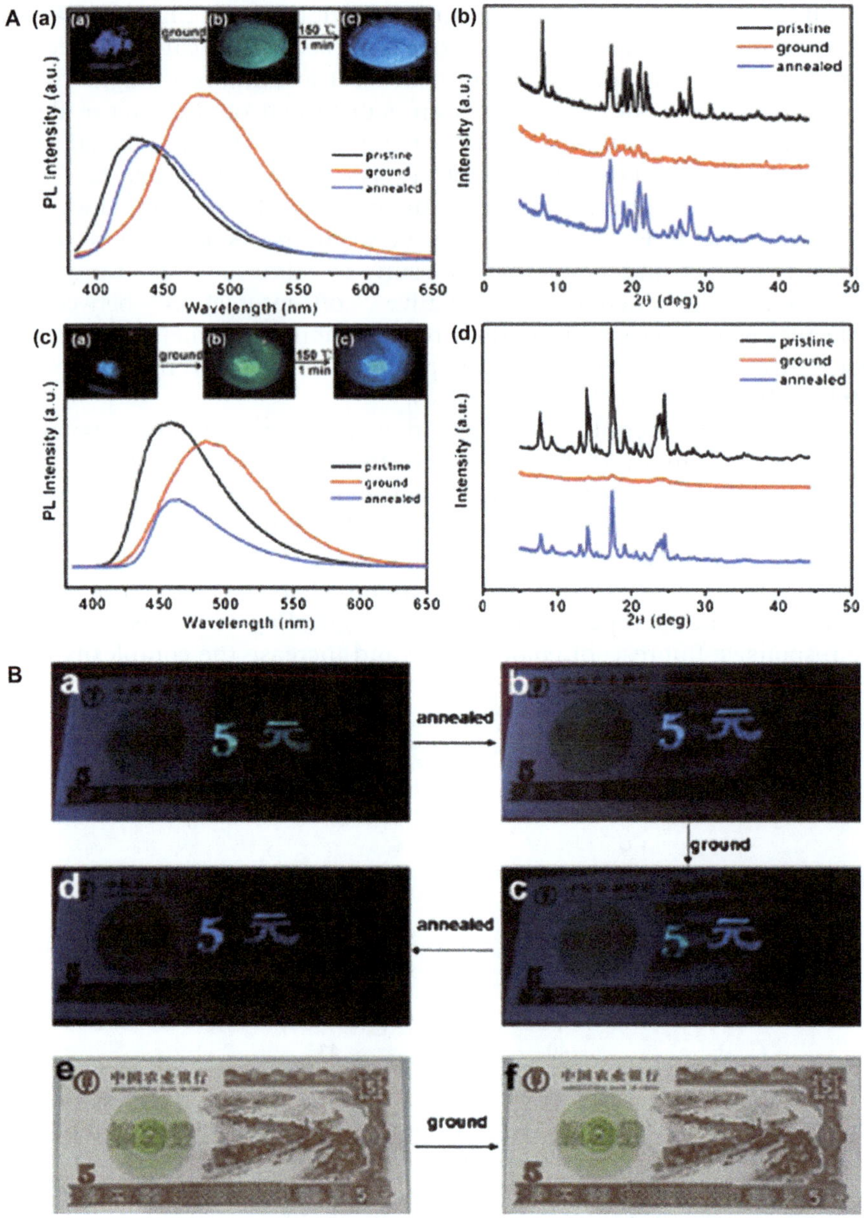

Figure 6.21 (A) PL spectra (a) and XRD patterns (b) of **40**: pristine, ground and annealed sample (150 °C for 1 min). PL spectra (c) and XRD patterns (d) of **41**: pristine, ground and annealed sample (150 °C for 1 min). Insets: real object illustration of the process. (B) Illustration of **40** as an anti-counterfeiting ink on a 5-yuan RMB practice note printed with Chinese characters of "5-yuan". Images are (a) the note immediately after "5-yuan" was printed, (b) after annealed; (c) when 5 was ground; (d) after annealed again under UV light irradiation; (e) and (f) are images of (b) and (c) under visible light, respectively.
Reproduced with permission from reference 43. Copyright 2013, The Royal Society of Chemistry.

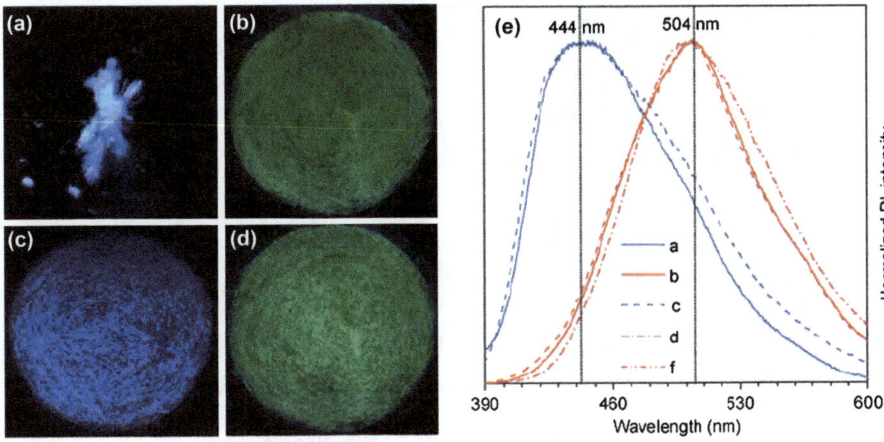

Figure 6.22 Images of **43**: (a) crystals; ground solid (b) before, (c) after annealing and (d) regrinding. (e) PL spectra of samples in the images. PL spectrum of (line f) amorphous solid of 2 is also listed for comparison. Annealing details: 175 °C, 15 min. Excitation wavelength: 350 nm. Photographs were taken under UV illumination.

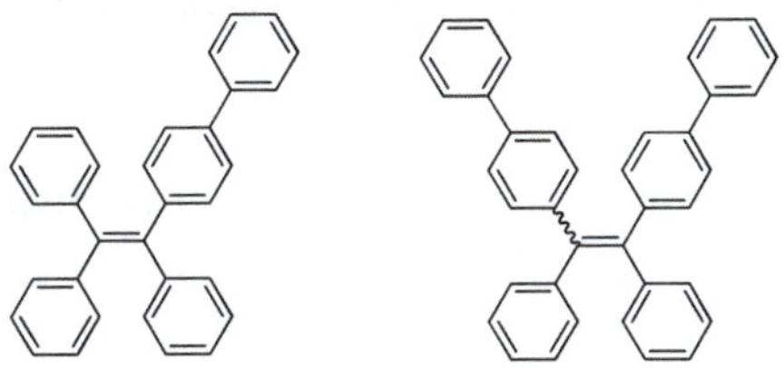

42 **43**

Dong *et al.* (2013)[45] designed and synthesized an analog of tetraphenylethylene (**44**), which was obtained in three colored crystals. As is shown in Figure 6.23, the emission of luminogen **44** could be tuned reversibly among three different colors through repeating grinding and heating. The emission of the ground powder can also be tuned reversibly between bright green and dark orange through repeated grinding and solvent fuming, and this process is independent of the macroscopic scale aggregate pattern.

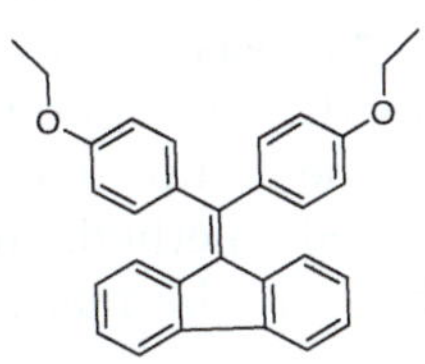

44

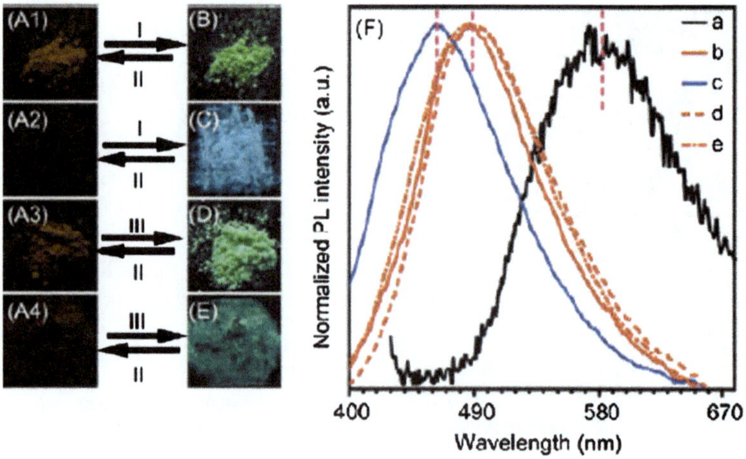

Figure 6.23 Photos of ground solid of luminogen **44** in (A1 and A3) pile or dispersed powder on quartz plate (A2 and A4); (B–E) photos of ground solid after treating are placed in corresponding places; (F) PL spectra of samples in the photos. Excitation wavelength: 400 nm for A, and 350 nm for others. Photos were taken under UV illumination. I, heating at 120 °C, 10 min, under ambient condition; II, grinding; III, fuming with chloroform, 5 min.
Reproduced with permission from reference 45. Copyright 2013, Science China Press and Springer-Verlag Berlin Heidelberg.

The dye **45** containing also the tetraphenylethylene AIE unit was reported by Tang *et al.* (2012).[46] The crystals of dye **45** showed strong yellow emission at 565 nm. After gentle grinding using a mortar and pestle, red powders were formed with red PL at 650 nm. After fuming with acetone vapor for 10 min, the sample reverted to its original appearance (yellow). The conversion between yellow and red emission colors could be repeated several times without fatigue because these stimuli were nondestructive in nature. On the other hand, heating the ground sample at 150 °C for 10 min changed its color from red to orange. Such switching was reversible and suffered little wavelength shift after several cycles. The unground sample was crystalline, whereas the ground sample was amorphous, as confirmed by XRD measurements. The samples treated thermally and fumigated with solvent vapor were also crystalline, suggesting that the amorphous powders crystallized upon solvent fumigation or thermal treatment. The mechanochromism observed for dye **45** was associated with the morphology change from the crystalline to the amorphous state and vice versa. The red amorphous powders cannot be completely recovered as yellow emissive crystals by the thermally treated method. It is indicative of stronger effect by solvent fumigation than the thermal process on the crystallization of dye **45**.

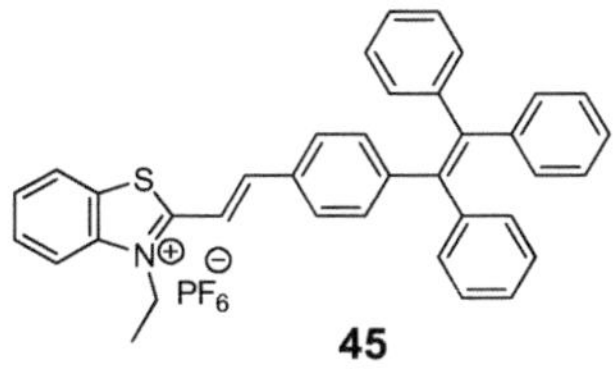

45

Tang *et al.* (2013)[47] reported another tetraphenylethylene ionic compound **46** containing pyridinium unit. The compound exhibited multi-functional properties including morpho- and mechano-chromism, optical waveguide and mitochondrion imaging. The pale-yellow crystals of **46** were readily obtained by slow evaporation of its dichloromethane-hexane mixture at room temperature. UV irradiation of the **46** crystals gives a strong green emission at 515 nm with a Φ_F value of 31.8% (Figure 6.24A and B). After gently grinding the crystals using a metal spatula, yellow powders are formed. The powders emit at 600 nm, giving a large emission contrast of *ca.* 85 nm. At the same time, the Φ_F value drops to 20.4%. Such a transformation is reversible, aided by fuming with acetone vapor for 10 min or heating at 150 °C for 10 min (Figure 6.24C and D). The powder X-ray diffraction analysis results revealed that the **46** crystals have well-ordered structure and its ground sample possesses low crystallinity or is even amorphous.

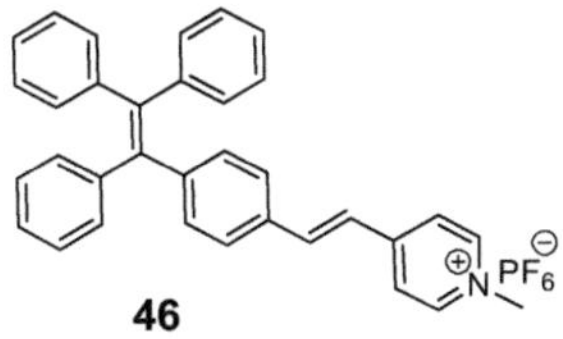

46

A new multi-responsive tetraphenylethene functionalized oxazoline molecular switch (**47**) was synthesized by Zhang *et al.* (2013).[48] In its molecular structure, relatively weak intermolecular interactions such as C–H···N, C–H···π and C–H···O were introduced after the oxazoline motif modification, enabling **47** to have mechanofluorochromic performance, whereas its parent compound, tetraphenylethylene (TPE), does not have mechanofluorochromic property because in general TPE cannot be realized the crystalline to amorphous phase transition under pressure. The multi-emission performance of **47** was realized in the solid state with various external stimuli, including mechanical grinding and acid/base treatment (Figure 6.25). The emission color of **47** changed from blue (λem: 450 nm, line a) to cyan (λem: 484 nm, line b) after the crystalline to amorphous phase transition by grinding, and it changed back to blue (line c) after wetting the ground sample with ethyl acetate to convert the aggregation state from amorphous to crystalline again. When both the crystals and the ground amorphous sample of **47** were fumed with HCl, their emission color changed to deep red (λem: 664 nm, line d and e), which suggests a structural

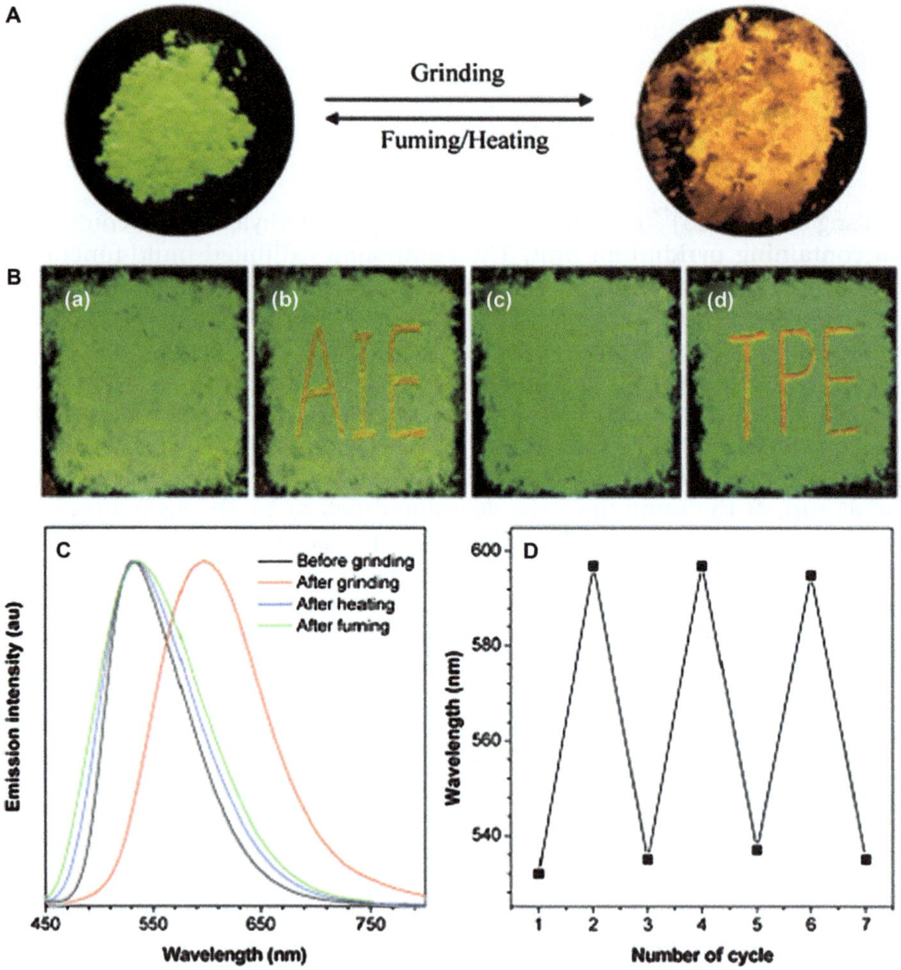

Figure 6.24 (A) Switching the solid-state emission of **46** by the grinding-fuming/heating process. (B) Fluorescent images of **46** thin films on filter papers (a) without and (b) with letters of "AIE" being written by using a metal spatula. The photograph in (c) was obtained by fuming the film in (b) with acetone vapor for 10 min, while that in (d) was obtained by writing the letters of "TPE" on the **46** film in (c) using a metal spatula. All the photos were taken under 365 nm UV irradiation. (C) Change in the emission spectrum of **46** crystals by the grinding-fuming/heating process. (D) Repeated switching of the solid-state fluorescence of **46** by repeated grinding and fuming/heating cycles. Reproduced with permission from reference 47. Copyright 2013, The Royal Society of Chemistry.

transformation from the ring-closed form to the protonated ring-open form. The protonated ring-open form powder could be deprotonated by alkali vapor, and, after being wetted by triethylamine (TEA), the red color turned

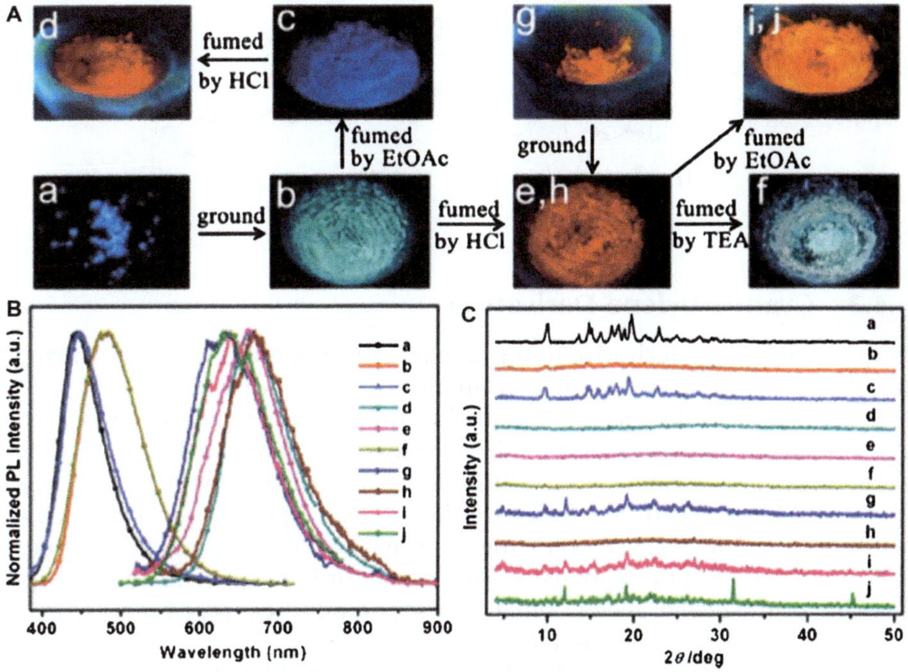

Figure 6.25 Fluorescent images and corresponding PL spectra of **47** and **48** under various external stimuli in solid state. Sample "a" is **47** crystals; samples "b, c, d, e, f and j" are obtained from sample "a" by various treatments as labelled in the figure. Sample "g" is **48** crystals; samples "h" and "i" are obtained from sample "g" by various treatments as labelled in the figure.
Reproduced with permission from reference 48. Copyright 2013, The Royal Society of Chemistry.

back to cyan with the emission peak blue-shifting from 664 nm to 484 nm (line f), indicating TEA didn't induce the amorphous to crystalline phase transfer. The amorphous to crystalline phase transition of HCl protonated **47** was achieved by fuming with EtOAc, which then exhibited an orange emission (λem: 630 nm, line j). Therefore, four emission colors (peaks at 450 nm, 484 nm, 630 nm and 664 nm) were obtained from the **47** crystals, and they could convert to each other through appropriate stimuli. In the same way, the crystals of **48** also showed mechanofluorochromic property after weak intermolecular interactions such as C–H···Br, O–H···Br, C–H···π and C–H···O were introduced. Its emission color changed from orange to red with the peak red shifted from 630 nm (line g) to 664 nm (line h) after phase transfer from crystalline to amorphous by grinding, and the emission color was recovered to orange (λem: 633 nm, line i) after re-crystallization by wetting with solvents, such as EtOAc.

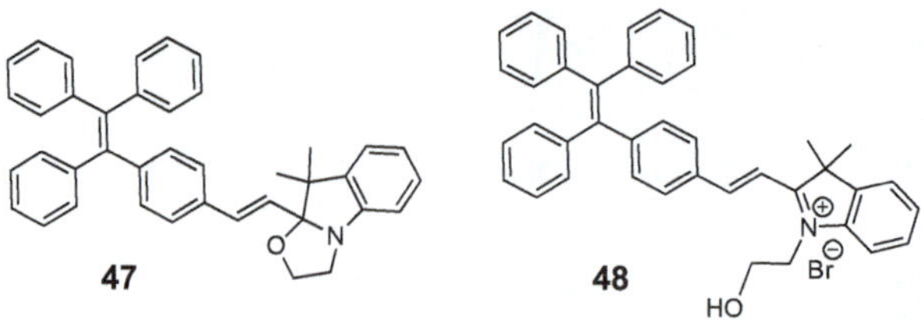

6.4.3 Cyanoethylene Derivatives

The cyanoethylene derivative found to have AIE properties is only following the silole derivative, thus, it is an important class of AIE materials.

In section 6.1.2, the mechanofluorochromic mechanism of dye **3** has been described. Park *et al.* (2010)[10] also successfully fabricated a rewritable fluorescent optical recording medium, that is, a poly(methyl methacrylate)-(PMMA)/dye **3** blend film, which exhibited a fast-responding and reversible multi-stimuli fluorescence switching (Figure 6.26). The authors pointed out that this multi-stimuli responsive system was unique in terms of the slip-stacking of molecular sheets, which provides a novel concept for rewritable fluorescent optical recording media.

Park *et al.* (2011)[49] designed and synthesized dyes **49** and **50** to investigate the change in the mechanochromic phase *via* direct crystalline structure analysis. The two dyes possess simpler molecular structure than dye **3** and were expected to possess highly crystalline property, which was

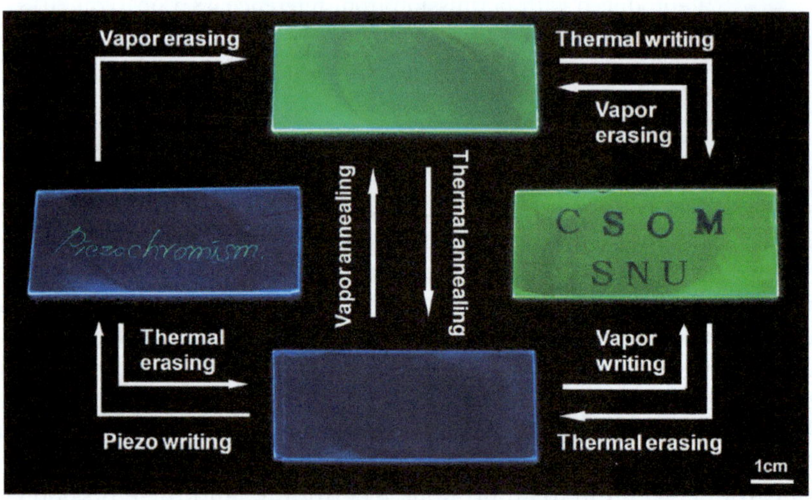

Figure 6.26 Photos of the fluorescence writing/erasing cycle of dye 3/PMMA film. Reproduced with permission from reference 10. Copyright 2010, American Chemical Society.

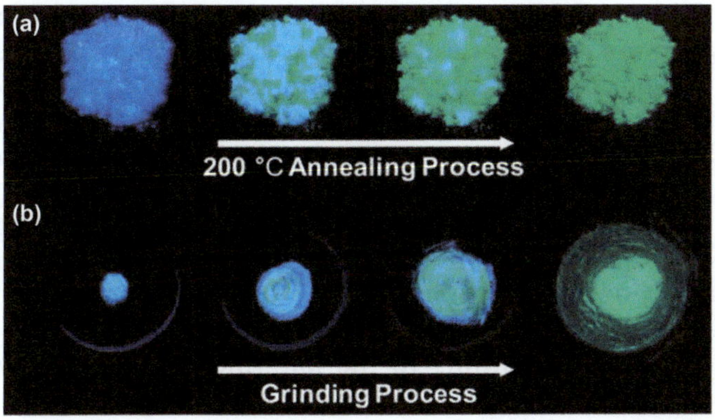

Figure 6.27 Dyes **49** phase transition processes from phase I to phase II: (a) thermal annealing process without melting, and (b) mechanical grinding process.
Reproduced with permission from reference 49. Copyright 2011, The Royal Society of Chemistry.

advantageous in preparing single crystals. Both dyes exhibited very high solid-state fluorescence quantum yields because of the characteristics of AIE properties as well as their capacity for polymorphic two-color emission. The polymorphic fluorescent phases were obtained using various sample preparation methods, such as suspension preparation, drop-casting, solution-recrystallization, melt-solidification, thermal annealing and mechanical grinding. The phase transition processes of dye **49** from phase I with a short emission wavelength to phase II with a long emission wavelength are shown in Figure 6.27. Based on the analysis of single crystal structures, the head-to-tail coupling of the local dipoles as well as the multiple C–H···π and C–H···N interactions in phase I were responsible for the specific molecular stacking architecture with weak excited state dimeric coupling. In phase II, the crystals exhibited an efficient excited-state dimeric coupling that was attributed to the substantial π–π overlap, which includes the antiparallel coupling of local dipoles. The different molecular packing structures produced differently colored fluorescence emission. The reciprocal transformation between phases I and II with the applications of thermal and mechanical stimuli resulted in reversible changes in the emission colors.

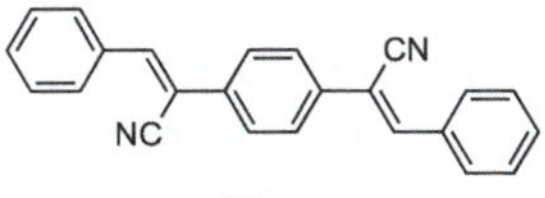

49

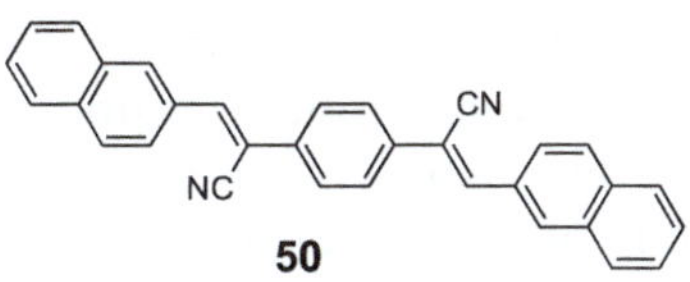

50

Zhang *et al.* (2012)[50] reported a series of cyanostilbene derivatives and found that the recrystallization of dye **51** from ethanol resulted in a light-blue powder with a faint luminescence (Φ_F<0.1%). When this powder was briefly ground using a pestle and mortar, the emission spectrum exhibited an obvious red shift of 14 nm (from 498 nm to 512 nm) with the Φ_F value as high as 24.1%. The results indicated that the compound had mechanochromic property. The ground sample emitted a green fluorescence when irradiated with UV. Dye **52** also exhibited a mechanically induced change in its luminescence. The emission color changed from green (531 nm) to greenish yellow (542 nm) by grinding. The excited state of the dye **51** crystals had a single-exponential decay ($\tau = 20.8$ ns). On the contrary, the ground sample decayed bi-exponentially, and its lifetime was obviously shortened to 4.0 ns. According to the long lifetime and low Φ_F in their crystalline states, the authors considered it to form excimers by strong dipole-dipole interactions or/and effective intermolecular π-stacking. The XRD measurement results showed well-defined microcrystalline-like and amorphous structures before and after grinding, respectively. The authors also thought that the mechanochromism might originate from the alternating mode of molecular packing from high-order to disorder by grinding. Dye **51** in the amorphous state exhibited higher λ_{max} and Φ_F values than those in the crystalline states. The authors suggested that the increase in Φ_F might be attributed to the damage of intermolecular π–π interactions when an external mechanical stimulus was applied. The red shift of λ_{max} was generally attributed to the formation of excimers. However, the fluorescence lifetime of the ground sample was obviously shorter than that in the crystalline state, which did not correlate with the characteristics of excimers. Thus, the mechanochromic mechanism of dye **51** remains unclear and needs further investigation.

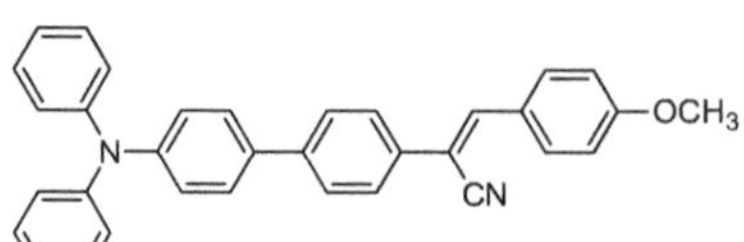

51

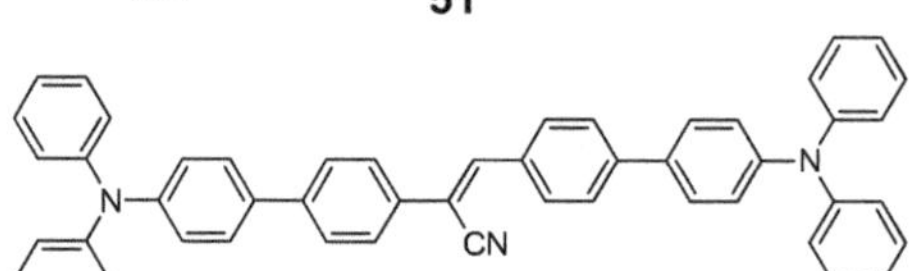

52

Zhang *et al.* (2013)[51] designed and synthesized two isomers (**53** and **54**), consisting of twisted triphenylamine and cyanostilbene. The fluorescent colors of the two isomers in the crystalline state were sky-blue for **53** (Figure 6.28a) and green for **54** (Figure 6.28b) with quantum yields (Φ_F) of 44.9% and 7.7%, respectively. The single crystal X-ray diffractometry results revealed the very different molecular conformations and packing modes of **53** (planar) and **54** (twisting), the authors thought, which might be the reason for different Φ_F. The reversible mechanochromic and thermochromic

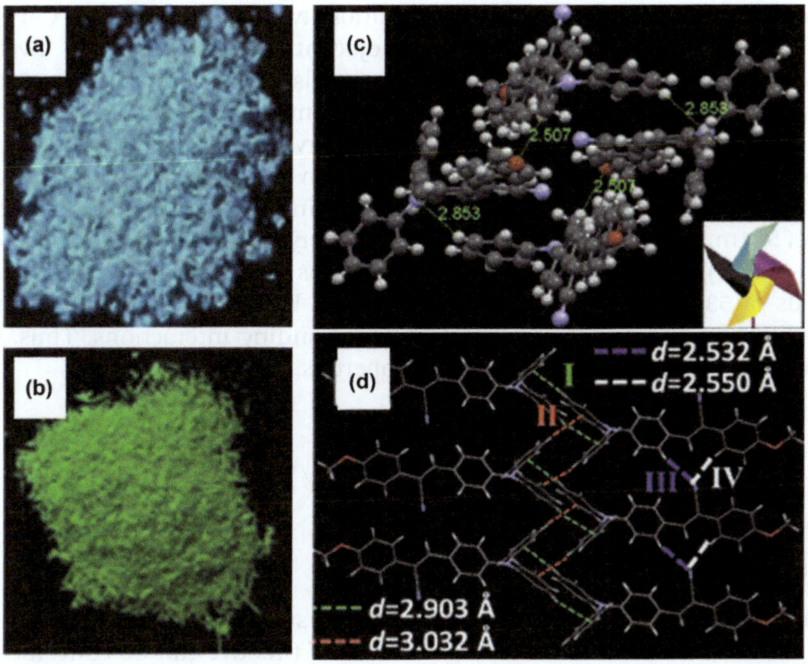

Figure 6.28 Compounds **53** (a) and **54** (b) crystal images under UV irradiation (λex = 365 nm). Molecular stacking structure with secondary bonding interactions in the crystal of **53** (c) and **54** (d).
Reproduced with permission from reference 51. Copyright 2013, Elsevier.

fluorescence switching was observed in the crystalline **53** powders. The sky-blue crystals were changed into the green-emissive solids after grinding, and they were recovered to their original state upon heating at 60 °C over 2 min. The powder X-ray diffractometry results demonstrated that the mechanochromic and thermochromic behaviors were also attributed to the crystalline-to-amorphisation phase transition. In addition to the photophysical properties, the different molecular packing motifs were also observed in their single crystals. The crystal structure of **53** containing one discrete molecule was monoclinic with space group $P2_1/c$. The two-dimensional structure with edge-to-face interactions was an achiral "pinwheel" pattern on the *bc* plane. The good coplanarity of cyanostilbene and absence of face-to-face π–π interaction was considered to result in the high Φ_F in the crystalline state. In the "pinwheel" stacking, Figure 6. 28c showed that the C–H···N interaction, which was longer than the Van der Waals radii of 2.75Å, existed between two adjacent molecules. The strong C–H···O interaction with short distance of 2.507 Å and angle of 151.7° were also observed. As for dye **54**, the single crystal consisting of two molecules belonged to $P\bar{1}$ triclinic space group and there was a face-to-face slipped π-stacking motif with a roll angle of 58.7° in the crystal. Another notable feature was that the C–H···N distance in **54**, 2.54 Å fell well within the sum of

Van der Waals radii (Figure 6.28d). In addition, two types of aromatic C–H···π interactions (I and II), making the adjacent molecules of the twisted triphenylamine connect to each other, were also observed in Figure 6.28d. However, such aromatic C–H···π and C–H···N interactions did not exist in the crystal of **53**. These various multiple secondary bonding interactions in **54** enhanced the molecular rigidity and stability in crystal lattice. This proposal was further supported by a higher melting point of **54** (m.p. 139.7 °C) than that of **53** (m.p. 131.2 °C). Moreover, the single crystal with a "pinwheel" motif (edge-to-face), relative to parallel stacking, was quite unstable. Briefly, the crystals of **53** with edge-to-face stacking could be readily damaged by force stimuli in the absence of various secondary bonding interactions. Thus, dye **53** presented mechanofluorochromic properties, rather than **54**.

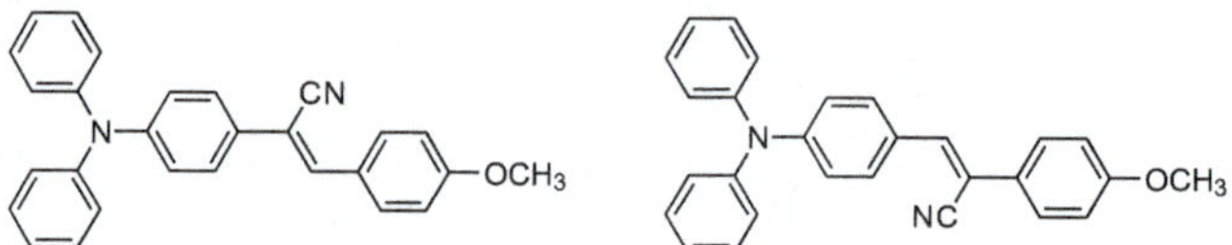

53 **54**

The AIE compound **55** with a D-A electron structure was synthesized by Wang *et al.* (2011).[52] The fluorescent colors of this dye can be conveniently switched using various environmental stimuli, which include mechanical force, organic vapor, heat, acid and base (Figure 6.29). Grinding and heating treatments effectively induced the fluorescence changes from red-orange to yellow by tuning the molecular packing in the solid states. The XRD measurements confirmed that the unground sample was a well-ordered microcrystalline-like structure, whereas the ground sample was amorphous.

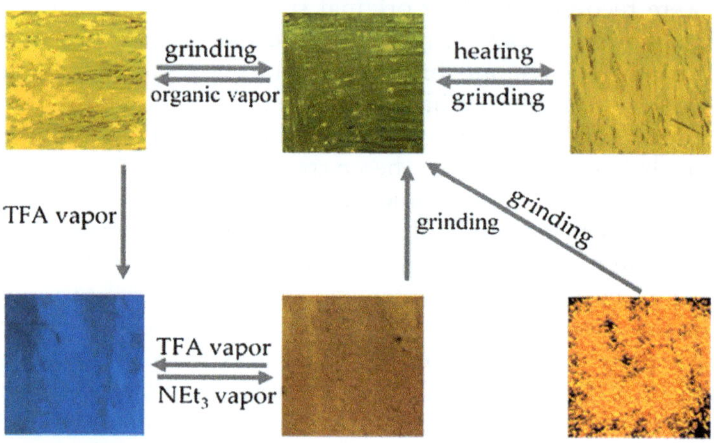

Figure 6.29 Stimuli-responsive behaviors of dye **55** upon grinding, heating and acid and base vapor treatments.
Reproduced with permission from reference 52. Copyright 2011, American Chemical Society.

Upon exposure to trifluoroacetic acid (TFA) vapor, the sublimated or heated sample (orange) was converted into a blue emissive state, which could be recovered by triethylamine (NEt_3) vapor. The chloroform solution of dye **55** emitted nearly unobservable fluorescence at room temperature and intense yellowish green fluorescence in the frozen condition (77 K). After addition of TFA, the solution changed into a weak green emissive. This phenomenon is considered to be the effect of protonation.[52]

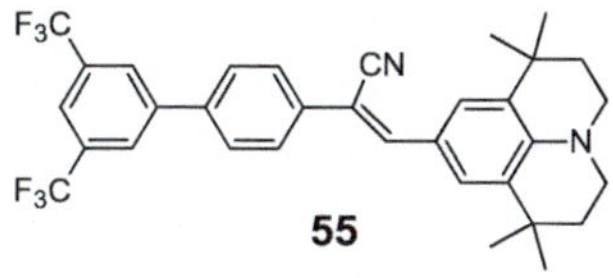

55

Another D-A electron structure compound **56** was also studied by Zhang *et al.* (2013)[53] It shows intramolecular charge transfer and aggregation-induced emission characteristics with high solid-state efficiency of 65.3%. Moreover, it exhibits reversible mechanochromic behavior between the crystalline and amorphous states with remarkable emission color change from green (504 nm) to yellow (545 nm) (Figure 6.30a and b). The XRD profile of the as-prepared solid clearly displays intensive and sharp diffraction peaks, reflecting a well-ordered microcrystalline structure. However,

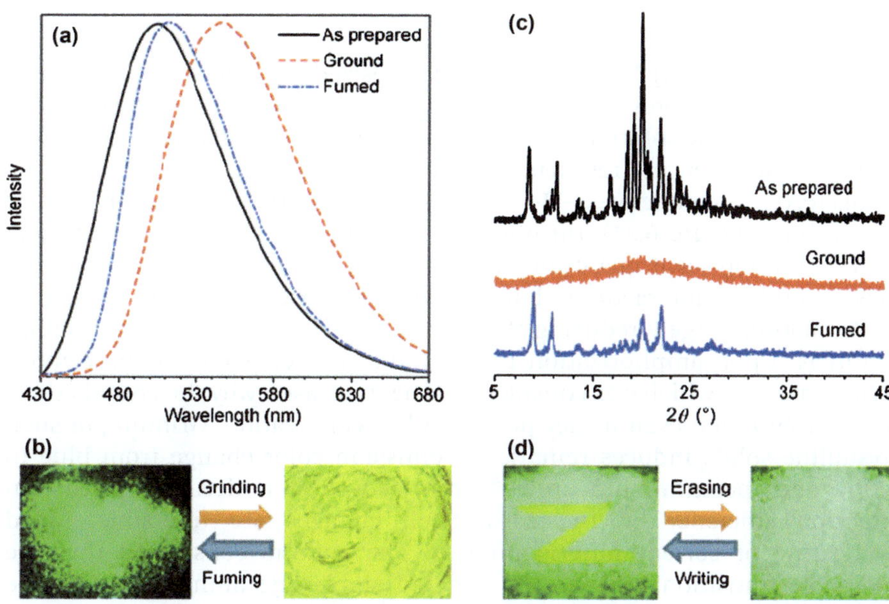

Figure 6.30 (a) Normalized PL spectra, (b) photographs under 365 nm UV light and (c) XRD patterns of **56** at different states as indicated. (d) Writing and erasing of letter "Z" on the filter paper with sprayed **56** under UV light. Reproduced with permission from reference 53. Copyright 2013, The Author(s).

the diffractogram of the ground solid shows only a weak and broad diffusion halo, suggesting its disordered amorphous stacking. After being fumed with solvent vapor, noticeable diffractions resembling those of the as-prepared solid emerge out, indicating the recovery of ordered crystalline lattice (Figure 6.30c). According to these results, it was reasonable to ascribe the mechanochromism as a reversible phase transition process between the ordered crystalline and disordered amorphous states. The authors also considered that the red-shifted emission upon mechanical stimuli should be attributed to the planarization of highly twisted conformations of **56**, which induces much longer effective conjugation length and thus much redder emission. Switchable mechanochromic luminescence with a remarkable shift of 41 nm endows the high efficiency **56** solid applicable in varying fields, for example, in the rewritable optical storage (Figure 6.30d) with no chemical structure destruction in the pressing and fuming processes.

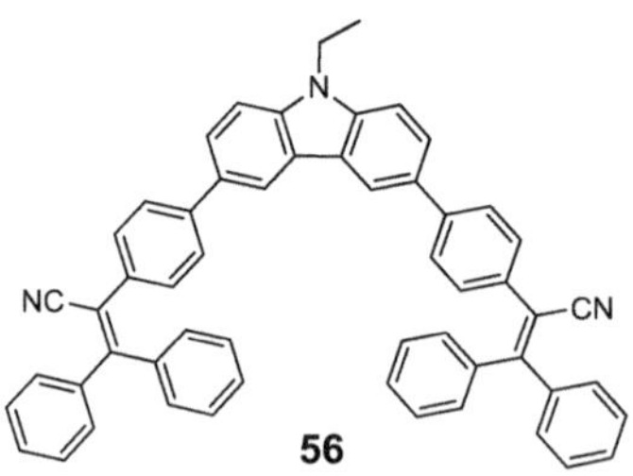

56

Zhang *et al.* (2013)[54] designed and synthesized a group of triphenylacrylonitrile luminophores, namely **57**, **58** and **59**. They are practically non-luminescent in the solutions due to the active intramolecular rotations. Such rotations, however, are restricted in the crystalline state due to the van de Waals forces and effective C–H···N, C–H···π and C–H···H–C intermolecular interactions (Figure 6.31), turning them into highly efficient blue emitters with high fluorescent quantum yields up to 99%. The authors believed that these multiple interactions have locked and rigidified the molecular conformation, largely reducing the energy loss *via* non-radiative relaxation channels. Upon amphorization by mechanical force, some of these interactions are lost, which was considered to be the reason why the crystals show stronger emission even if they possess a low conjugation. Grinding of such crystalline solids induces remarkable emission color change from blue to yellow and spectrum shift up to 78 nm, as well as noticeable decrease in emission efficiency by 10.5–24%. Such changes are reversible, aided by heating or solvent fuming process. The authors considered that the reversible transition between the ordered stable crystalline state and the disordered metastable amorphous state is crucial to the switchable emission color and intensity of the luminogens. The authors believed that in the crystalline state, the luminogens may adopt a more twisted conformation in order to fit into the crystalline lattice. This shortens their effective conjugation length and hence makes them emit bluer light. On the other hand,

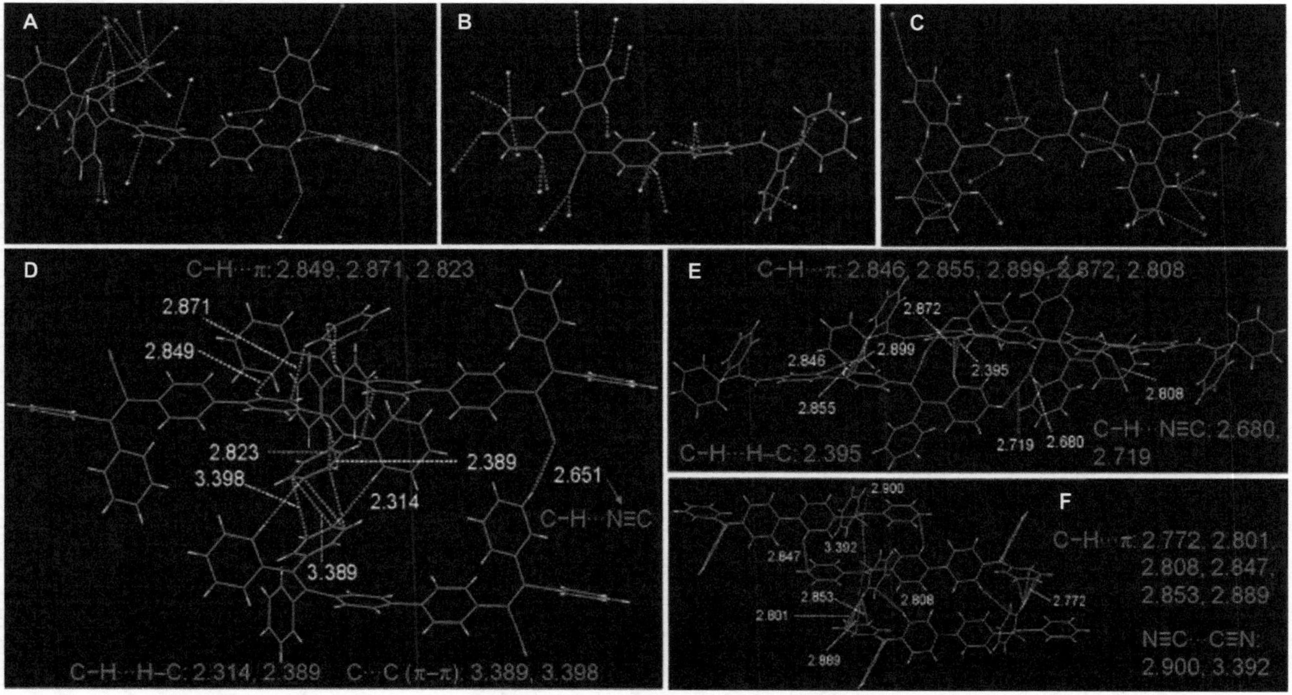

Figure 6.31 Crystal structure of: (A) **57**, (B) **58** and (C) **59**, and multiple C–H···π, π-π, C–H···N, C–H···H-C and N≡C···C≡N intermolecular interactions in crystals of: (D) **57**, (E) **58** and (F) **59**.
Reproduced with permission from reference 54. Copyright 2013, Wiley-VCH.

once the crystalline lattice collapses by external stimuli, the dye molecules may relax to a more planar conformation and thus emit redder light. The contrasts in emission color and intensity are so high that they can be readily detectable by the naked eye, thus rendering such multi-stimuli-responsive solid emitters promising as smart materials.

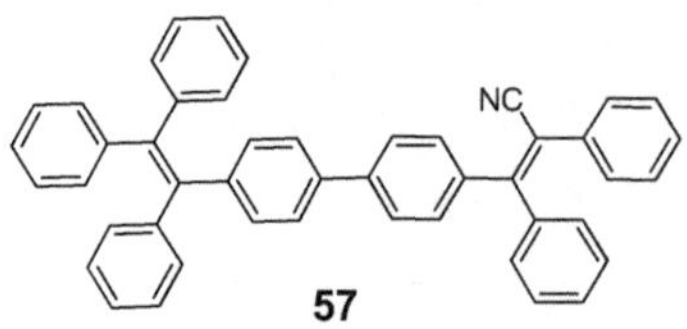

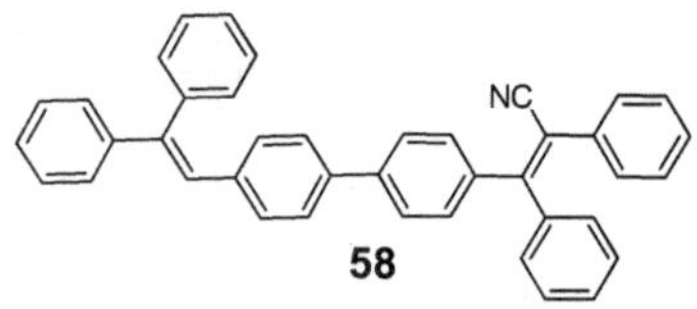

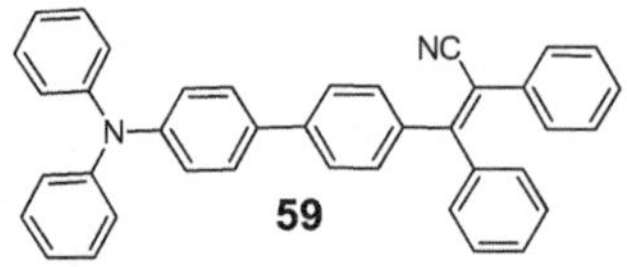

Song *et al.* (2013)[55] reported a liquid crystalline cyanoethylene derivative **60** which presents excellent mechanochromism behaviors with a significant red shift from 485 nm as high as 23 nm after grinding. The recrystallized blue-green **60** powder was changed into relatively yellow-green with the fluorescence quantum efficiency (Φ_F) obviously converting from original 52.7% to 38.7% after being ground. The un-ground dye **60** displays indicative of well-defined microcrystalline-like structures, and the grinding dye is in the amorphous state. Observing the morphological structure of **60** under different aggregative state by using scanning electron microscopy revealed that the damage on surface morphology which is caused by external stimuli can be restored by exposing it to ethanol vapor or heat treatment. The authors also suggested that the mechanochromism originated from the altering mode of molecular packing from the high-ordered to disordered by grinding.

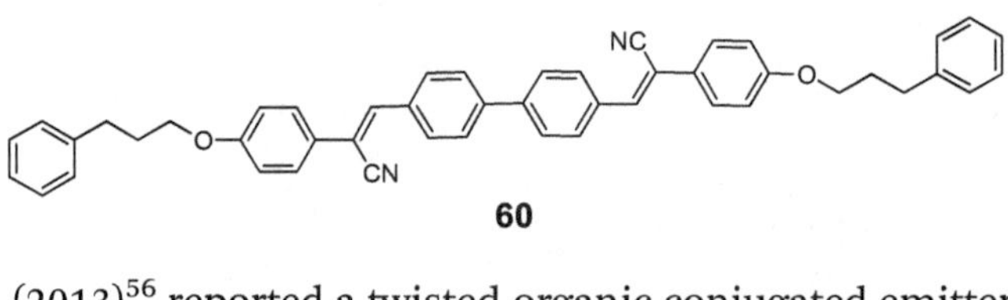

60

Liu *et al.* (2013)[56] reported a twisted organic conjugated emitters molecule (**61**, BAPN) containing 9,9′-bianthracene (BA) group. The BAPN solid possesses emission-tunable behaviors by transforming molecular aggregation patterns to switch performance of the two intramolecular emitting systems alternately. The three crystalline states of BAPN with diverse emitting properties are correlated to the different molecular stacking assemblies, which provide an insight into the source of fluorescence from the BAPN molecular architecture. Emission of one type of crystalline state remains "dark" at room temperature, while it turns on at 77 K, revealing that some intramolecular low-rotation motion still survives in this BAPN crystalline state at room temperature. Most importantly, the authors successfully demonstrate an extraordinarily efficient strategy to establish the accurate information about structure–property relationship through adopting modularization of BAPN molecule.

The pristine powder (Crystal I) and Crystal II of BAPN possess mechanofluorochromic characteristic (Figure 6.32B). All of their ground solid states emit the same green-yellow fluorescence. Furthermore, the emission turns on as grinding the BAPN Crystal II. It is interesting to note that the blue emission can be recovered by treating the ground powder with a poor solvent (ethyl acetate, EA), indicating the repeatability of blue and green-yellow emission by grinding and treating with EA process. Most of the XRD diffraction peaks disappeared in the ground solid-state, suggesting that the crystal lattice is partly altered. In addition, the fluorescence lifetime (2.67 ns) of BAPN in the ground solid-state is longer than that in the pristine solid-state (1.26 ns), revealing that the more condensed packing formation results in less conformational freedom or altering the excited state in the ground solid-state. The blue emission source in BAPN is mainly from Module I because of the analogous spectra of BAPN and BA in their pristine state, while the green-yellow emission of BAPN originates from Module II due to the

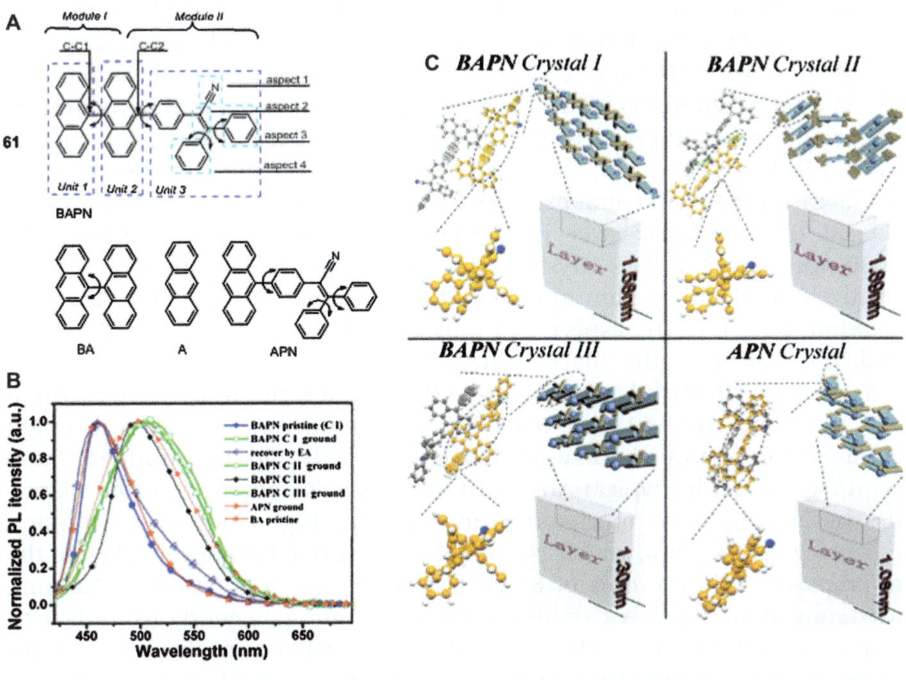

Figure 6.32 (A) The chemical structure of compound **61**. (B) The PL spectra of BA, APN and BAPN in solid-states at room temperature. (C) Schematic drawing of molecules, dimers, and layers in BAPN and APN crystals. Reproduced with permission from reference 56. Copyright 2013, Wiley-VCH.

more or less similarity of emitting performance between BAPN and APN in their ground solid-state. Some clues based on modules will be shown to explain the mechanofluorochromic nature of BAPN. On the one hand, APN does not show mechanofluorochromic performance, corresponding to be identical in the XRD character of the pristine and ground solid-state. However, the aggregation should be looser in the ground solid state of APN for two reasons: (1) the Φ_F value (0.42) of APN in the ground solid-state is far below that (0.91) in the pristine solid-state, suggesting that the aspect 3 or 4 of the APN molecule in the ground solid-state with a low-rotation motion dissipate some of the excited energy; (2) the ground solid-state has a shorter lifetime (3.19 ns) than that of the pristine state (3.92 ns). Thus, the self-assembly of APN molecule favors condensed aggregation. On the other hand, the mechanofluorochromic performance of BA is not available. Remarkably, the emission of BA in the ground solid-state exhibits a new shoulder peak at around 425 nm that resembles the spectra of pristine powder at 77 K, suggesting the aggregation of BA in the pristine solid-state seems to be loose. Indeed, BA seems to be a steric group that may keep the chromophores from approaching and interacting. Consequently, the steric hindrance of Module I will plays an important role in influencing the assembly of BAPN molecules. BAPN forms three different types of crystals: Crystal I (blue emitting with $\Phi_F = 0.18$), Crystal II ("dark" state with $\Phi_F < 0.01$) and Crystal III (green emitting with $\Phi_F = 0.15$). The crystals consist of layers by weak sparse π–π stacking in Crystal I (or Crystal II) and strong N···H interaction in Crystal III (or APN crystal). The mechanofluorochromic behavior can be observed in Crystal I and Crystal II obviously, but not in Crystal III and APN crystal. The thickness of the layer directly illustrates the contact from loose to condensed between molecules (Figure 6.32C), layer in Crystal II (1.89 nm) > layer in Crystal I (1.58 nm) > layer in Crystal III (1.30 nm) > layer in APN Crystal (1.08 nm). In particular, since presenting relatively weak interaction among BAPN molecules in Crystal II, the low-rotation motion of the aspect 4 can dissipate the excited energy efficiently, and thus leading to the "dark" state.

Yuan *et al.* (2013)[57] designed and synthesized a series of D–π–A structured AIE molecules (**62**, **63** and **64**) with high solid-state efficiency through combination of propeller-like triphenylacrylonitrile and diarylamines. These luminogens exhibit typical AIE and intramolecular charge transfer characteristics, remarkable mechanochromism, and high solid-state emission efficiency up to 100%, which is rarely found in ICT luminogens. While the as-prepared solid powders of **62** exhibit strong green emission at 500 nm, they emit yellow lights with a maximum at 551 nm upon gently grinding with a spatula, shifting 51 nm (Figure 6.33). Upon further annealing at 80 °C for 10 min or fuming with DCM vapor for 3 min, the original green emission is restored. Such mechanochromic conversions between green and yellow emission colors can be repeated many times without fatigue, due to the nondestructive nature of the mechanical stimuli. The mechanochromism is highly associated with the molecular arrangement that greatly influences the

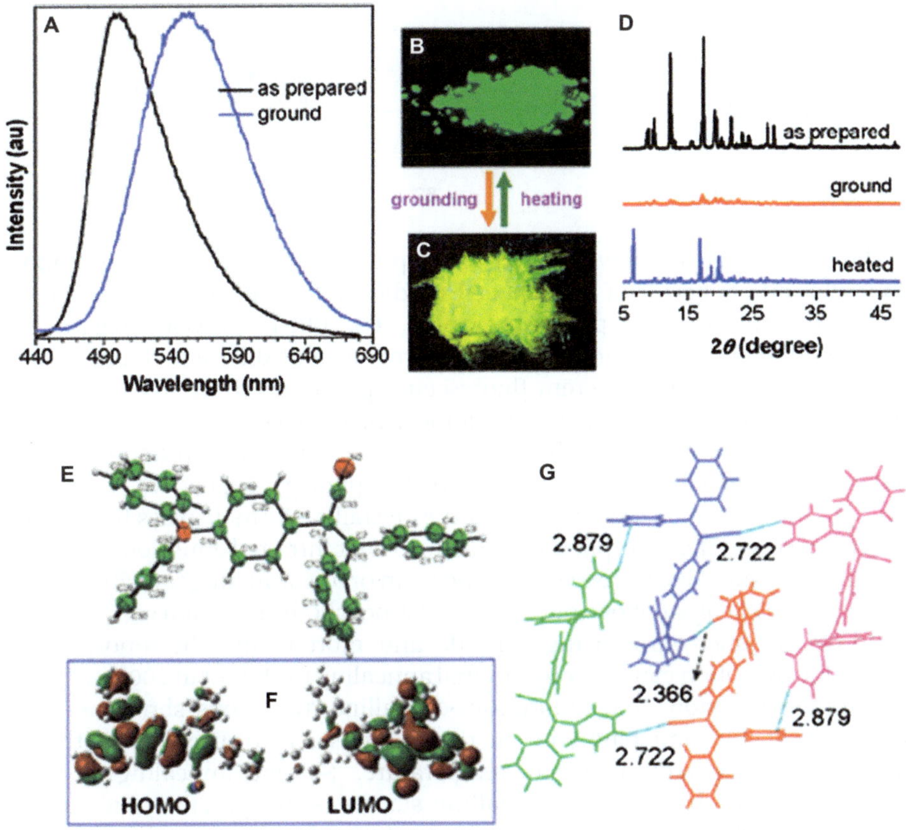

Figure 6.33 (A) Emission spectra of as prepared and ground **62** solids and their (B and C) photographs taken under UV illumination and (D) XRD patterns. (E) ORTEP drawing, (F) HOMO and LUMO energy levels and (G) molecular packing of **62** in crystals with C–H···π (2.879 Å), C–H···N (2.722 Å), and C–H···H–C (2.366 Å) hydrogen bonds indicated. Reproduced with permission from reference 57. Copyright 2013, The Royal Society of Chemistry.

photophysical properties. Similar phenomena are also observed for **63** and **64**, whose as-prepared/ground solid emission maxima are 503/549 and 496/534 nm, giving remarkable mechanochromic red shifts of 46 and 38 nm, respectively. In this case, single crystal structures of the compounds provide more insights into the mechanochromic mechanism. The authors also considered that the origin of the luminescence switching can also be ascribed to the highly twisted conformations, which can be planarized through facile mechanical stimuli and restored upon thermal annealing or solvent fuming. While the twisted stacking architectures result in much shorter effective conjugation lengths and bluer emissions, the planarized conformations generate increasingly effective conjugation lengths, thus producing much redder emissions.

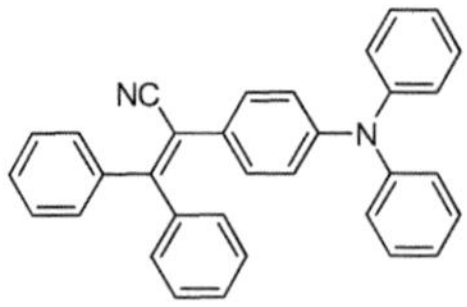

61

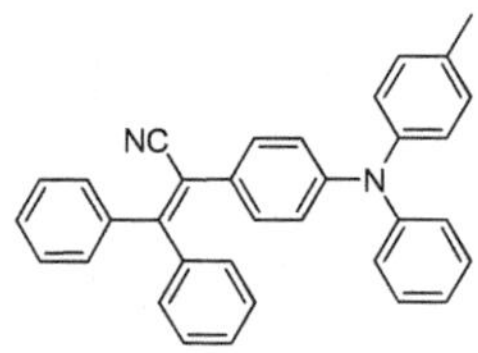

62

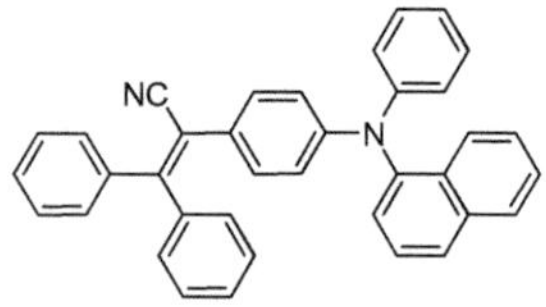

63

Zhu *et al.* (2013)[58] investigated the optical properties and mechanofluorochromic behavior of N,N′-dioctyl-1,7-di(4-(1,2,2-triphenyl)vinyl) phenyl-3,4:9,10- tetracarboxylic perylenebisimide (**65**, PBI-TPE). With regard to PBI-TPE, the crystalline and amorphous forms exhibit different colors under normal room light and different fluorescence properties under UV light. The amorphous form is a very dark red (almost black) color and the crystalline form is red, showing λmax values in emission around 744 nm and 665 nm, respectively. Therefore, PBI-TPE exhibits morphology dependent emission ($\Delta\lambda = 79$ nm). It can be seen by eye that the crystalline form displays bright red fluorescence and the amorphous form, due to near infra-red emission, displays virtually no visible fluorescence, resulting in an on-off switching transition. The results of XRD analysis show that the crystal-amorphous transition can occur by grinding the microwires using a pestle and mortar, and the amorphous-crystalline transition can occur by thermal annealing for 1 hour at 200 °C. These crystalline-amorphous and amorphous-crystalline transitions show the subsequent red shift and blue shift in the PL emission, respectively, with no spontaneous recovery at room temperature. Solvent annealing of the amorphous state back to the crystalline state was not possible due to the inherent stability of the amorphous state after solvent evaporation. In this case, the authors pointed out that the twisted and non-planar structure of tetraphenylethylene is the most likely reason for its ability to undergo amorphization after grinding. It is possible that the more planar PBI-TPE molecule in the amorphous state results in closer contact between molecules and perhaps excimer formation.

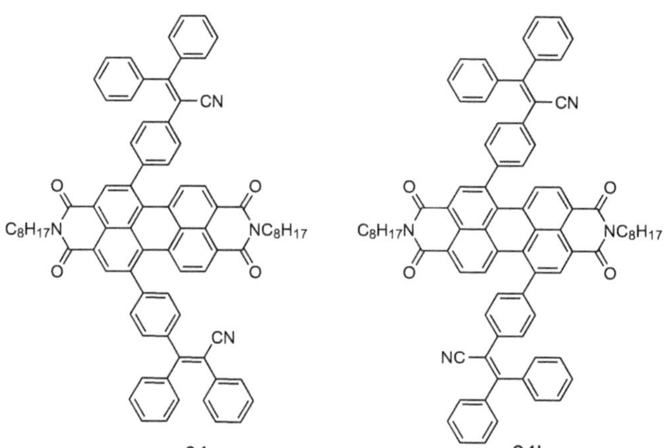

64a 64b

64

Zhao *et al.* (2013)[59] reported a novel solid-state enhanced red emission system based on quinolinemalononitrile (**66**), with an impressive alternative approach to modify the typical π-electron acceptor in the dicyanomethylene-4H-pyran (DCM) derivatives. Specifically, in distinct contrast with the reference compound **67** with severe ACQ characteristic, the stimuli-responsive behavior of **66** was realized by grinding/fuming or pressing/heating (Figure 6.34a). The pristine powder of **66** emitted a strong orange light with a peak centered at 605 nm (upon excitation with 365 nm UV light). After being pressed or ground, a bright red emission with a fluorescence peak at 645 nm occurred, a red shift of ~40 nm, arising from more tight stacking with the increased π–π interaction. The ground sample could be almost recovered to the original state by annealing or fuming. The DSC (Figure 6.34b) and XRD results confirmed that during the pressing procedure, the transformation of crystalline state to amorphous state to some extent occurred and formed a metastable state.

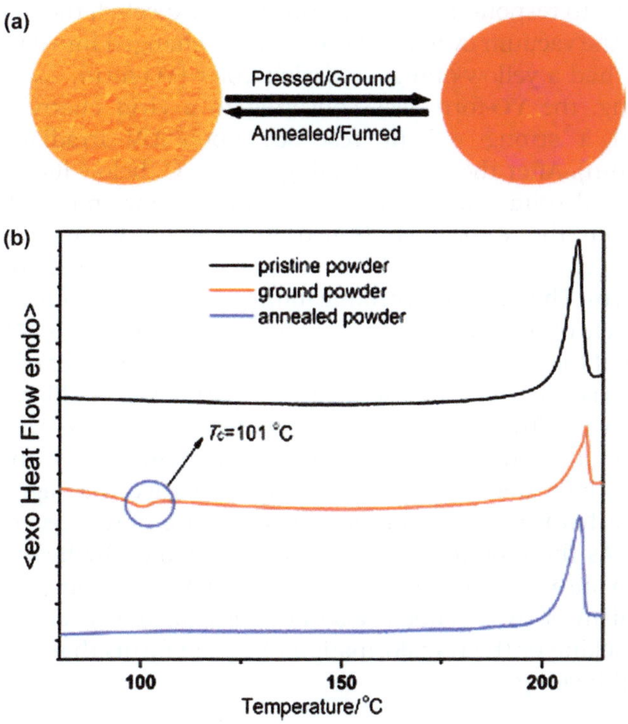

Figure 6.34 (a) Photographic images of **66** powder under 365 nm UV light: pristine, ground (grinding pristine powder with pestle), annealed (ground powder annealed at 100 °C for 5 min), and fumed (reground powder in dichloromethane vapor for 5 min). (b) DSC curves of pristine, ground, annealed powder of **66** under nitrogen. (Scan rate = 20 °C min^{-1}.) Reproduced with permission from reference 59 Copyright 2013, American Chemical Society.

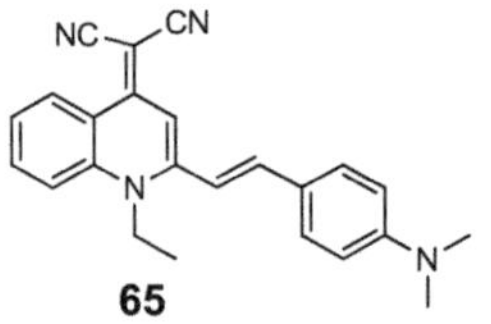

65

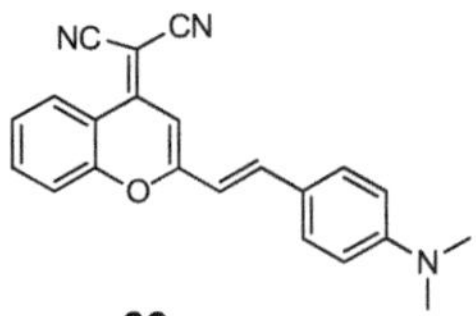

66

6.4.4 Silole Derivative

Although the AIE concept was proposed based on a silole derivative by Tang *et al.* in 2001,[2] up to now, the only reported silole derivative is compound 7.[25] Compound 7, with AIE characteristics is a multi-substituted silole that contains 1-phenyl-2,2-dicyanoethene moieties. Such a strategy is helpful in designing and constructing organic soft porous crystals with other conjugated building blocks and in developing novel smart and stimuli-responsive photo/electronic materials. The authors recently pointed out (2012)[25] that the multi-stable crystalline states with different fluorescent colors increased the potential of dye 7 to respond to external stimuli. As expected, the orange powder was dried in the vacuum oven and subsequently changed into a yellow-orange solid, which had a yellow-orange emission color (YO-form, $\lambda_{em} = 576.5$ nm). After grinding, the YO-form was converted to a vivid cherry-colored solid that showed a strong red fluorescence upon UV irradiation (R-form, $\lambda_{em} = 600.5$ nm). After thermal annealing at 150 °C (below its melting point, 228 °C), the red solid immediately turned into yellow-orange. The YO-solid recovered from the R-form continued to emit an intense yellow-orange fluorescence ($\lambda_{em} = 566$ nm), which is approximately equal to that of the original O-form. The mechanochromic luminescence was readily reproducible during the cycle of the grinding-annealing operation (Figure 6.35). The typical XRD pattern of the pristine YO-form solid exhibited a number of recognizable diffraction peaks that can be ascribed to a partially crystallized solid. After grinding, most diffraction peaks disappeared or became vague, while a number of weak new ones at different diffraction angles emerged. This phenomenon indicated that the initial crystal lattice was significantly disrupted by mechanical force, and the crystalline size of the new polymorph is quite small. As the ground R-form crystal was treated by heating, the solid turned again into YO with a recovered XRD pattern. Thus, the mechanofluorochromism of dye 7 should also be ascribed to a reversible change of molecular packing or the transformation of crystal forms that was induced by grinding and heating.

6.4.5 Containing Multi-AIE Units

As described, a lot of mechanofluorochromic compounds containing two or more AIE units have been reported. Here, two other examples are provided to demonstrate this kind of mechanofluorochromic compound. Dye **68** was designed and synthesized by Chi and Xu *et al.* (2011).[11] The dye is AIE-active

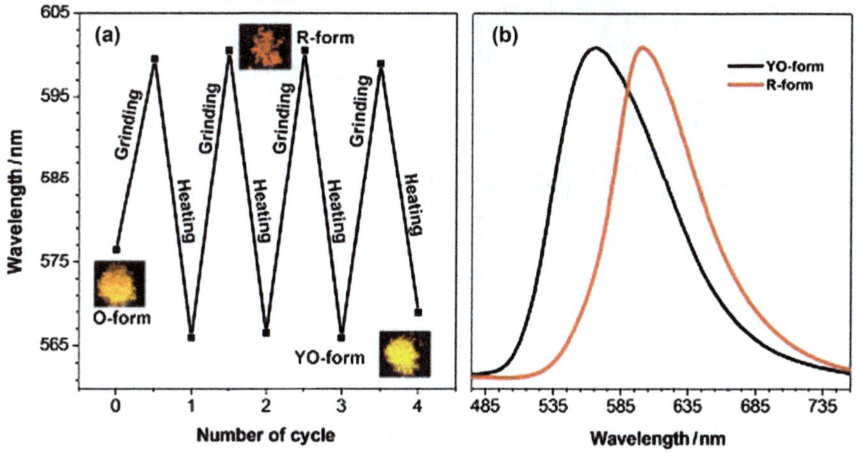

Figure 6.35 (a) Switching of the solid-state fluorescence of molecule 7 by repeated grinding and heating ($\lambda_{ex} = 429$ nm). Insets show the fluorescence photographs of the R-form (after grinding), O-form (before grinding) and heating at 150 °C for 30 s; (b) The fluorescence spectra of the YO-form and R-form upon excitation at 429 nm.
Reproduced with permission from reference 25. Copyright 2012, The Royal Society of Chemistry.

because it contains two AIE units of tetraphenylethylene and distyrylanthracene. It was found that the dye exhibited significantly mechanofluorochromic activity (Figure 6.36A and B). After grinding, the emission wavelength increased from 506 nm to 574 nm (red shifted by 68 nm). The XRD results indicated that mechanofluorochromism resulted from the reversible morphological change between the crystalline and amorphous structures (Figure 6.36C). The DSC results showed that the pressed sample had a significant cold-crystallization peak at approximately 336 °C. This peak indicated the existence of a metastable-state aggregation in the pressed sample, which could be converted into a more stable state through annealing. The cold-crystallization transition of the pressed sample seems to be a common feature for numerous mechanofluorochromic compounds.

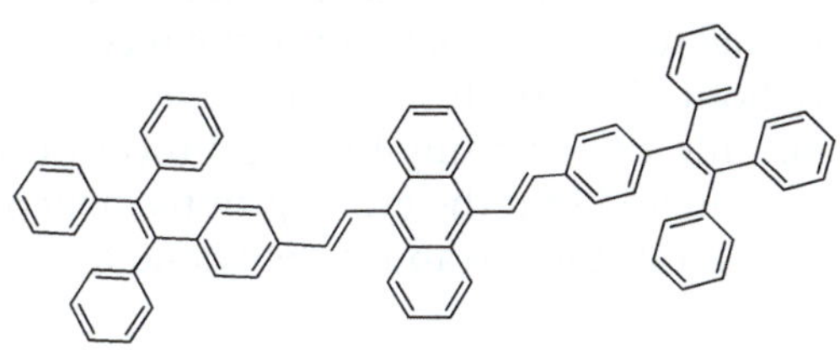

67

Another example is that cyanoethylene and tetraphenylethylene AIE units were combined by Sun *et al.* (2013)[60] to synthesize compound **69**, and tetraphenylethylene trimer, **70**, was used as the reference. The

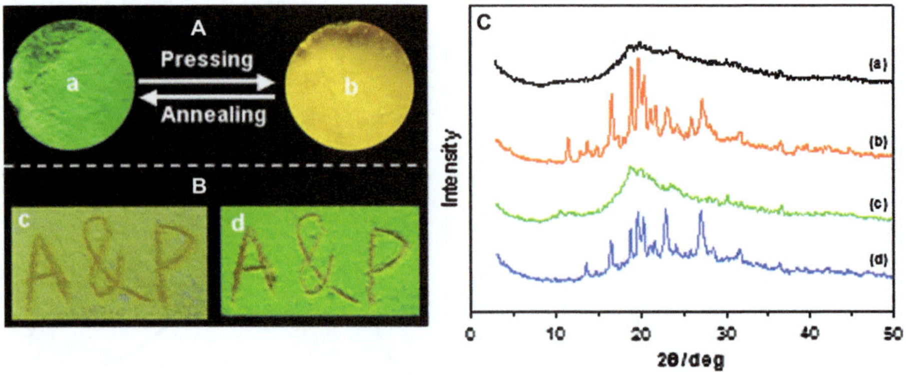

Figure 6.36 The images of dye **68** taken at room temperature. (A): annealing sample (a) and pressing sample (b) under 365 nm UV light; (B): cast on a filter paper after writing "A&P" with a metal spatula (a) natural light and (b) 365 nm UV light; (C) XRD curves of the dye **68** samples: (a) pressing; (b) annealing the (a) sample at 340 °C for 1 min; (c) pressing the (b) sample; (d) annealing the (c) sample at 340 °C for 1 min.

"fresh-prepared" compound **69** is a yellow solid with bright yellow emission (λem = 541 nm) after running from the column. After being aged for months, the solid turns a light green and its emission blue shifts from yellow to yellowish-green. By simply grinding the as-prepared solid of **69** with a mortar and pestle, an orange powder with orange emission (λem = 563 nm) was obtained, red shifted 22 nm. After thermal annealing at 160 °C for 10 min, the color and emission reverted to the original states. The thermal annealing progress can be replaced by solvent vapor fuming, such as DCM or THF. The reversible process is realized by the change between polymorphous and amorphous states. The polymorphs of the solvent fumed sample are similar to that of the as-prepared solid but dissimilar to the thermally treated sample. Compound **70** also has mechanofluorochromic properties, however, comparing the mechanochromic behaviors for **69** and **70** reveals two different characteristics (Figure 6.37). First, the initial state of **69** is polymorphous, whereas **70** is amorphous, suggesting a stronger crystallizability for **69** in normal temperature. Second, the mechanochromism of **69** can be repeated by both the grinding-heating and grinding-fuming processes, while for **70**, only the grinding-fuming process can be taken into effect. The authors considered that the cyano groups in **69** play a crucial role in the mechanochromism of the tetraphenylethylene derivatives.

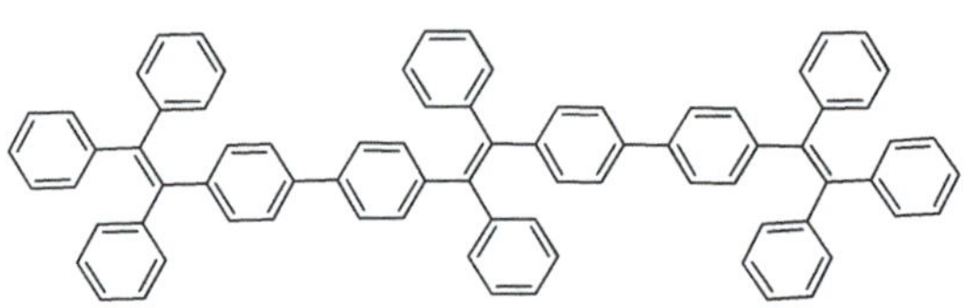

68

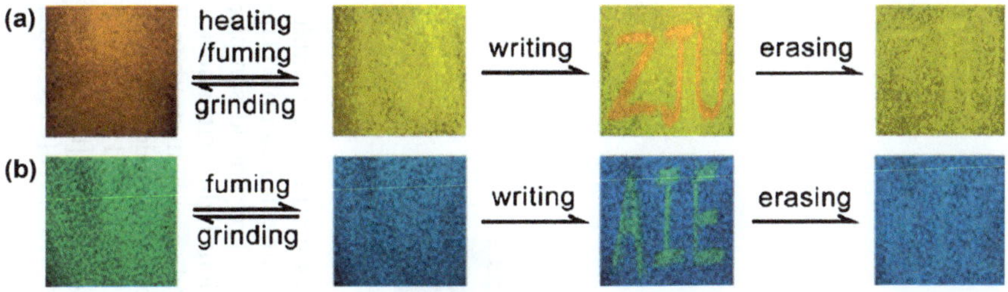

Figure 6.37 Photos taken under UV light (λex = 365 nm): (a) **69** and (b) **70** are ground on filter papers, after initializing by heating or fuming process, the letters of "ZJU" were written on the **69** "paper" and "AIE" on the TTPE "paper" with a spatula, and subsequently the papers were erased by thermal annealing (only **69**) or vapor fuming (both **69** and **70**) (the letters "ZJU" and "AIE" becoming invisible under UV light).
Reproduced with permission from reference 60. Copyright 2013, American Chemical Society.

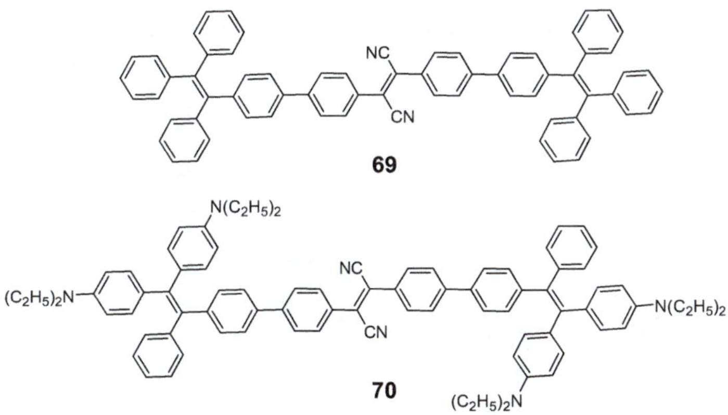

6.4.6 Metal Complexes

Ligand dye **71** that contains tetraphenylethylene and terpyridine moieties as well as the zinc ion complex of dye **72** was reported by Chi and Xu *et al.* (2011).[61] Dye **72** was the first mechanofluorochromic AIE complex reported at that time. Both the ligand and the complex exhibited AIE effects. Their colors and emissions could be efficiently switched using various external stimuli, which include grinding, heating, solvent-fuming (Figure 6.38A and B) and exposure to acid and base (Figure 6.38C). The results of XRD, DSC and time-resolved emission measurements demonstrated that the grinding-, heating- and solvent fuming-induced fluorescence changes could be all obtained by tuning the molecular packing in the solid states. The variation scopes of dyes **71** and **72** before and after grinding were 38 and 81 nm, respectively. The complexation could magnify the wavelength variation range in mechanofluorochromism. Upon exposure to acid and base vapors, such as TFA and NEt_3, the fluorescence emissions and colors of dyes **71** and **72**

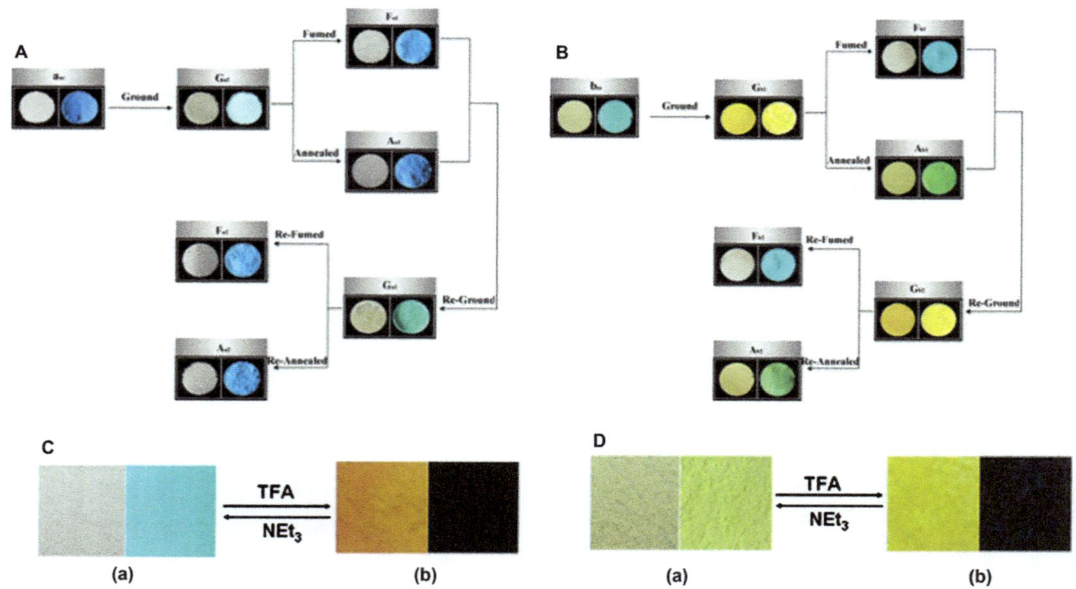

Figure 6.38 (A) Dye **71** taken at room temperature under ambient light (left) and UV light (right). Samples: (a_{as}) as-synthesized sample; (G_{a1}) ground sample; (F_{a1}) fumed sample (ground sample in methanol vapor for five minutes); (A_{a1}) annealed sample (The ground sample was annealed at 160 °C for five minutes and cooled down at room temperature.); (G_{a2}) re-ground sample; (F_{a2}) re-fumed sample; (A_{a2}) re-annealed sample. (B) Dye **72** taken at room temperature under ambient light (left) and UV light (right). Samples: (b_{as}) as-synthesized sample; (G_{b1}) ground sample; (F_{b1}) fumed sample (ground sample in methanol vapor for five minutes); (A_{b1}) annealed sample. (The ground sample was annealed at 300 °C for 1 hour and cooled down at room temperature.); (G_{b2}) re-ground sample; (F_{b2}) re-fumed sample; (A_{b2}) re-annealed sample. (C) Dye **71** and (D) dye **72** taken at room temperature under ambient light (left) and UV light (right). Samples: (a) as-synthesized sample [filter paper wetted by the solution of the dye **71** (in THF) or the dye **72** (in acetone) and then dried]; (b) fumed with TFA (CF_3COOH) vapor for 5 s.
Adapted with permission from references 61. Copyright 2011, The Royal Society of Chemistry.

exhibited an off/on switching effect (Figure 6.38C). To the best of our knowledge, this study was the first one in the literature to report on a metal ion complex that shows both AIE effect and multi-stimuli-responsive fluorescence features.

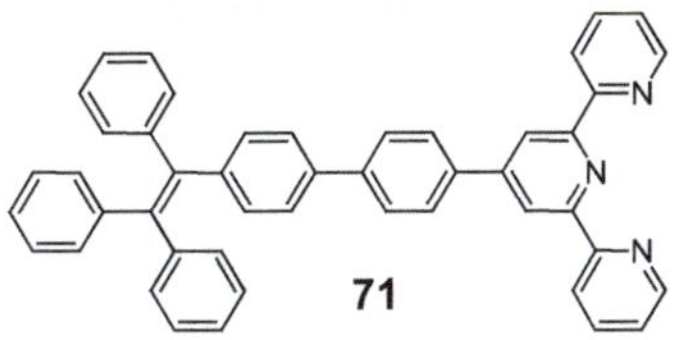

71

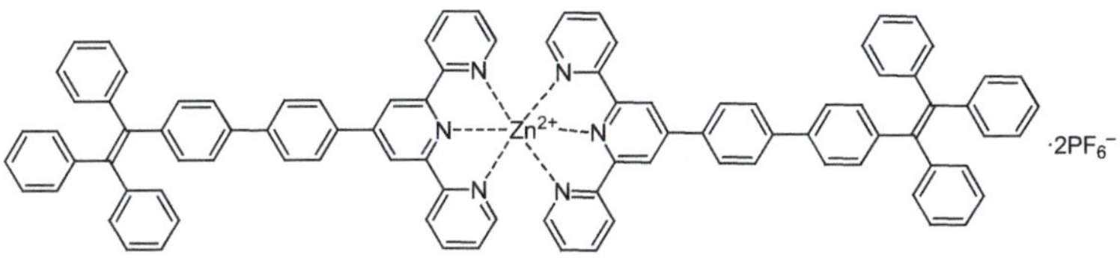

72

Yi *et al.* (2012)[62] reported a novel cationic Ir(III) complex (dye **73**) prepared by employing a dendritic ancillary ligand. The emission color of its as-prepared powder changed upon grinding and reverted to the original emission *via* heating and/or recrystallization. This novel study was the first reported example of a cationic Ir(III) complex that simultaneously exhibited mechanofluorochromism (Figure 6.39) and AIE phosphorescent properties. However, its molecular structure contained no common AIE unit. The close

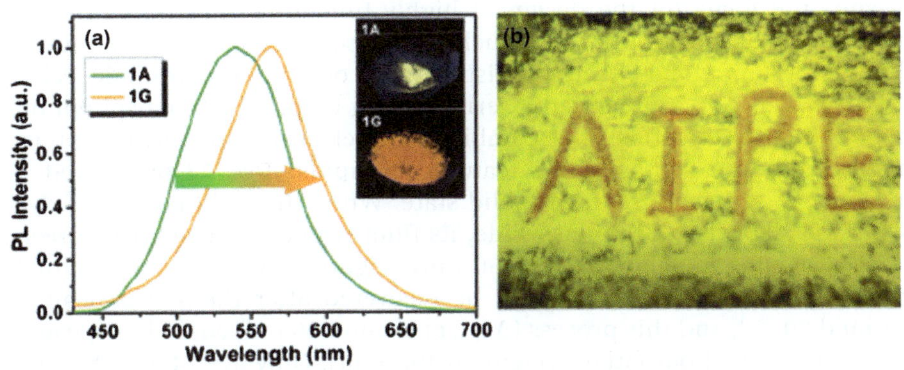

Figure 6.39 (a) Emission spectra of the **73** samples 1A (as-synthesized) and 1G (ground). (b) The powder 1A was cast on the filter paper and the letters "AIPE" were written with a spatula under UV light at room temperature. Reproduced with permission from reference 62. Copyright 2012, The Royal Society of Chemistry.

investigation of the molecular packing structure of dye **73** through single crystal analysis then revealed that the multiple intermolecular C–H···π interactions existed in the single crystal structure of dye **73**. The crystal structure of dye **73** could be easily destroyed because of its weak interactions, which might be triggered by crystalline-amorphous transition when an external pressure was applied, which consequently resulted in mechanofluorochromism.

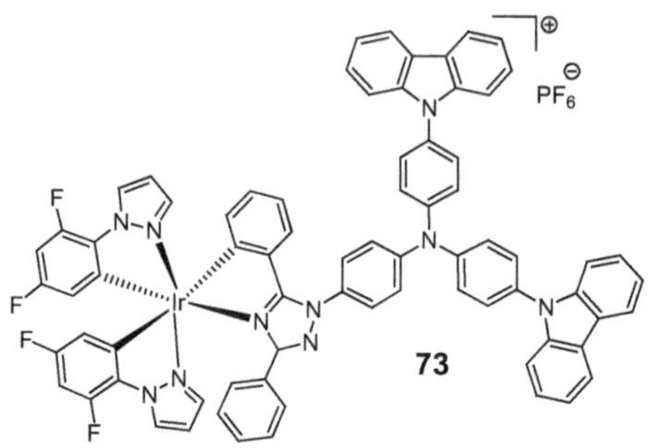

73

6.4.7 Atypical AIE Mechanofluorochromic Compounds

In this section, some unique mechanofluorochromic compounds will be described. In the chemical structures of these compounds, there are no any common AIE unit contained, but they exhibit AIE properties, and thus, could be called "atypical AIE mechanofluorochromic compounds".

The inherent electronic nature and geometric parameters of the heteroatoms (such as B, Si, P and S) caused the π-conjugated systems to exhibit intriguing properties that cannot be observed in pure carbon-based conjugated derivatives. Consequently, these "heteroatom-doped" systems became the focus for the design of highly functional materials in organic/bioelectronics applications, such as sensors, organic light-emitting diodes, organic field-effect transistors and organic photovoltaic cells. Ren *et al.* (2011)[63] synthesized a series of new amphiphilic phosphonium materials (dye **74** series) that combined the electronic features of phospholes with the self-assembly features of lipids. Dye **74-4a** showed a mechanical PL response in the solid state. When the solid powder of dye **74-4a** was deposited on a glass slide, its fluorescence changed from green ($\lambda_{em} = 538$ nm) to yellow ($\lambda_{em} = 560$ nm) (Figure 6.40, left). The original green fluorescence features could be recovered after thermal annealing (around 80 °C), and this process ($\Delta\lambda_{em}$ at around 20 nm) could be repeated several times without fatigue (Figure 6.40, right). Only dye **74-4a** exhibited the mechanochromic behavior. The other combined systems, such as dyes **1a, 1b, 2a, 2b, 3a** and **3b,** with smaller conjugated cores and counteranions did not exhibit any significant mechanically responsive emission in similar conditions. A possible eximer of dye **74-4a** was formed during mechanical grinding.

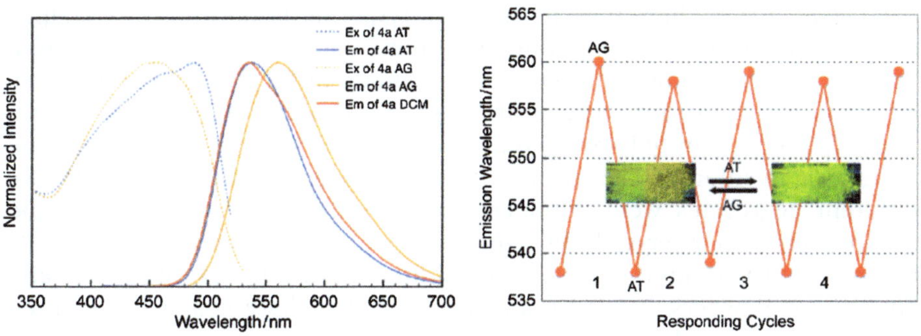

Figure 6.40 (Left) Excitation (ex, dots) and emission (em, solid) spectra of 1b; (right) responding cycles; AT = after thermal annealing; AG = after grinding. Reproduced with permission from reference 63 Copyright 2011, American Chemical Society.

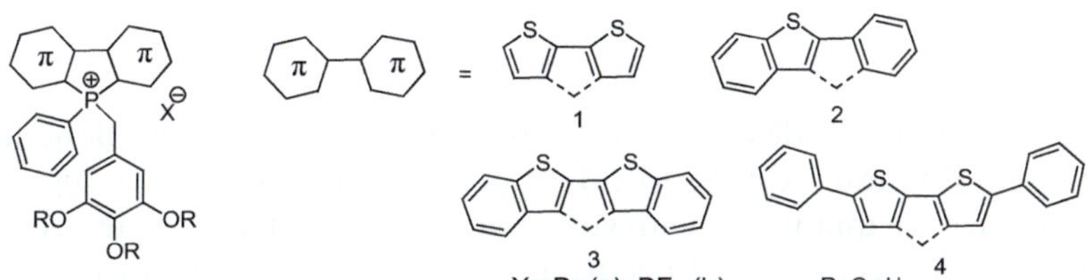

X= Br (a); BF_4 (b); BPh_4 (c); OTf (d) R=$C_{12}H_{25}$

74

Ren and Baumgartner (2012)[64] synthesized a series of new phospholium acenes, which were quaternized with various benzyl groups (dye **75** series). This class of phosphole system also exhibited enhanced emission in the solid state and at low temperature in solution, showing AIE nature. The crystals of the new phospholium compounds also exhibited mechanically responsive features. The powders obtained from mechanical shearing of the crystals between two glass slides displayed a blue shift and broadened emission (1a, $\Delta\lambda_{em} = -3$ nm; 1b, $\Delta\lambda_{em} = -3$ nm; 1c, $\Delta\lambda_{em} = -2$ nm; 2b, $\Delta\lambda_{em} = -4$ nm; 2c, $\Delta\lambda_{em} = -2$ nm; 3a, $\Delta\lambda_{em} = -2$ nm; 3b, $\Delta\lambda_{em} = -9$ nm; 3c, $\Delta\lambda_{em} = -11$ nm). This mechanically responsive emission shift was the result of the formation of different conformational isomers upon application of a mechanical force. Only 3b and 3c with a larger conjugated head exhibited a relatively strong mechanically responsive emission, which indicated that the intermolecular interactions have a significant effect on the mechanically responsive emission because of the restricted conformation changes.

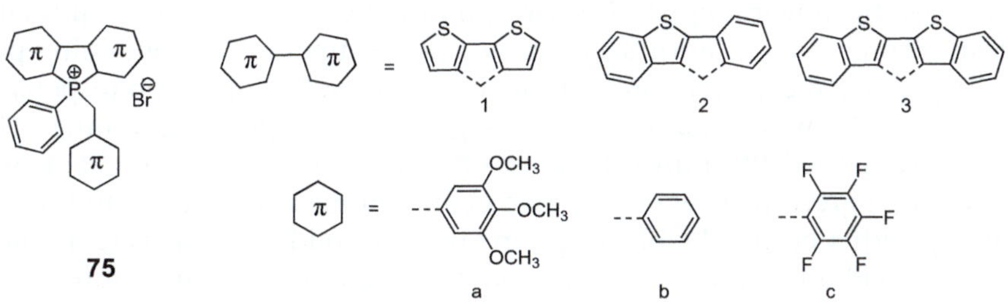

75

Xu *et al.* (2013)[65] have tried to integrate 1,3,4-oxadiazole and the V-shaped structural characteristic of a sulfone into one compound (**76**). It was found that **76** is also an AIE compound. The crystals of **76** show a strong green emission at $\lambda = 510$ nm. After grinding with a mortar and pestle, the obtained powders emitted strong yellow light at $\lambda = 570$ nm. When the solutions of **76** in chloroform were evaporated to dryness at different speeds, two different aggregates were obtained: an amorphous aggregate with rapid evaporation and a crystalline aggregate with slow evaporation. The PL emission of the ground sample is red shifted relative to that of crystals and amorphous powders. The authors suggested that there is a stronger interaction in the ground sample than those in crystals and amorphous samples.

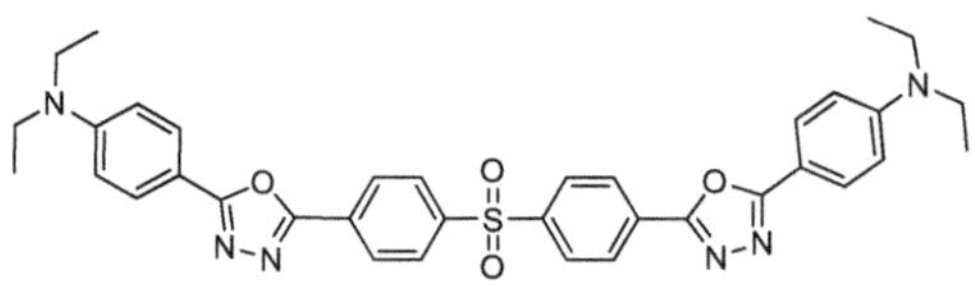

76

Dong *et al.* (2013)[66] designed and synthesized (diphenylbutadienediyl)dibenzoic acid (**77**) and it ester derivative (**78**), both of which exhibit high solid-state fluorescence quantum yield with characteristic AIE properties. A drastic change in fluorescence of **77** occurred upon grinding and can be fully restored to its original color upon exposure to solvents or their vapors (Figure 6.41). However, the mechanofluorochromic change of **78** is not as significant as that of **77**. The authors considered that the destruction of intermolecular hydrogen bonding (H-bonding) interactions by applying high pressure, leading to the state where close packing is the overriding factor governing molecular packing, is responsible for the altered emission of the ground samples.

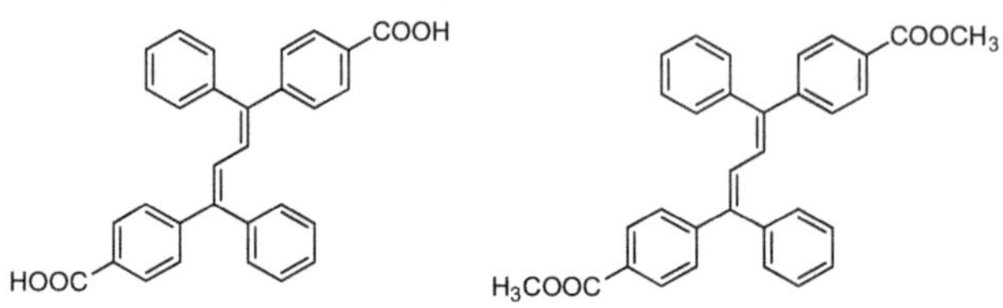

77 **78**

Sun *et al.* (2013)[67] designed and synthesized a series of AEE-active D–A–D molecules by integrating a diphenylquinoxaline core with two aromatic chromophores such as anthracene or pyrene. Compound **79** containing pyrene displayed a mechanofluorochromism. The reversible mechano-responsive behavior of **79** was realized by grinding/heating or pressing/fuming. The freshly crystallized **80** from a mixture of chloroform and hexane (1 : 1, v/v) resulted in a green powder with bluish green luminescence at 511 nm. Upon grinding with a pestle, the bluish-green emissive solid transforms to a yellow emissive solid and showed an emission maximum at 526 nm (Figure 6.42a)

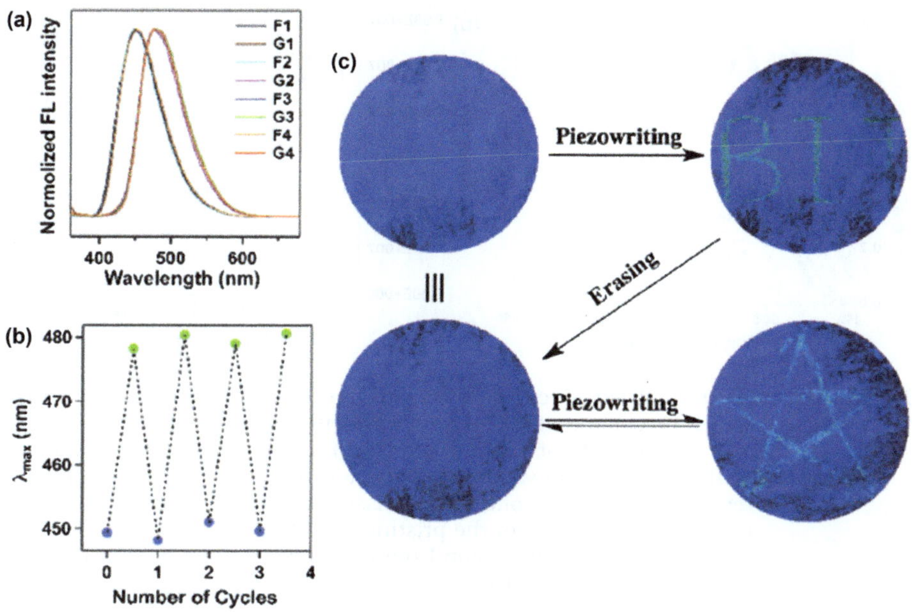

Figure 6.41 (a) Normalized fluorescence (FL) spectra of the **77** powder upon fuming–grinding. (b) Maximum emission wavelength change versus repeating cycles. (c) Photographs of **77** cast filter paper upon piezowriting-erasing under UV-light (365 nm).
Reproduced with permission from reference 66. Copyright 2013, The Royal Society of Chemistry.

that was believed to arise from the more tight stacking with increased π–π interactions. The authors pointed out that if the red shift in the PL spectrum of the ground sample is due to the increased conjugation induced by the molecular planarization, then a similar change should be observed in the absorption spectra but should remain the same in the case of excimer emission. The experimental results did show a clear difference in the absorption spectra between the as-synthesized sample (λmax = 351 nm) and the ground sample (λmax = 355 nm). In particular, a notable absorption shoulder appeared near 450 nm in the ground sample. Consequently, the colors of the pristine and ground samples are different which rules out the possibility of the red shifted emission in the ground sample coming from the excimer emission. However, the anthracene containing compound **80** exhibited thermofluorochromism (Figure 6.42b) but no mechanofluorochromism.

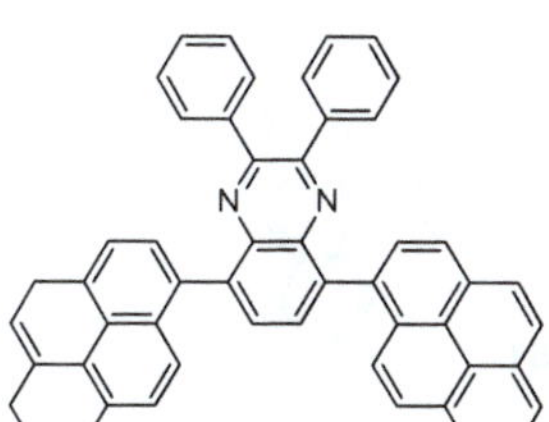

79

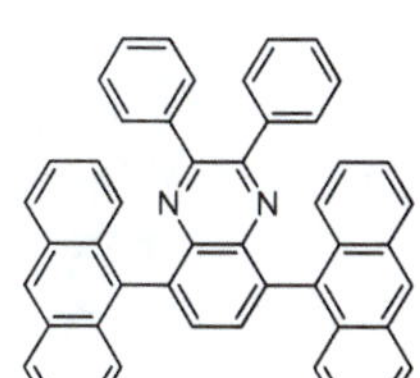

80

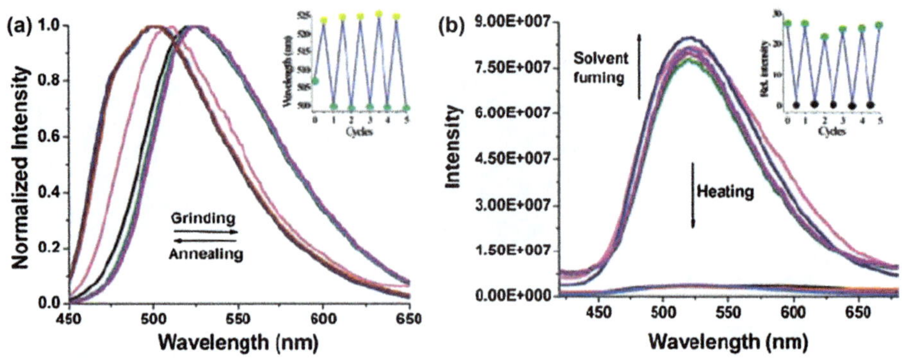

Figure 6.42 (a) Normalized PL curves of **79** with reversible grinding and heating states. Inset shows the emission wavelength of the repeated grinding and annealing treatments of **79**; (b) PL curves of **80** with reversible heating and solvent vapor fuming states. Inset shows the quenching–recovering cycles of solid-state emission against the heating–fuming procedure starting from the pristine solid (green).
Reproduced with permission from reference 67. Copyright 2013, The Royal Society of Chemistry.

Sun *et al.* (2013)[68] reported three compounds containing anthracene (**81** and **82**) and pyrene (**83**) moieties. Among the three **TDAn** compounds, **81** and **82** exhibited prominent AIE and mechanofluorochromic properties, which also demonstrates that AIE and mechanofluorochromism have certain correlation. Upon grinding the as-synthesized compounds with a pestle, the green (507 nm) and bluish green (491 nm) emissive solids of **81** and **82** changed to yellow emissive solids (Figure 6.43a) and showed emission maxima at 548 and 540 nm, respectively, while for **81** and **82** the quantum yields (Φ_F) of the ground samples were found to be 4.1% and 5.3%, respectively. The reversible switching of the emission wavelengths can be realized by the repeated process of grinding and heating without decomposition of the samples (Figure 6.43b and c). The reversible transformation between green or blue-green fluorescent and orange emission was also imputed to phase transition. The corresponding ground samples exhibited a clear exothermic peak at 134 and 250 °C for **81** and **82**, respectively (Figure 6.43). These thermal behaviors were ascribed to the meta-stable state of the ground samples to a more stable state transformation by annealing. The molecular conformation planarization mechanism was used to explain the mechanofluorochromism of **81** and **82**.

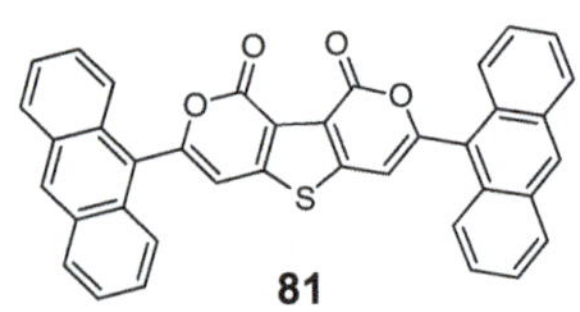

81

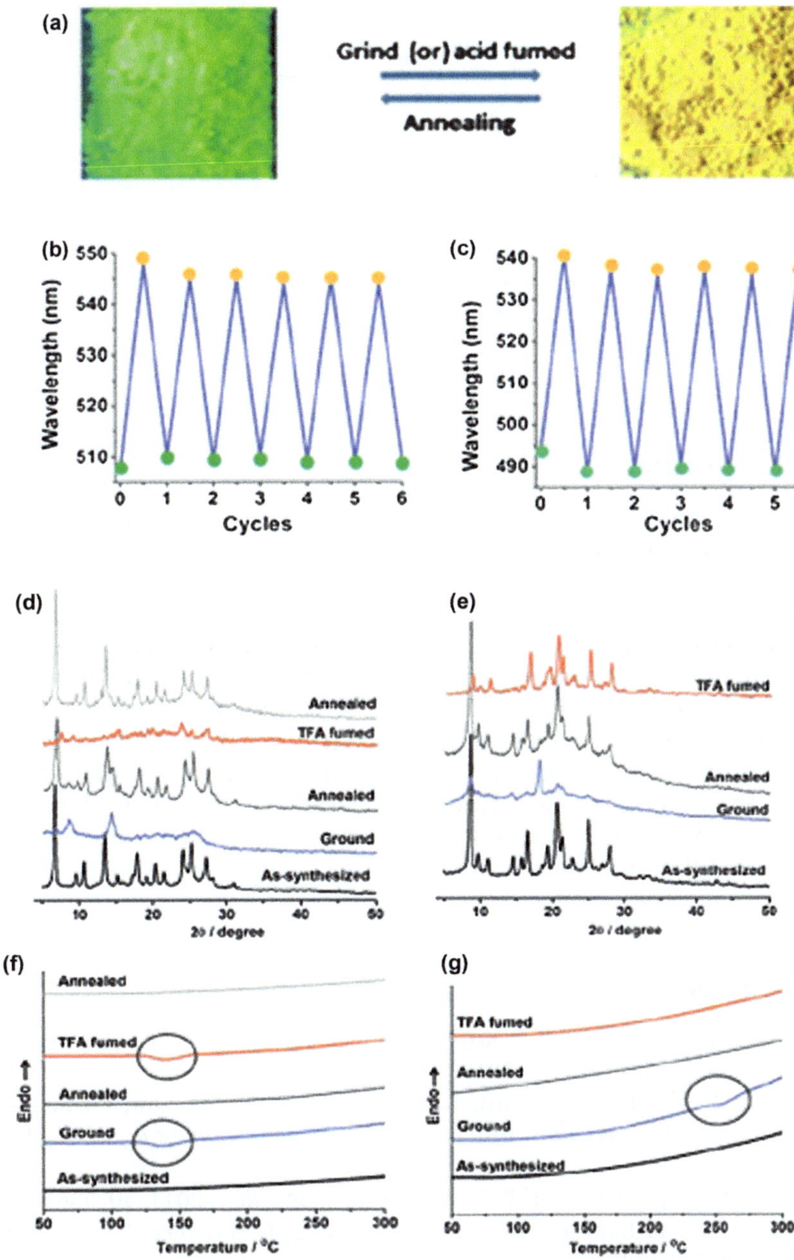

Figure 6.43 Photographs of the ground (or acid-fumed) and the annealed samples of **TDAn** under UV light (365 nm) (a); emission wavelengths of the repeated grinding and annealing treatments of **81** (b) and **82** (c). XRD (d and e) and DSC (f and g) profiles of **81** (a and c) and **82** (b and d). Reproduced with permission from reference 68. Copyright 2013, The Royal Society of Chemistry.

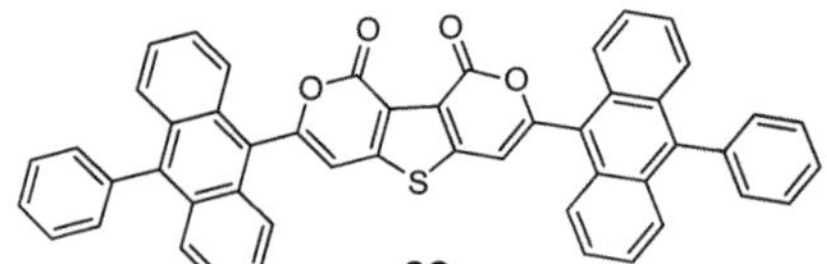

82

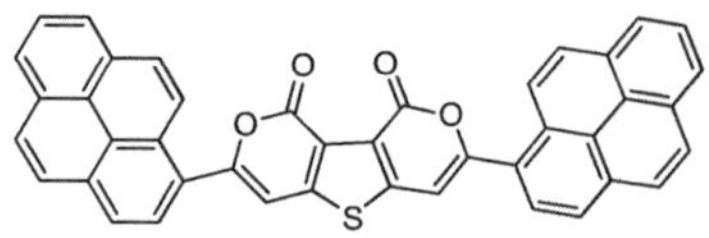

83

6.5 Conclusion

The recent research on mechanofluorochromic materials based on aggregation-induced emission molecules dependent on changes in physical molecular packing modes has been presented in this chapter. The structure-property relationships and mechanisms of the AIE mechanofluorochromic materials have been described. The structural relationship between the aggregation-induced emission compounds and mechanofluorochromism is recognized. Most mechanofluorochromic AIE compounds were found to exhibit multi-stimuli-responsive fluorescence switching behaviors, and should be promising candidates for optical date storage systems, stress sensors, chemosensors, light emitting devices *etc.* The molecular design strategy based on AIE structure may be widely applicable to the mining of new mechanofluorochromic materials.

References

1. Z. G. Chi, X. Q. Zhang, B. J. Xu, X. Zhou, C. P. Ma, Y. Zhang, S. W. Liu and J. R. Xu, *Chem. Soc. Rev.*, 2012, **41**, 3878–3896.
2. J. Luo, Z. Xie, J. W. Y. Lam, L. Cheng, H. Chen, C. Qiu, H. S. Kwok, X. Zhan, Y. Liu, D. Zhu and B. Z. Tang, *Chem. Commun.*, 2001, 1740–1741.
3. Y. Hong, J. W. Y. Lam and B. Z. Tang, *Chem. Soc. Rev.*, 2011, **40**, 5361–5388.
4. J. Liu, J. W. Y. Lam and B. Z. Tang, *J. Inorg. Organomet. Polym. Mater.*, 2009, **19**, 249–285.
5. L. Qian, B. Tong, J. Shen, J. Shi, J. Zhi, Y. Dong, F. Yang, Y. Dong, J. W. Y. Lam, Y. Liu and B. Z. Tang, *J. Phys. Chem. B*, 2009, **113**, 9098–9103.
6. Z. Y. Yang, Z. G. Chi, T. Yu, X. Q. Zhang, M. N. Chen, B. J. Xu, S. W. Liu, Y. Zhang and J. R. Xu, *J. Mater. Chem.*, 2009, **19**, 5541–5546.
7. J. W. Chung, B. K. An, F. Hirato, J. H. Kim, H. Jinnai and S. Y. Park, *J. Mater. Chem.*, 2010, **20**, 7715–7720.

8. Y. Q. Dong, J. W. Y. Lam, Z. Li, A. J. Qin, H. Tong, Y. P. Dong, X. D. Feng and B. Z. Tang, *J. Inorg. Organomet. Polym. Mater.*, 2005, **15**, 287–291.
9. Y. Q. Dong, J. W. Y. Lam, A. J. Qin, Z. Li, J. Z. Sun, H. H. Y. Sung, I. D. Williams and B. Z. Tang, *Chem. Commun.*, 2007, 40–42.
10. S. J. Yoon, J. W. Chung, J. Gierschner, K. S. Kim, M. G. Choi, D. Kim and S. Y. Park, *J. Am. Chem. Soc.*, 2010, **132**, 13675–13683.
11. X. Q. Zhang, Z. G. Chi, H. Y. Li, B. J. Xu, X. F. Li, W. Zhou, S. W. Liu, Y. Zhang and J. R. Xu, *Chem.-Asian J.*, 2011, **6**, 808–811.
12. B. K. An, D. S. Lee, J. S. Lee, Y. S. Park, H. S. Song and S. Y. Park, *J. Am. Chem. Soc.*, 2004, **126**, 10232–10233.
13. J. W. Chen, B. Xu, X. Y. Ouyang, B. Z. Tang and Y. Cao, *J. Phys. Chem. A*, 2004, **108**, 7522–7526.
14. F. Wang, M. Y. Han, K. Y. Mya, Y. Wang and Y. H. Lai, *J. Am. Chem. Soc.*, 2005, **127**, 10350–10355.
15. B. J. Xu, Z. G. Chi, H. Y. Li, X. Q. Zhang, X. F. Li, S. W. Liu, Y. Zhang and J. R. Xu, *J. Phys. Chem. C*, 2011, **115**, 17574–17581.
16. B. J. Xu, Z. G. Chi, X. F. Li, H. Y. Li, W. Zhou, X. Q. Zhang, C. C. Wang, Y. Zhang, S. W. Liu and J. R. Xu, *J. Fluoresc.*, 2011, **21**, 433–441.
17. J. T. He, B. Xu, F. P. Chen, H. J. Xia, K P. Li, L. Ye and W. J. Tian, *J. Phys. Chem. C*, 2009, **113**, 9892–9899.
18. Y. Dong, B. Xu, J. Zhang, X. Tan, L. Wang, J. Chen, H. Lv, S. Wen, B. Li, L. Ye, B. Zou and W. J. Tian, *Angew. Chem. In. Ed.*, 2012, **51**, 10782–10785.
19. Y. Ooyama, N. Yamaguchi, S. Inoue, T. Nagano, E. Miyazaki, H. Fukuoka, I. Imae, K. Komaguchi, J. Ohshita and Y. Harima, *Tetrahedron*, 2012, **68**, 529–533.
20. C. D. Dou, L. Han, S. S. Zhao, H. Y. Zhang and Y. Wang, *J. Phys. Chem. Lett.*, 2011, **2**, 666–670.
21. W. Z. Yuan, Y. Q. Tan, Y. Y. Gong, P. Lu, J. W. Y. Lam, X. Y. Shen, C. F. Feng, H. H.-Y. Sung, Y. W. Lu, I. D. Williams, J. Z. Sun, Y. M. Zhang and B. Z. Tang, *Adv. Mater.*, 2013, **25**, 2837–2843.
22. L. Y. Bu, M. X. Sun, D. T. Zhang, W. Liu, Y. L. Wang, M. Zheng, S. F. Xue and W. J. Yang, *J. Mater. Chem. C*, 2013, **1**, 2028–2035.
23. Y. L. Wang, W. Liu, L. Y. Bu, J. F. Li, M. Zheng, D. T. Zhang, M. X. Sun, Y. Tao, S. F. Xue and W. J. Yang, *J. Mater. Chem. C*, 2013, **1**, 856–862.
24. X. Zhou, H. Y. Li, Z. G. Chi, X. Q. Zhang, J. Y. Zhang, B. J. Xu, Y. Zhang, S. W. Liu and J. R. Xu, *New J. Chem.*, 2012, **36**, 685–693.
25. J. Mei, J. Wang, A. Qin, H. Zhao, W. Yuan, Z. Zhao, H. H. Y. Sung, C. Deng, S. Zhang, I. D. Williams, J. Z. Sun and B. Z. Tang, *J. Mater. Chem.*, 2012, **22**, 4290–4298.
26. X. Q. Zhang, Z. G. Chi, X. Zhou, S. W. Liu, Y. Zhang and J. R. Xu, *J. Phys. Chem. C*, 2012, **116**, 23629–23638.
27. H. Y. Li, X. Q. Zhang, Z. G. Chi, B. J. Xu, W. Zhou, S. W. Liu, Y. Zhang and J. R. Xu, *Org. Lett.*, 2011, **13**, 556–559.
28. H. Y. Li, Z. G. Chi, B. J. Xu, X. Q. Zhang, X. F. Li, S. W. Liu, Y. Zhang and J. R. Xu, *J. Mater. Chem.*, 2011, **21**, 3760–3767.

29. X. Q. Zhang, Z. G. Chi, B. J. Xu, C. J. Chen, X. Zhou, Y. Zhang, S. W. Liu and J. R. Xu, *J. Mater. Chem.*, 2012, **22**, 18505–18513.
30. B. J. Xu, Z. G. Chi, J. Y. Zhang, X. Q. Zhang, H. Y. Li, X. F. Li, S. W. Liu, Y. Zhang and J. R. Xu, *Chem.–Asian J*, 2011, **6**, 1470–1478.
31. J. Wang, J. Mei, R. Hu, J. Sun, A. Qin and B. Z. Tang, *J. Am. Chem. Soc.*, 2012, **134**, 9956–9966.
32. J. Q. Shi, N. Chang, C. H. Li, J. Mei, C. M. Deng, X. L. Luo, Z. P. Liu, Z. S. Bo, Y. Q. Dong and B. Z. Tang, *Chem. Commun.*, 2012, **48**, 10675–10677.
33. X. Luo, J. Li, C. Li, L. Heng, Y. Q. Dong, Z. Liu, Z. Bo and B. Z. Tang, *Adv. Mater.*, 2011, **23**, 3261–3265.
34. S. F. Xu, W. Liu, M. X. Sun and W. J. Yang, *Sci. Sin. Chim.* (in Chinese), 2013, **43**, 1105–1120.
35. X. G. Gu, J. J. Yao, G. X. Zhang, Y. L. Yan, C. Zhang, Q. Peng, Q. Liao, Y. S. Wu, Z. Z. Xu, Y. S. Zhao, H. B. Fu and D. Q. Zhang, *Adv. Funct. Mater.*, 2012, **22**, 4862–4872.
36. X. Q. Zhang, Z. G. Chi, B. J. Xu, J. Jiang, X. Zhou, Y. Zhang, S. W. Liu and J. R. Xu, *Chem. Commun.*, 2012, **48**, 10895–10897.
37. W. Liu, Y. L. Wang, L. Y. Bu, J. F. Li, M. X. Sun, D. T. Zhang, M. Zheng, C. Yang, S. F. Xue and W. J. Yang, *J. Luminesc.*, 2013, **143**, 50–55.
38. X. L. Luo, W. J. Zhao, J. Q. Shi, C. H. Li, Z. P. Liu, Z. S. Bo, Y. Q. Dong and B. Z. Tang, *J. Phys. Chem. C*, 2012, **116**, 21967–21972.
39. L. Y. Bu, Y. P. Li, J. F. Wang, M. X. Sun, M. Zheng, W. Liu, S. F. Xue and W. J. Yang, *Dyes Pigments*, 2013, **99**, 833–838.
40. W. Liu, Y. L. Wang, M. X. Sun, D. T. Zhang, M. Zheng and W. J. Yang, *Chem. Commun.*, 2013, **49**, 6042–6044.
41. J. He, B. Xu, F. Chen, H. Xia, K. Li, L. Ye and W. Tian, *J. Phys. Chem. C*, 2009, **113**, 9892–9899.
42. X. Q. Zhang, Z. G. Chi, J. Y. Zhang, H. Y. Li, B. J. Xu, X. F. Li, S. W. Liu, Y. Zhang and J. R. Xu, *J. Phys. Chem. B*, 2011, **115**, 7606–7611.
43. Q. K. Qi, Y. F. Liu, X. F. Fang, Y. M. Zhang, P. Chen, Y. Wang, B. Yang, B. Xu, W. J. Tian and S. X.–A. Zhang, *RSC Adv.*, 2013, **3**, 7996–8002.
44. J. Q. Shi, W. J. Zhao, C. H. Li, Z. P. Liu, Z. S. Bo, Y. P. Dong, Y. Q. Dong and B. Z. Tang, *Chin. Sci. Bull.*, 2013, **58**, 2723–2727.
45. C. Y. Li, X. L. Luo, W. J. Zhao, Z. Huang, Z. P. Liu, B. Tong and Y. Q. Dong, *Sci. China Chem.*, 2013, **56**, 1173–1177.
46. N. Zhao, Z. Yang, J. W. Y. Lam, H. H. Y. Sung, N. Xie, S. Chen, H. Su, M. Gao, I. D. Williams, K. S. Wong and B. Z. Tang, *Chem. Commun.*, 2012, **48**, 8637–8639.
47. N. Zhao, M. Li, Y. L. Yan, J. W. Y. Lam, Y. L. Zhang, Y. S. Zhao, K. S. Wong and B. Z. Tang, *J. Mater. Chem. C*, 2013, **1**, 4640–4646.
48. Q. K. Qi, X. F. Fang, Y. F. Liu, P. Zhou, Y. M. Zhang, B. Yang, W. J. Tian and S. X.–A Zhang, *RSC Adv.*, 2013, **3**, 16986–16989.
49. S. J. Yoon and S. Y. Park, *J. Mater. Chem.*, 2011, **21**, 8338–8346.
50. Y. Zhang, J. Sun, G. Bian, Y. Chen, M. Ouyang, B. Hua and C. Zhang, *Photochem. Photobiol. Sci.*, 2012, **11**, 1414–1421.

51. Y. J. Zhang, G. L. Zhuang, M. Ouyang, B. Hu, Q. B. Song, J. W. Sun, C. Zhang, C. Gu, Y. X. Xu and Y. G. Ma, *Dyes Pigments*, 2013, **98**, 486–492.
52. C. Dou, L. Han, S. Zhao, H. Zhang and Y. Wang, *J. Phys. Chem. Lett.*, 2011, **2**, 666–670.
53. W. Y. Lu, Y. Q. Tan, Y. Y. Gong, H. Li, W. Z. Yuan, Y. M. Zhang and B. Z. Tang, *Chin. Sci. Bull.*, 2013, **58**, 2719–2722.
54. W. Z. Yuan, Y. Q. Tan, Y. Y. Gong, P. Lu, J. W. Y. Lam, X. Y. Shen, C. F. Feng, H. H.-Y. Sung, Y. W. Lu, I. D. Williams, J. Z. Sun, Y. M. Zhang and B. Z. Tang, *Adv. Mater.*, 2013, **25**, 2837–2843.
55. W. G. Mao, K. Chen, M. Ouyang, J. W. Sun, Y. B. Zhou, Q. B. Song and C. Zhang, *Acta Chim. Sinica*, 2013, **71**, 613–618.
56. P. Zhang, W. Dou, Z. H. Ju, X. L. Tang, W. Liu, C. Y. Chen, B. Wang and W. S. Liu, *Adv. Mater.*, 2013, **25**, 6112–6116.
57. Y. Y. Gong, Y. Q. Tan, J. Liu, P. Lu, C. F. Feng, W. Z. Yuan, Y. W. Lu, J. Z. Sun, G. F. He and Y. M. Zhang, *Chem. Commun.*, 2013, **49**, 4009–4011.
58. M. P. Aldred, G. F. Zhang, C. Li, G. Chen, T. Chen and M. Q. Zhu, *J. Mater. Chem. C*, 2013, **1**, 6709–6718.
59. C. X. Shi, Z. Q. Guo, Y. L. Yan, S. Q. Zhu, Y. S. Xie, Y. S. Zhao, W. H. Zhu and H. Tian, *ACS Appl. Mater. Interfaces*, 2013, **5**, 192–198.
60. X. Y. Shen, Y. J. Wang, E. G. Zhao, W. Z. Yuan, Y. Liu, P. Lu, A. J. Qin, Y. G. Ma, J. Z. Sun and B. Z. Tang, *J. Phys. Chem. C*, 2013, **117**, 7334–7347.
61. B. J. Xu, Z. G. Chi, X. Q. Zhang, H. Y. Li, C. J. Chen, S. W. Liu, Y. Zhang and J. R. Xu, *Chem. Commun.*, 2011, **47**, 11080–11082.
62. G. Shan, H. Li, J. Qin, D. Zhu, Y. Liao and Z. Su, *Dalton Trans.*, 2012, **41**, 9590–9593.
63. Y. Ren, W. H. Kan, M. A. Henderson, P. G. Bomben and C. P. Berlinguette, *J. Am. Chem. Soc.*, 2011, **133**, 17014–17026.
64. Y. Ren and T. Baumgartner, *Inorg. Chem.*, 2012, **51**, 2669–2678.
65. P. Y. Gu, Y. H. Zhang, G. Y. Liu, J. F. Ge, Q. F. Xu, Q. C. Zhang and J. M. Lu, *Chem.-Asian J.*, 2013, **8**, 2161–2166.
66. T. Han, Y. J. Zhang, X. Feng, Z. G. Lin, B. Tong, J. B. Shi, J. G. Zhi and Y. P. Dong, *Chem. Commun.*, 2013, **49**, 7049–7051.
67. R. Rao M, C. W. Liao, W. L. Su and S. S. Sun, *J. Mater. Chem. C*, 2013, **1**, 5491–5501.
68. R. Rao M, C. W. Liao and S. S. Sun, *J. Mater. Chem. C*, 2013, **1**, 6386–6394.

CHAPTER 7

Mechanofluorochromic Mechanism

BIN XU, QINGKAI QI, JIBO ZHANG AND WENJING TIAN*

State Key Laboratory of Supramolecular Structure and Materials,
Jilin University, Changchun 130012, China
*Email: wjtian@jlu.edu.cn

7.1 Introduction

Mechanofluorochromic luminescent materials are attracting considerable attention for their potential applications in mechano-sensors, indicators of mechano-history, security papers, optoelectronic devices and data storage.[1] A number of mechanochromic systems based on organic molecules, metal complexes and polymers, which switch their fluorescence colors in response to mechanical stimuli such as shearing, grinding and rubbing, have been rapidly developed and well investigated, leading to increased understanding of the luminescence behavior of condensed solid materials.[2–5]

The photophyscial properties of a given system in the condensed phase, such as fluorescence emission and vibrational relaxation, undergo significant variation according to the mode of molecular packing and molecular configuration, because the intermolecular interactions or intramolecular effects invariably alter the photophyscial processes of the molecules. In particular, dynamic solid state fluorescence switching under mechanical stimuli will lead to the subtle interplay on the structures and properties in molecular and supramolecular level. In this respect, understanding and establishing the meaningful structure-property relationships in the solid state, which are usually related to the correlations of intermolecular

RSC Smart Materials No. 8
Mechanochromic Fluorescent Materials: Phenomena, Materials and Applications
Edited by Jiarui Xu and Zhenguo Chi

Published by the Royal Society of Chemistry, www.rsc.org

interactions and intramolecular effects during the mechanofluorochromic processes, are very important for obtaining mechanofluorochromic material systems with desired chemical and physical properties. Up to now, several possible mechanisms of mechanofluorochromic behavior, such as excimer forming, phase structural transition, J- or H-aggregation and intramolecular coplanarity, have been assigned by researchers. Herein, we focus on the mechanofluorochromic mechanism in the molecular systems that switch their emission by responding to a mechanical stimulus such as shearing, grinding or pressing.

The intention of this chapter is to emphasize the current aspects of the mechanofluorochromic mechanism based on organic systems, paying particular attention to the intermolecular and intramolecular effects.

7.2 Intermolecular Effects on Mechanofluorochromism

7.2.1 Excimer Formation

In 2002, Lowe and Weder *et al.* published pioneering work about organic mechanofluorochromism based on two blended films using linear low-density polyethylene (LLDPE) as the matrix and two cyano group substituted oligo(*p*-phenylene vinylene) derivative as the dopants.[6,7] They demonstrated that the solid-state tensile deformation could facilitate substantial changes in the emission characteristics of the LLDPE-dye blend, as shown in Figure 7.1. In the host polymers, the dyes form aggregates in which stacked structures of at least two molecules are constructed, leading to the emission from excimers. These aggregating molecules dissociate as a result of the tensile deformation, and the dyes are consequently dispersed at molecular level in the polymers. For example, a polyethylene film doped with dyes exhibits luminescence before stretching. On the tensile deformation, the luminescent color changes from orange to green, owing to an increase in the ratio of monomer to excimer emission. When the polyethylene film was doped with different dyes, a different change of luminescent color, from green to orange, could be achieved. Moreover, the monomer-to-excimer ratio can be modified by changing the dye structure, the polymer composition and the processing conditions, so the luminescent response of polymers can thus be tunable. In addition, a derivative of oligo(*p*-phenylene vinylene), which has cyano groups and long alkyl chains also shows mechanochromic luminescence.[8] It exhibits blue emission both in the solution (465 nm) and in the solid state (481 nm), which is the characteristic of monomer emission. When compressed, the crystalline powder emits a green-yellow color (548 nm), which shows a broad and lacks structure emission spectrum. This observed green-yellow emission is attributed to the excimer formation of the luminescent π-conjugated moieties. Heating the compressed sample enables recovery of the original luminescent color without forming the

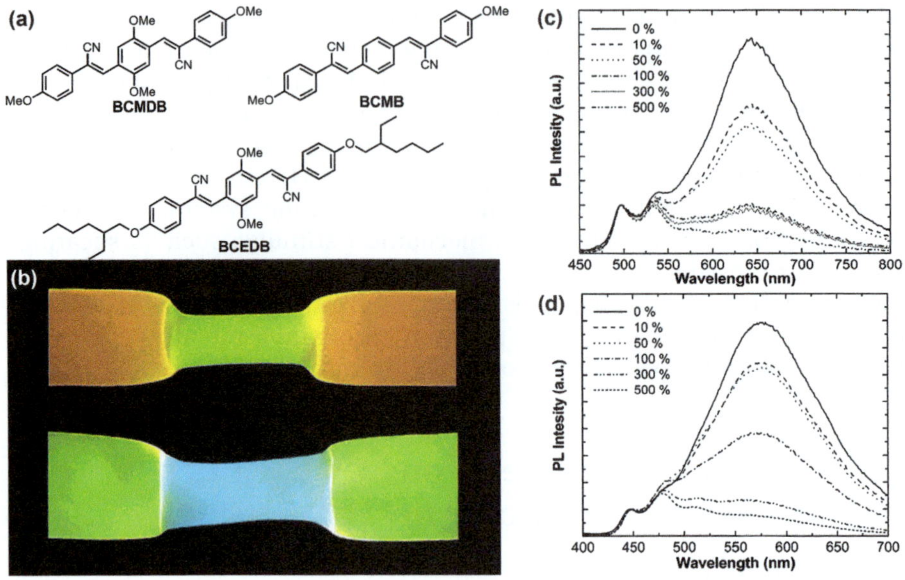

Figure 7.1 (a) Chemical structures of BCMDB, BCMB and BCEDB. (b) Pictures (under 365 nm illumination) of blend films of LLDPE and 0.18% w/w BCMDB (top) and 0.20% w/w BCMB (bottom) stretched at room temperature to a draw ratio of 500%. And PL emission spectra of blend films of LLDPE and 0.20% w/w BCMDB (c) and 0.20% w/w BCMB (d) as a function of draw ratio. All spectra were normalized to the intensity of the monomer peak.
Reproduced with permission from reference 7. Copyright 2003, The American Chemical Society.

isotropic phase. When the crystalline powder is heated the compound melts and exhibits smectic and nematic liquid crystalline (LC) phases, with the aligned molecules only arranging in ordered layers in the smectic phase (in the nematic phase the molecules are still aligned but do not form layers). Therefore, the compound shows blue emission in the smectic phase, that is a monomer emission, and after the transition from the smectic to nematic phase, green-yellow emission, an excimer emission is demonstrated.

The mechanochromic luminescence of LC molecules, which have particularly complex structures, has also been observed.[9,10] For example, the compounds having two dendritic moieties with long alkyl chains to induce LC properties, and pyrene and anthracene moieties as luminescent cores can form micelles arranged in a cubic lattice - micellear cubic LC phases in a wide range of temperature. Yellow emission is observed under UV irradiation, and is ascribed to the formation of excimer. When mechanically sheared, the yellow emission of the LC states switches to a shorter wavelength. Thermal analysis of the compounds indicates that the structures have been converted from a cubic, metastable LC phase to a columnar, stable one. The metastable cubic phases are formed for these bulky compounds as the result of hydrogen bonding interactions and π–π interactions. After the phase transitions induced by mechanical force, the metastable cubic form can be obtained again when the samples are annealed in their isotropic states. Among these LC molecules, the thin film of the molecule containing pyrene prepared by solution coating on a glass substrate shows a change in luminescence from yellow to blue-green on rubbing. Figure 7.2 shows the shear-induced structural change of the assembled structures of the pyrene molecule. In the cubic phase, around 20 molecules form a segmented column. The pyrene groups are arranged in a π-stacked structure, which is enough to form excimers, and leads to the emission of a yellow color (579 nm). These segmented columns consisting of the π-conjugated moieties are surrounded by the flexible alkyl chains, forming a micellar structure. The columns form disordered stacked structures since the distance of the hydrogen bond between the amide chains is around 5.0 Å, whereas the distance between the aromatic rings forming stable π-stacked structures is approximately 3.5 Å. It is worth mentioning that mechanical shearing can induce a change in molecular assembly from a segmented columnar, cubic phase, to a non-segmented columnar structure, columnar phase, where a more linear hydrogen-bonded structure is formed. In these assembled structures, luminescent cores are fixed in a molecular arrangement in which no excimer formation occurs. Therefore, the luminescent color changes from yellow to blue-green; in other words, from excimer to monomer emission.

Different from the monomer-to-excimer mechanism, another example based on excimer-to-excimer was found in the mechanochromic process of dendritic molecule.[11,12] The glycine and aspartic acid (Gly-Asp) based dendron compound with pyrene group as the focal luminophore displays a reversible mechanochromic luminescent property, which arises from the switchable self-assemblies between the hexagonal packing and lamellar arrangement, leading to different color emission. The dynamic recovery of the emission colors between bright green and blue was evident upon multiple cycles of grinding and treating the sample with solvent, or heating and cooling. The emission band at 390 nm in the dilute solution is assigned to the monomeric emission of the pyrenyl moiety. The two broader and less structured emission bands at longer wavelength are suggestive of two different excimer-like components of pyrenyl group, which is 420 nm (**E1**) for the drop cast aggregate state, and 470 nm (**E2**) for the ground solid state, respectively. The **E1** excimer is less stable, with a partially overlapped packing, and exhibits a short lifetime. On the other hand, the **E2** excimer in a sandwich packing is more stable due to the enhancement of π–π interaction, and shows a long lifetime. This unusual stimuli-responsive behavior can be understood by the molecular interactions and arrangement of dendritic branches and lumenophores, which finally determine the apparent phase structures. The dendritic block tends to self-assemble when the molecules crystallize from the solution or are cooled from the melt. In this way, the pyrenyl groups packs in a confined environment, which may force them

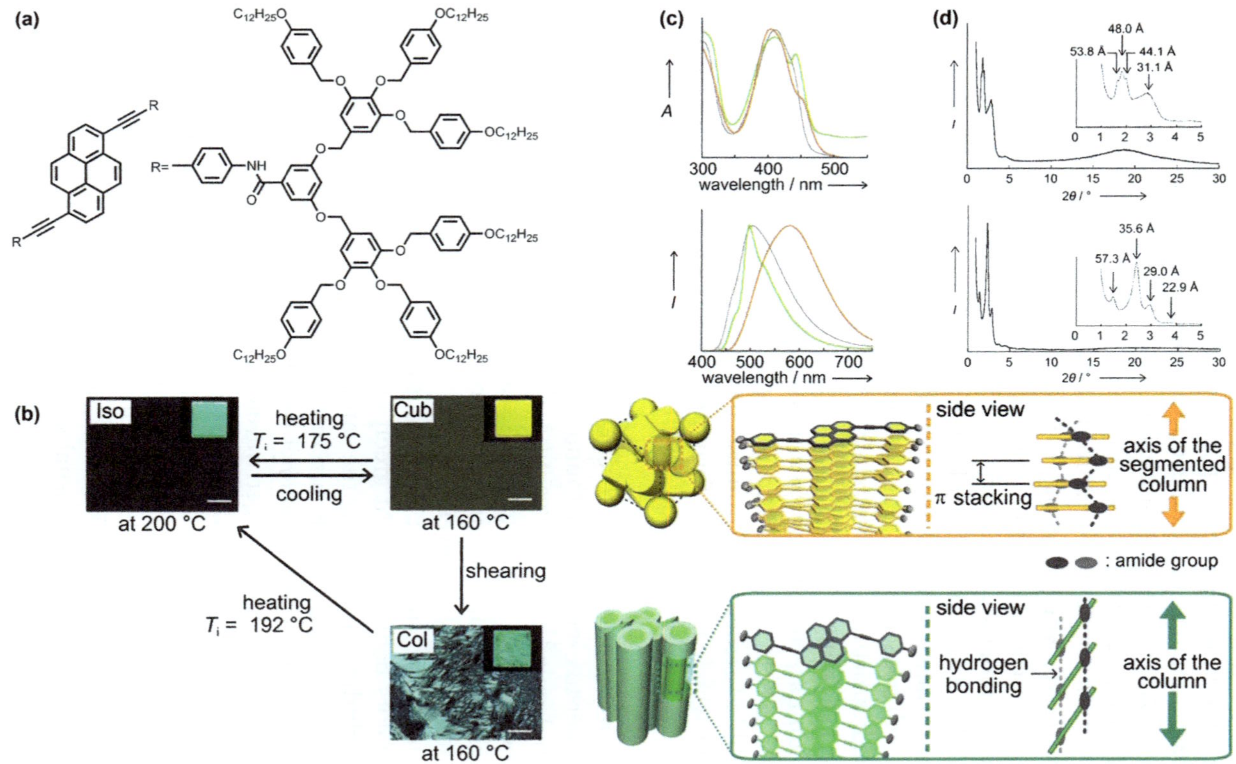

Figure 7.2 (a) Molecular structure. (b) The change of photoluminescent color (Ti: LC–isotropic phase-transition temperature) and illustrations of the assembly in the cubic phase and in the shear-induced columnar phase. (c) Absorption and emission spectra in the cubic phase at 160 °C (orange line), in the shear-induced columnar phase at 160 °C (green line), and in the isotropic phase at 200 °C (gray line). (d) X-ray diffraction patterns of the compound: (a) in the cubic phase and (b) in the shear-induced columnar phase.

to be partially overlapped, resulting in the blue emission. The alteration of hydrogen bonding under the stimulus of the external force contributes to the variation of self-assembled structures in which the pyrenyl groups tend to a sandwich-like arrangement, and results in a bright green **E2** emission. This is a rarely reported mechanochromic example based on the mechanism of excimer-to-excimer transition for tuning the luminescent property.

Very recently, direct evidence of excimer formation based on a high-pressure single-crystal X-ray diffraction analysis was reported.[13] 2,3,4,5-Tetra(2-thiazolyl)thiophene exhibits the distinct luminescence response under the mechanical grinding or hydrostatic pressure, as shown in Figure 7.3. In the single crystal structure, the thiazole rings A and D take almost coplanar arrangements with the central thiophene ring with the dihedral angles of 18.5° and 3.0°, respectively, indicating an effective π-conjugation in the 2,5-dithiazolylthiophene skeleton. The thiazole rings B and C adopt twisted conformations, with the large dihedral angles of 69.0° and 78.6°, respectively. The crystals exhibited a yellow emission with the λ_{max} of 556 nm. By grinding the crystals, the emission color was largely blue shifted to green (490 nm).

The crystal packing structure will give insights into the origins of the long-wavelength emission in the crystalline state and its mechanochromism. This compound forms weak intermolecular C–H···N hydrogen bonds at several positions. A relatively strong interaction was observed between the H atoms at the 5-position of D or A ring and the N atom of B or C ring in the adjacent molecules. The hydrogen bonds play an important role in fixing the orientation of the two π-conjugated 2,5-dithiazolylthiophene skeletons in a face-to-face fashion. These multiple intermolecular hydrogen bonds construct a 3D hydrogen-bond network in which the face-to-face teraryl dimers are integrated. This structural analysis suggests that the broad and structureless spectrum of the yellow emission band in the crystalline state is attributed to the preformed excimer, which is consistent with the slightly longer fluorescence lifetime of the yellow emission in the crystalline state (0.42 ns) as compared to those of the green emission in the ground powder (0.32 ns) and the blue emission in solution (0.14 ns). Upon grinding the crystals, the weak 3D hydrogen-bond network is disordered without forming another long-range ordered phase. From the powder XRD and solid-state ^{1}H NMR analysis, the yellow-emissive excimer is no longer formed. Instead, the degree of the intermolecular distances between the neighboring fluorophores varies to some extent, resulting in a broad fluorescence band with a green color.

Another noteworthy result is that this compound showed a significant red shift in the fluorescence maximum under hydrostatic pressure, which exhibits a gradually red-shifted emission during the compression process, and turns back to the original emission when the hydrostatic pressure returned to the ambient pressure. The high-pressure IR spectra of the compound revealed the continuous change with an increase in pressure, which implied that no significant phase transition occurs. A high-pressure single-crystal

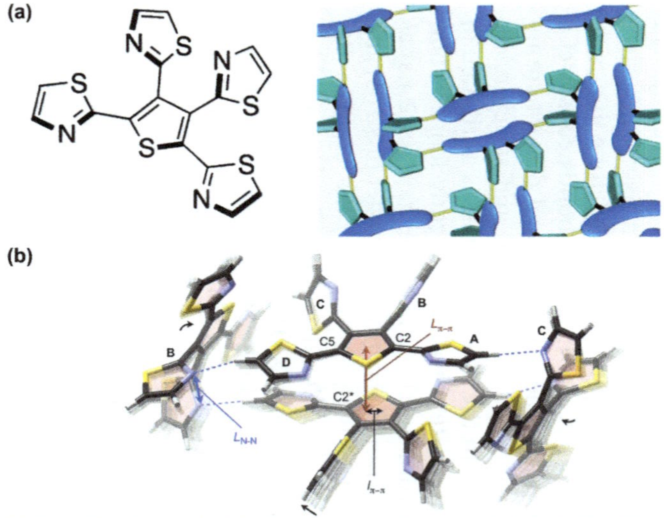

Pressure / GPa	π-π Distance $L_{\pi\text{-}\pi}$ / Å	π-π Angle ∠C2-C5-C2* / °	π-π Slipping [b] $l_{\pi\text{-}\pi}$ / Å	N···N Distance $L_{N\text{-}N}$ / Å
0	3.910	81.65	0.574	3.758
1.0	3.752	83.52	0.426	3.742
2.0	3.659	85.15	0.311	3.739
3.0	3.588	86.59	0.214	3.714
4.0	3.529	87.60	0.148	3.713

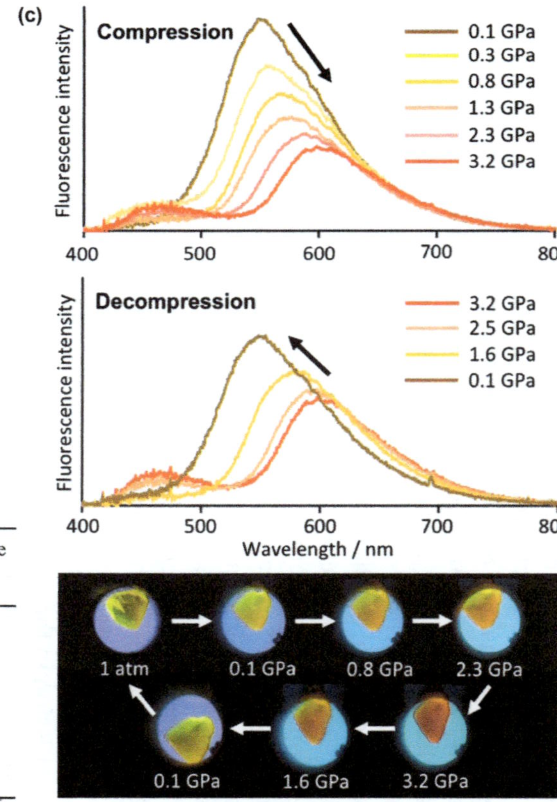

Figure 7.3 (a) Molecular structure and the representation of crystal packing. (b) Structural parameters of the optimized packing from 0 GPa to 4.0 GPa. (c) Fluorescence microspectroscopy of crystal under high pressure in the range of 0.1–3.2 GPa and the micrographs of the crystal under high pressure. A UV LED lamp was used for the excitation (365 nm).
Reproduced with permission from reference 13. Copyright 2013, The American Chemical Society.

XRD analysis of the compound (up to 2.8 GPa) shows that the optimized geometries under high pressure possessed similar compression behaviors in the packing structure to each other. The hydrogen-bonded lattice is deformed, which reduces the void space between the face-to-face 2,5-dithiazolylthiophene moieties. The interfacial distance between their mean planes becomes much shorter from 3.96 Å at the ambient pressure to 3.69 Å at 2.8 GPa. In addition, the two teraryl skeletons have slipped along the long axis, so that these units overlap with each other to a greater extent. The density functional theory (DFT) calculation of the closely stacked dimer in the optimized geometry indicated the greater orbital interaction between the two teraryl units at 3.0 GPa than that at ambient pressure. As a result of the closer proximity, a highly overlapped excimer is formed, which is likely the origin of the red-shifted emission under hydrostatic pressure.

As a result, the different mechanisms of a blue shift by the grinding crystals and of a red shift under hydrostatic pressure are fully investigated. The anisotropic and isotropic modes of mechanical loading in the 3D hydrogen-bond network suppress and enhance the excimer formation, respectively.

7.2.2 Transition of Aggregation Mode

Crystalline organic solid easily forms variation polymorphs under different conditions, which shows it to be capable of mechanochromic luminescence. For example, K. Araki *et al.* reported a pyrene based compound consisting of four hexyl amide groups at the *para* position of phenyl units.[14,15] As shown in Figure 7.4, the addition of methanol to the chloroform solution of the compounds gives the precipitation of a white powder ("B-form") that

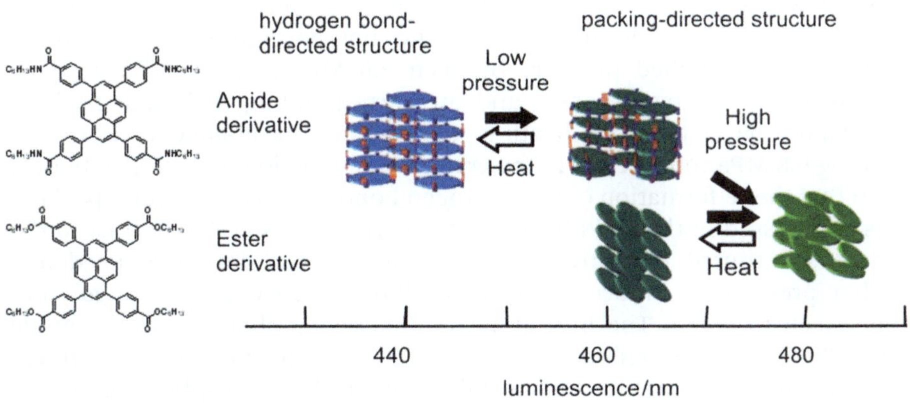

Figure 7.4 Schematic illustration of mechanofluorochromic response of the derivatives.
Reproduced with permission from reference 15. Copyright 2011, The Royal Society of Chemistry.

exhibits strong blue luminescence under UV irradiation. Once the precipitate is formed, the B-form solid does not dissolve in chloroform or in other common organic solvents, but it slowly dissolved in warm 1-methyl-2-pyrrolidinone. When pressed with a spatula, the B-form transformed to a yellow solid ("G-form") that exhibits green luminescence, which can be converted back into B-form by a heating process or re-precipitated from the dissolved solution. The B-form was found to be stable with ordered hydrogen-bonding arrays, whereas the G-form was in a metastable state in which the hydrogen bonding was more random. Under the applied pressure, the tight hydrogen bonding interactions were destroyed, resulting in the change of the twisting angles between the phenyl rings and pyrene unit and pyrene packing mode, which inhibits the formation of H-aggregates in the columnar-assembled structures. These factors suggested that the transition of aggregation mode facilitated the change in luminescent color. In order to further study the role of the amide hydrogen bonds and side chains, a series of pyrene derivatives modified by different amide substituted or ester substituted were designed and synthesized. Among them, two molecules with short chain lengths were found to have no mechanofluorochromism properties. However, the other molecules with relatively long chain lengths exhibited evident mechanofluorochromic responses. The formation of the molecular packing required appropriately sized side chains. The molecular size and shape of the amide substituted derivatives and ester substituted derivative were quite similar. However, the mode of molecular packing was definitely different, that is, the tetraphenylpyrene cores of the molecule with amide substituent were firmly connected by four amide hydrogen bonds to form the H-type columnar aggregate, whereas the ester derivative did not form this type of molecular assembly. It is worth mentioning that the molecules with amide substituent can fully return to its initial emission after heating above 112 °C, when increasing the pressure from 2.0 MPa to 9.8 MPa. However, at a pressure of 9.8 MPa, the emission of the sample further red shifted, and the luminescence could not be restored to the initial emission. The applied pressure between 2.0 MPa and 9.8 MPa partially disrupted the hydrogen bonds, and the corresponding molecular packing could only be restored by heating. Meanwhile, the applied pressure exceeding 9.8 MPa could result in the excessive destruction of hydrogen bonds, such that the re-formation of the hydrogen bond-directed molecular packing was suppressed in the solid form by the thermal process. In contrast to the amide substituted derivative, the ester substituted derivative needed a higher pressure to induce a noticeable luminescence color change from blue-green to green. The blue-green luminescence of the pressed sample could be fully recovered by heating to 40 °C, indicating that the green-emitting pressed sample was unstable under the atmospheric pressure. Therefore, the presence of the amide hydrogen bonds was essential for a sensitive and repeatable mechanofluorochromic response and that the excessive destruction of the hydrogen bonds restricted the re-formation of the hydrogen bond-directed structure.

The detailed structural information on the phase transition, and the comprehensive optical and photophysical characterization would allow for a deep insight into the structure-property relationships in π-conjugated organic materials. Cyano stilbene derivatives are very important optoelectronic materials, which often show the extremely large fluorescence enhancement in the solid state, commonly known as aggregation-induced enhanced emission (AIEE).[16] An AIEE-active cyano distyrylbenzene derivative, DBDCS, exhibits very high solid-state fluorescence quantum yield due to the AIEE process, as well as multi-stimuli two-color luminescence switching. The single crystal of DBDCS emits green light under UV illumination, which is to be called the G-phase of DBDCS crystal. When heated to 125 °C, however, the crystals change from G-phase to B-phase and the crystal appearance changes from transparent to opaque, while the emission changes from green to blue. The molecular packing structure of the G-phase single crystal reveals that DBDCS form the planar "molecular sheets" assisted by the multiple C–H···N and C–H···O hydrogen bonds combined with the appropriate length of alkyl substituents. It is shown that the molecular sheets in this G-phase crystal are arranged in the slip-stacks along the molecular axis, where the driving force may result from the antiparallel coupling between the local dipoles in the adjacent molecular sheets. Since the outer phenyl rings with butoxy-substituents are electron-rich, while the central phenyl ring with cyano group is electron-poor, DBDCS is a D-A-D molecule comprising two local dipoles, which add to a zero net dipole moment. Antiparallel dipole coupling places the central "A" ring of the upper sheet just above the "D" ring of the lower sheet, bringing about the efficient excitonic and excimeric coupling between the DBDCS molecules (Figure 7.5).

Thermal annealing of the G-phase crystal to the B-phase must accompany the specific stacking change. Although the B-phase could not be determined by single crystal X-ray analysis due to the poor crystal quality, the small-angle X-ray scattering (SAXS) pattern of the blue-emitting crystal shows (100) and (200) reflection peaks of the lamellar structure. It suggests that the principal slip direction of the molecular sheets is along the shorter axis of DBDCS molecule to effectively offset the aromatic rings and establish efficient head-to-tail coupling of the local dipoles in the adjacent molecular sheets. When grinding the blue B-phase powder in a mortar or rubbing it with a spatula, it was observed that the fluorescence color changed immediately to green. This mechanochromic transition thus represents the reversal of the annealing process and is likely to be driven by the stacking mode change as well. It should be noted that the mechano-process is not a complete conversion process, since the shearing force applies unevenly to the bulk pristine powder. However, reannealing of the ground green-emitting powder completely recovers the initial state from the SAXS pattern, which confirms the same nature of blue emission in the pristine and reannealed powder.

On the other hand, H-aggregation behavior is observed in the absorption spectrum of the G-phase crystal. The small displacement along the short axis of the adjacent molecules in the G-phase leads to a substantial overlap of the

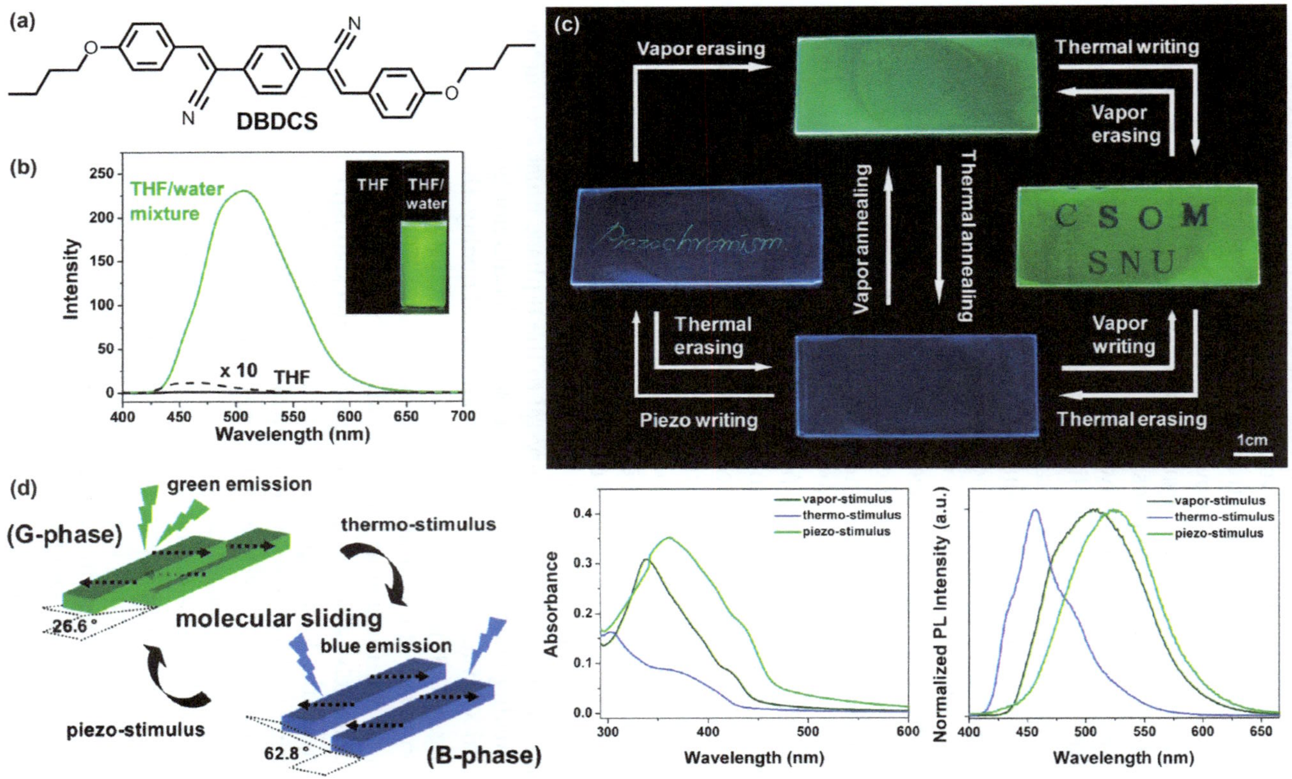

Figure 7.5 (a) Molecular structure and (b) PL spectra of DBDCS in THF solution and THF/water mixture. (c) Photos of the luminescence writing/erasing cycle of DBDCS/PMMA film as well as the UV-visible absorption and PL spectra under different stimulus. (d) Illustration of two different modes of slip-stacking in DBDCS molecular sheets.
Reproduced with permission from reference 16. Copyright 2010, The American Chemical Society.

π-systems and, thus to considerable excited state delocalization. The latter allows for the efficient vibronic coupling of the interchrophore breathing modes, which is responsible for the significantly red-shifted and unstructured "excimer-like" emission. Upon thermal treatment, the metastable G-phase transforms into the thermodynamically favored B-phase, aided by the efficient slip along the molecular sheets. Both the absorption and emission spectra point to a pronounced different coupling situation as compared with G-phase. In the B-phase, the excimer formation is diminished, while the excitonic interaction substantially increases.

On the basis of structural, optical, photo-physical and computational studies, two different phases, *i.e.* the metastable green-emitting G-phase and the thermodynamically stable blue-emitting B-phase, were identified. In G-phase, antiparallel coupling of the local dipoles kinetically stabilizes a structure with very weak excitonic coupling, but the excimer formation is favored by the pronounced overlap of the π-systems. Upon annealing, a smooth slip of the molecular sheets with a low activation barrier forms the B-phase with a head-to-tail arrangement of the local dipoles. Here the excimer formation is diminished, while the excitonic interaction substantially increases.

Another example of the transition of the aggregation mode was reported in two anthracene derivatives BP2VA and BP4VA, which N atom substitutes on different sites of the benzene ring.[17] For example, BP2VA exhibited unusual spectacular luminescence characteristics: grinding and exertion of external pressure on the powder led to a change in its photoluminescence color from green to red. The initial BP2VA powder exhibited a strong green emission which peaked at 528 nm. After grinding, BP2VA powder showed a large red shift with a yellow emission (λ_{max} at 561 nm), and after being heated above 160 °C, the ground powder recovered to its initial green emission (λ_{max} at 528 nm). The two emission colors are completely reversible through grinding and heating. The red shift of 33 nm in fluorescence emission under grinding and the recovery of initial state upon heating, indicated that the BP2VA powder performs a significant mechanochromic effect.

To further understand the mechanochromic effect, the influence of the applied pressure on the luminescence of BP2VA powder was investigated. As the applied pressure increased, the fluorescence emission of the BP2VA powder clearly showed a gradual red shift. It shows that the applied pressure from 0–8 GPa caused a more noticeable luminescence color change of the BP2VA powder from green (528 nm) to red (652 nm) compared to that upon grinding, which is the largest shift in the mechanochromic effect ever reported at present. This suggests apparently that the grinding method is not enough to cause a stronger mechanochromic effect in the case of BP2VA.

Single crystal analysis can provide a powerful tool to understand the relationship between the molecular aggregation state and the luminescence properties directly and definitely. As shown in Figure 7.6, the π–π interactions in the three crystalline states of BP2VA with different stacking modes

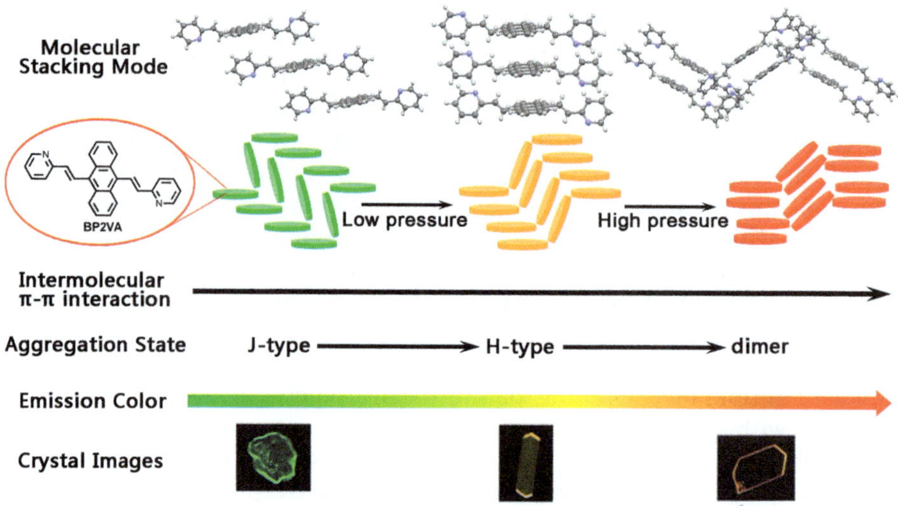

Figure 7.6 Stacking modes and corresponding emission colors for the various molecular aggregation states in BP2VA powder.
From reference 17, reproduced by permission of Wiley-VCH.

enhanced gradually, and the emission of three crystals located at 527 nm (**C1**), 579 nm (**C2**) and 618 nm (**C3**), respectively. In the case of **C1**, molecules adopt a stack mode of J-type aggregation along the Y axis. While in **C2**, H-type aggregation is formed along the X axis. In particular, dimers with tight face-to-face stack are found in **C3** along the X axis. It is worth noting that the overlap of the anthracene planes between the adjacent molecules is increased from **C1** to **C3**. These different stack modes in **C1**, **C2** and **C3**, reasonably result in various molecular aggregation state, leading to the diverse fluorescence colors, which obviously indicates that BP2VA possesses three or even more different aggregation structures in the solid state. Such a strong π–π interaction induced the red emission of **C3** with λ_{max} at 618 nm to show a red shift relative to those of **C1** and **C2**. These red shifts thus can result from the reduced band gap of the BP2VA molecule owing to the enhanced π–π interaction in terms of tight-binding model, which supposes that the band gap depends on the degree of π–π interaction of chromophores. Nonetheless, another important factor for the red-shifted fluorescence should be taken into consideration, that is, the increase of the exciton coupling and orbital overlap between the neighboring molecules from **C1** over **C2** to **C3**, which could lead to a strong red shift of the emission of the lowest state of the coupled chromophores. Therefore, these suggest that the PL emission of BP2VA in the aggregation/solid state could be changed by altering its molecular stacking mode.

To gain more insight into the origin of the mechanochromic effect of the BP2VA powder with the red shift of PL spectrum under grinding, the phase characteristics of the BP2VA powder were investigated by powder X-ray diffraction (PXRD) analysis, with the simulated patterns from single crystal data as reference. The PXRD pattern of the BP2VA unground powder agreed well with the simulated XRD pattern from the crystal data of **C1**. It suggested that the initial samples should adopt the same molecular arrangement as that of **C1** polymorph, with J-type aggregation along the molecular long axis and no effective π–π interactions of the central anthracene planes. Although some resolvable peaks of the ground sample were consistent with those of the unground sample, the intensity became weaker, indicating that the initial aggregation state was changed by grinding. The orange emission of the ground powder with λ_{max} at 561 nm is similar to the emission of **C2** (λ_{max} at 579 nm), which showed the red-shifted fluorescence and blue-shifted absorption as comparing to those of the unground powder. It suggested the H-type aggregation, similar to the molecular packing of **C2** was possibly formed during the grinding process, although the partial initial aggregation state remained after grinding. Besides, a red shift from the unground to the ground powder possibly originated from the enhanced exciton coupling between the pairs of neighboring chromophores, which already occurred over a long distance in the unground powder, although the orbital overlap is still very small. This can also be confirmed by the blue shift from the emission of the unground powder to that of 77 K solution, where the molecule is absolutely in the single molecular state.

On the basis of single crystal structural, photophysical and computational studies, it was identified that the enhanced π–π interaction between the adjacent anthracene planes of the three crystals, as well as the increased exciton coupling and orbital overlap between the neighboring molecules, induced their fluorescence emission from green to red. The pressure-PL spectra demonstrate that, under grinding or the external pressure, the molecular aggregation state in the BP2VA powder changed as an enhanced intermolecular π–π interaction, and induced the mechanochromic luminescence.

7.2.3 Other Intermolecular Effects

Except for the excimer formation and aggregation mode transition effect, there are also some special intermolecular effects for mechanofluorochromic luminescence, such as fluorescence resonance energy transfer (FRET), photo-induced electron transfer (PET), intermolecular proton transfer, intermolecular cycloaddition reaction (2 + 2, 4 + 2) and so on.

The mechanofluorochromic response based on the FRET mechanism was reported in a blend of donor and accepter systems, which phosphole derivatives were used as a donors and Rhodamine B as an acceptor.[18] Interestingly, upon mechanical grinding of the film, the donor emission was further quenched significantly. Concurrently, the emission decay of the donor in the mechanically ground film was shortened, indicating the presence of a non-radiative energy-transfer process. Subsequent thermal annealing resulted in the recovery of the original emission spectrum.

Encouraged by this mechanically responsive FRET, the mechanically responsive emission of the phosphole-lipid system could be further amplified through the energy transfer from the donor to a similar acceptor species. It is well known that FRET can occur over distances of up to 100 Å between donor and acceptor. But more importantly, the FRET efficiency is highly dependent on the donor-acceptor distance, which makes such intermolecular process very useful for sensing the external stimuli, particularly in the application toward mechanical responses. In this case, under mechanical force, changes in the intramolecular conformation and phase separation likely decrease the intermolecular distance between donor and acceptor, further enhance the efficiency of the energy-transfer process. The PXRD pattern also suggests a disturbed lamellar organization of the donor upon mechanical grinding of the doped film. It is thus inherently plausible that both the efficient exciton migration (donor-donor) and the enhanced energy transfer (donor-acceptor) work together to enhance the orange emission of the acceptor upon mechanical grinding as a result of created trap sites. In addition, thermal annealing can provide sufficient energy for conformational changes that increase the spatial vicinity of the donor and acceptor because of the flexible structural features of the phosphole-lipids, which consequently prevent efficient energy transfer. Therefore, a new mechanochromic concept (FRET) for the amplification of the emission shift (blue to orange) responsive to the external stimuli (mechanical grinding and thermal annealing) was demonstrated for the donor-acceptor system.

In the donor-acceptor systems, the photo-induced electron transfer (PET) process often occurs, resulting in the change of photophysical properties.[19,20] 2,5-di(*E*)-distyrylfuran as an electron donor exhibited an intensive greenish-blue fluorescence both in the solution and in the crystal state, and the planar π-conjugated structure leads to strong tendency to aggregate by π–π stacking. Meanwhile, *N*-alkyl substituted maleimides as the acceptors to quench the fluorescence of donor *via* PET process. Interestingly, the fluorescence of this system can be recovered by applying mechanical force. In fact, once the complex (1 : 1) of donor and acceptor is deposited on the substrate, the fluorescence intensity of the donor will be dramatically quenched. However, the fluorescence can be recovered by the applying mechanical force. To reveal the mechanism, the emission spectra of donor and acceptor in solid state were investigated. The olivinic crystal of the donor has an emission maximum at 468 nm with a shoulder around 500 nm, which disappeared when 1 : 1 complex formed. Meanwhile, a new, weak, red-shifted charge transfer emission peak emerged at 545 nm with a shoulder at 587 nm that in accordance with the transition characteristics in PET. Because there is no reaction under the experimental conditions and no emission nature of the acceptor, the emission should come from the donor itself in the aggregation mode. The dominant of PET is also supported by the result of fluorescence kinetics and energy levels of the complex. From the XRD analysis, it is noted that the co-crystals of donor and acceptor formed during the process of solvent evaporation. Under pressure, the metastable

D-A cocrystals were decomposed then phase separation occurred and the donor crystallized itself with intensive fluorescence. It should be emphasized that the fluidity of the acceptor under pressure promotes the separation process. Therefore, it demonstrated that the balance between the individual self-assembly and the coassembly of donor-acceptor should be considered. Moderate coassembly interaction keeps the D-A complex at a metastable quenching state, but stronger self-assembly effect (self-crystallization) of the donor drives phase separation and recovers the fluorescence by applying mechanical forces. From this strategy, the concept of mechanochromic luminescence can be expanded from single molecules to co-crystallized complex systems, and various emission colors of mechanochromic luminescence can be obtained by changing the donor molecules.

Another important intermolecular effect of mechanochromism is the intermoelcular proton transfer, which was found in piroxicam.[21] When the proton transfer is induced by external stimuli, the tautomeric change is sometimes accompanied by a change in color. Piroxicam has three colorless polymorphs, forms **I**, **II** and **III**. However, the crystal and molecular structures of only two polymorphs, forms **I** and **II**, have been solved. For both polymorphs, crystallinity decreases as grinding time increases. After 60 min of cryogrinding, amorphous piroxicam is obtained. Polymorphic transformation due to grinding is not observed with either polymorph. From the XRD analysis, the decrease in crystallinity of forms **I** and **II** as a result of mechanical stress is accompanied by an increase in the intensity of yellow color with respect to grinding time. It suggests that the solid state of piroxicam changes from order (crystalline) to disorder (amorphous), and the molecules of piroxicam undergo a structural change. Because piroxicam was found not to undergo chemical degradation upon grinding, this structural change could result from either proton transfer and/or a change in the conformation of the piroxicam molecules.

^{13}C NMR provides direct evidence of deprotonation of the acidic enol group of the piroxicam molecules in the crystalline state to yield a population of the enolate anion. When the amorphization of the polymorphs is cryogrind, the changes in the relative integrated peak area indicate that the recrystallization accompanied by the formation of the neutral enol from the charged enolate. The NMR shoulder and the reduction in the relative peak area of the signal suggest that a relatively small population of the deprotonated species is formed in the amorphous piroxicam due to cryogrinding of crystalline piroxicam. Thus, most of the amorphous piroxicam consists of neutral piroxicam molecules; and the deprotonated species comprise only about 8% of the amorphous phase. However, a small population of charged species would be sufficient to impart color to a previously colorless sample. The diffuse-reflectance infrared Fourier transform spectroscopy spectra also give an evidence of the protonation of the pyridine N during cryogrinding. The band corresponding to C=O stretch for both polymorphs was found to remain unchanged during milling, suggesting that this carbonyl group remains intact during grinding. The new band appeared in the N–H stretch region corresponds to a second N–H bond in the amorphous piroxicam demonstrates protonation of the basic pyridine N.

The compounds exhibiting mechanochromic behavior accompanied by a change in their solid state during grinding appear to possess hydrogen-bonding ability, which causes the neutral molecules to order specifically in their crystalline state. Grinding, which disrupts the hydrogen bonding, causes the molecules to disorder, such that the disordered molecules are free to undergo an intermolecular proton transfer from the acidic functional group of one molecule to the basic functional group of another, resulting in a population of charged molecules in the disordered phase. As a result, crystalline piroxicam, under the mechanical stress, undergoes mechanochromism and transforms from the crystalline to the amorphous state, which accompanied by a change in the molecular structure and resulted in a color change from a colorless powder to a yellow mass. The intensity of the yellow color increases linearly with the increased time of the applied mechanical stress. While the solid-state change corresponds to the amorphization, the molecular change corresponds to the intermolecular proton transfer between the neutral piroxicam molecules in the crystalline phase.

Except for the intermolecular photophysical process, there is a special example based on solid-state organic reactions from asymmetric cyanostilbene derivatives [CN(L)-TrFMBE and CN(R)-TrFMBE], which shows shear- and UV-induced fluorescence switching by reversible $[2+2]$ cycloaddition.[22] These compounds exhibit virtually no fluorescence in both the crystalline and solution states. Their crystals are easily obtained with recrystallization techniques because of their strong intermolecular interactions. Surprisingly, these non-fluorescent crystals were found to gradually become highly fluorescent with strong sky-blue emissions when they are exposed to the prolonged irradiation from a UV light, as shown in Figure 7.7. From the optical microscopic observations and SEM images, there are distinct changes in the coloration and surface morphology of the irradiated crystals. In particular, the crystals became opaque with a color change from yellowish green to off-white, and the smooth and clear surfaces changed to extensive cracking and clefts after the UV irradiation. These suggest that the fluorescence enhancement upon the UV irradiation is probably caused by the reconstructive phase transition in either molecular conformation or stacking mode in the crystal state. Interestingly, these compounds can also be switched on by a small external shear-strain. The initially non-fluorescent powdered crystal of CN(L)-TrFMBE immediately became highly fluorescent when it was briefly smeared with a spatula. In order to understand the origin of this unusual behavior, a comprehensive crystal structure analysis and theoretical calculations were carried out.

CN(L)-TrFMBE and CN(R)-TrFMBE are structural isomers due to their significantly twisted conformations related to the different positions of the central CN group. The most important structural feature of CN(L)-TrFMBE and CN(R)-TrFMBE is the presence of antiparallel π-dimers, which comprise very close pairs of stilbenic C=C bonds with the center-to-center distances

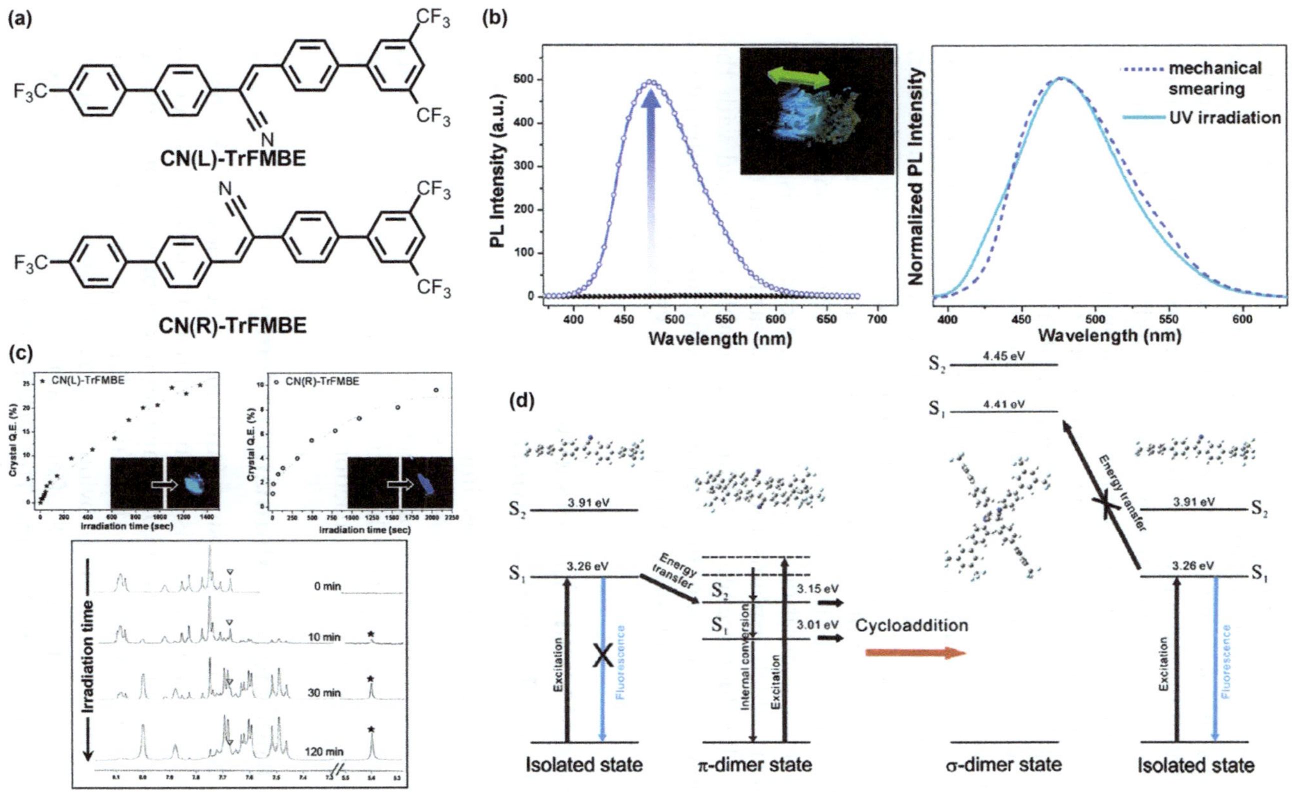

Figure 7.7 (a) Molecular structure. (b) Fluorescence turn-on of a powdered CN(L)-TrFMBE crystal on a quartz plate after smearing with a metal spatula and normalized PL spectra after mechanical smearing and UV irradiation. (c) Absolute PL quantum efficiencies of CN(L)-TrFMBE and CN(R)-TrFMBE as functions of irradiation time. And the ^1H NMR spectra after various irradiation times. (d) Proposed mechanism for the shear-induced emission of asymmetric cyano-stilbene in the solid state. Reproduced with permission from reference 22. Copyright 2009, The American Chemical Society.

of 3.785 and 3.889 Å, respectively. It is evident from the π-dimer geometries that the three asymmetrically substituted CF_3 units are the key elements of the antiparallel π-dimer formation because of the resulting C–F···H–C and C–F···π interactions in addition to the favorable π–π and dipolar interactions. By the comprehensive structural, optical and crystallographic analyses, these crystals contain π-dimer molecular pairs with a strong tendency to [2+2] cycloaddition, which results in the formation of volume-expanded σ-dimer. The pair of molecules in the antiparallel π-dimer is related through a center of symmetry and maintains close contact to enable [2+2] cycloaddition. The parameters of the antiparallel π-dimers of these compounds are suitable for the basic requirement of the topochemical [2+2] cycloaddition reaction to occur in the crystalline state. It thus suggests that the initial antiparallel π-dimers will be transformed into covalently bonded "σ-dimer ([2+2] cycloadducts)" *via* [2+2] cycloaddition when the crystals are illuminated with UV light. This is consistent with the initial non-fluorescent state of the single crystal and also the crystal cracking visible. The rationale for the former observation under UV irradiation is that the absorbed UV photon contributes exclusively to [2+2] cycloaddition rather than photoluminescence generation. The volume of the σ-dimers is larger than that of π-dimer, which explains the observed shear-induced emission. It suggests that the fluorescence switching is probably associated with the shear-induced displacement of the π-dimer pair, which increases the rate of photoluminescence at the cost of frustrated [2+2] cycloaddition.

7.3 Intramolecular Effects on Mechanofluorochromism

The intramolecular effect is another important factor which will effectively influence the absorption and emission spectra of the compounds under mechanical force stimuli. The intramolecular effects are mainly originated from the following two aspects: the intramolecular conformation change and the intramolecular chemical reaction.

7.3.1 Intramolecular Conformation Change

Many aggregation-induced emission (AIE) molecules possess mechanofluorochromic behavior, which is often ascribed to become better coplanarity than the initial twisted conformation under pressure or grinding. For example, an anthracene derivative (TPE-An) containing tetraphenylethylene moieties exhibits obvious mechanofluorochromic and AIE properties.[23] As shown in Figure 7.8, after grinding, the emission wavelength increased from 506 nm to 574 nm. The WAXD results indicated that the mechanofluorochromic resulted from the reversible morphological change between the crystalline and amorphous structures. DSC studies show that the pressed sample had a significant cold-crystallization peak, indicating the

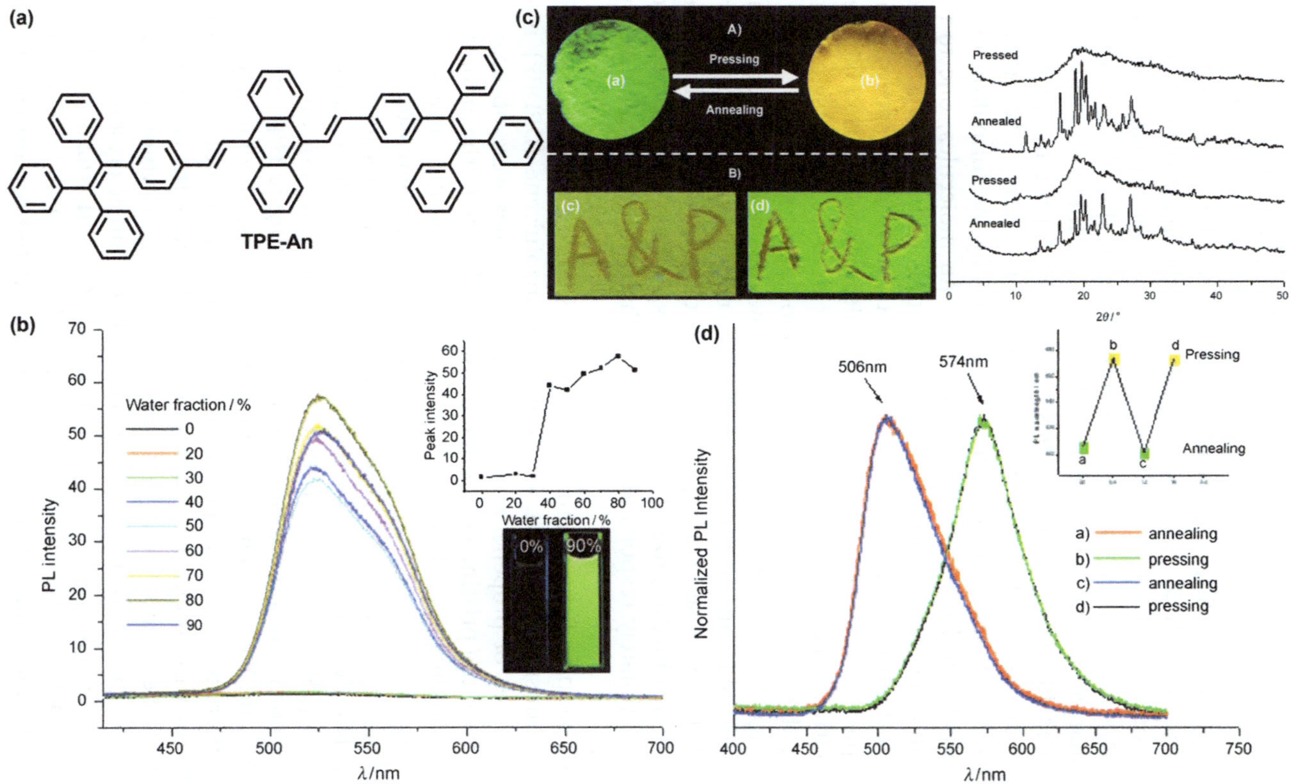

Figure 7.8 (a) Molecular structure of TPE-An. (b) PL spectra of TPE-An in water/THF mixtures. (c) The images of TPE-An taken at room temperature under stimuli and the WAXD pattern together with the normalized PL spectra. Reproduced with permission from reference 23. Copyright 2011, Wiley-VCH.

existence of a metastable-state aggregation in the pressed sample, which can be converted into a more stable state through annealing.

TPE-An molecule adopts a highly twisted conformation due to the existed steric hindrance between the aryl rings. Owing to the twisted conformation and weak π–π interactions, the molecular packing would be relatively loose with some defects (cavities) and the crystals are readily destroyed by the external pressure, which may be triggered by the planarization of the molecular conformation or the slip deformation under external pressure. The destruction of the crystalline structure leads to the planarization of the molecular conformation because of the release of the twist strain. This should be considered as one of the possible reasons for the increase of the conjugation degree, resulting in a red-shift in the PL spectrum. Whether the red-shift in the PL spectrum of TPE-An is caused by the formation of excimers from the anthracene cores between TPE-An molecules, due to the decreased intermolecular distance and destruction of highly twisted molecular conformation upon pressing or not, the UV absorption spectra is an important piece of evidence. If the increased molecular conjugation induced by the planarization of molecular conformation can cause such a significant red shift in the emission spectra, the absorption spectra should exhibit a similar change, whilst remaining the same in the case of excimer emission. Therefore, the red shift of the absorption spectra after grinding may originate from the increased conjugation with the decreased optical band gap of the molecules in the ground sample, that is, better intramolecular coplanarity is ultimately responsible for the mechanofluorochromism of the compound.

Another example of the intramolecular effect is reported on the unusual mechanochromism of a group of diphenyldibenzofulvene (DPDBF) molecules.[24] A series of propeller-shaped DPDBF molecules were observed to emit strongly in the pristine crystalline phase, but faintly in the amorphous phase after grinding. The decreased intensity after grinding was associated with a significant red shift. To investigate the origin of the emission change, the two single crystals of DPDBF with totally different emissions were investigated. The two polymorphs were green single crystal (GSC) and yellow single crystal (YSC), as shown in Figure 7.9. After ruling out any specific strong intermolecular interactions (such as π–π stacking or H/J-aggregates) in both GSC and YSC, it seems that the different emission color may result from the difference of the propeller-like conformation of molecules in the crystals. So the various emission colors may be ascribed to the conformation difference of the independent molecules in the two polymorphs. Closer examination of the conformation of molecules in the single crystals revealed that the torsion angles of the two phenyl rings in GSC ($\theta1 = 62.5^\circ$, 57.0° and $\theta2 = 51.1^\circ$, 51.4°) are larger than those in YSC ($\theta1 = 52.9^\circ$, $\theta2 = 47.8^\circ$), indicating that the molecules in GSC adopt a more twisted conformation than those in YSC. Thus a lower degree of conjugation of molecules in GSC induces the blue emission comparing to that of YSC. GSC also exhibits higher emission efficiency than that of YSC, which may be attributed to more

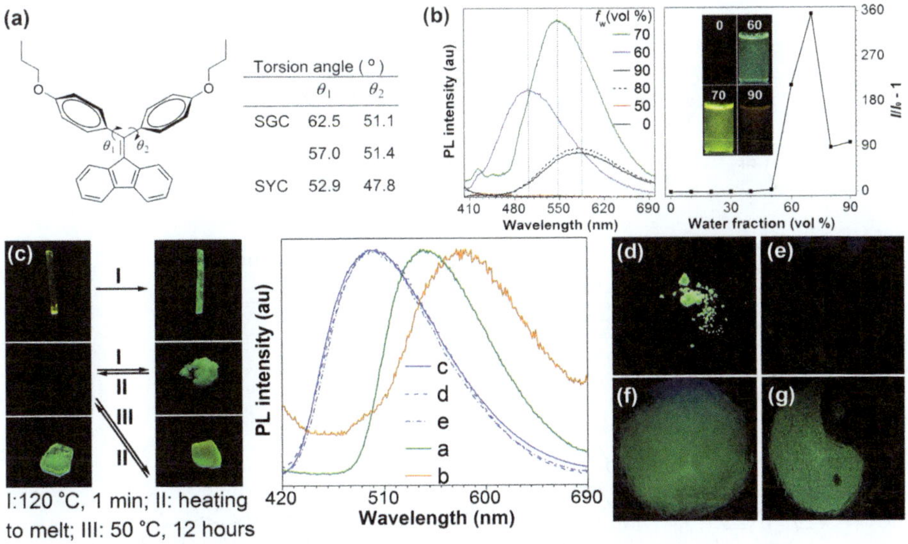

Figure 7.9 (a) Molecular structure and the selected torsion angle in crystals. (b) PL spectra in acetonitrile and acetonitrile/water mixtures. (c) Digital images under different stimulus (under UV illumination). Photographs of the dye (d) before and (e) grinding. And annealed (f) regrinding of selected areas (g).
Reproduced with permission from reference 24. Copyright 2011, Wiley-VCH.

C–H···π and C–H···O intermolecular interactions in GSC than in YSC. It will help to further solidify the molecular conformation and block the non-radioactive pathways.

Besides, some tetraphenylethene (TPE) derivatives also have obvious mechanochromic property.[25,26] One typical example is a benzothiazolium-functionalized tetraphenylethene molecule, in which it was found that its crystal can change the emission from yellow to red after gentle grinding using a pestle and a mortar. Interestingly, the three polymorphs of crystal with different emission wavelengths and efficiency can be prepared from different solvent mixtures.[27] To gain insight into the distinct emission behavior of the crystals, the geometric structures and packing arrangements were studied. Due to the propeller-shaped TPE unit, all the crystals adopt a highly twisted conformation. The torsion angles between the bridged phenyl ring and the vinyl core of TPE in the crystals of **1**, **1** · 2/3 THF and **1** · EtOAc are 70.42°, 70.37° and 67.94°, respectively, suggesting that the molecular conjugation is in order. This agrees well with their observed emission maximum, in which the crystals of **1** · EtOAc are the redder emitters. The energy band gaps of **1**, **1** · 2/3 THF and **1** · EtOAc are calculated to be 1.89, 1.83 and 1.79 eV, respectively, correlating with their different emission colors.

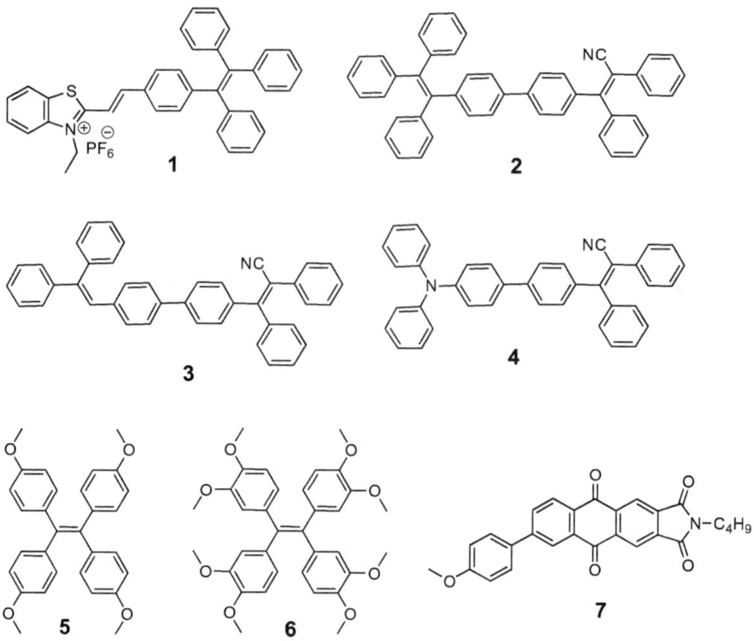

Another example is a group of cyano-containing triphenylacrylonitrile (TPAN) derivative luminophores, namely **2**, **3** and **4**. According to the conformation analysis, the specific C–H· · ·π, C–H· · ·O and C–H· · ·N interactions help lock and rigidify the propeller-like conformations of AIE molecules, which effectively restrict their intramolecular rotations and thus dramatically enhance their crystalline-state emissions. All the molecules adopt a highly twisted conformation in the crystalline state. Moreover, their propeller-like conformations prevent the formation of detrimental species such as excimers or exciplexes. The molecules may adopt a more twisted conformation in order to fit into the crystalline lattice. This shortens their effective conjugation length and hence makes them emit blue light. On the other hand, once the crystalline lattice collapses by the external stimuli, the dye molecules may relax to a more planar conformation and thus emit red PL.[28]

Two methoxy-substituted TPE derivatives, **5** and **6**, were reported to exhibit typical mechanofluorochromic behavior.[29] **5** has two different polymorphs, **5**-1 and **5**-2. The strong fluorescence of the **5**-1 pristine crystals can change from blue (420 nm) to cyan (480 nm) by grinding. While the emission from the **5**-2 crystal changes from 440 nm to 487 nm after grinding. And the emission from the **6** pristine crystal red shifts from 460 nm to 480 nm after being ground. The more planar conformation and better conjugation of a single molecule in the amorphous state are the inherent reason for the fluorescence red shift. The fluorescence can almost go back to the original again upon thermal treatment. The crystal structures and theoretical calculations both prove that their emissions are determined by the single molecular conformation (or conjugation state) rather than the intermolecular effects, such as H- or J-aggregation and π–π stacking. In the absence of the intermolecular effects, **5** and **6** have comparable high quantum yields in their crystalline and amorphous states.

D–A molecules containing anthraquinone imide (AQI) moieties with different electron-donating or -withdrawing substituents also possess mechanochromism properties, such as compound 7. The intermolecular interactions such as the enhanced π–π stacking have been excluded by the experimental results from XRD analyses of these compound powders and optical spectroscopic characterization of the molecularly dispersed polymer blends of these compounds. The intramolecular interactions thus might be responsible for the mechanochromism, because pressure would induce a decrease in the dihedral angle between the D and A moieties, and consequently, result in a better conjugation. Photoluminescence (PL) experiments of the solids illustrated that the emission spectra were also red-shifted to a similar extent as that of the absorption spectra under the pressure. Therefore, the origin of the mechanochromism in aryl-substituted AQIs should due to the enhanced ICT caused by external pressure, which results from the decreased torsion angles between the electron-donating and -withdrawing segments in the D–A molecules. This provides a new insight into the mechanochromic mechanism that is different from those described above.[30]

7.3.2 Intramolecular Chemical Reaction

The mechanically induced changes in color and/or fluorescence, due to the formation or break up of excimers, transition of aggregation mode and changes in conjugation or in charge state, have been introduced. In these systems, the mechanically active species has generally been in response to their physical properties. However, a special example has been reported on the covalently linked spiropyrans (SP) as highly effective color-generating mechanophores that can provide visible detection and mapping of mechanical stresses through their mechanically induced transformation to the merocyanine (MC) conformation in glassy and elastomeric chain growth polymers.[31,32] The spiropyrane molecule is well characterized and exhibits both thermo- and photochromic responses that can reversibly transform a closed, colorless spiropyran form to a highly colored, planar merocyanine structure through the rupture of the spiro carbon-oxygen (C–O) bond. The increased conjugation of the merocyanine shifts the absorption to longer wavelengths, thus gives visible color and fluorescence of molecule. In addition, the spiro C–O bond is the critical point for the mechanically induced activation, and there are many possible attachment points on the spiropyran that differ in their ability to transmit force efficiently across the junction. The simple mechanical models indicate that the attachment on the opposing sides of the spiro-junction, specifically, positions of the indole side and positions of the benzopyran side, preferentially stresses the spiro C–O bond over the spiro C–C bond. The results of the dynamics simulation suggest that the selective spiro C–O bond rupture under a variety of applied forces, ranging from 2 to 3 nN.

Experimentally, an elastomeric mechanophore-linked poly(methyl acrylate) (PMA) and a glassy mechanophore cross-linked poly(methyl methacrylate) (PMMA) mechanochemical transduction revealed the mechanochemical transduction with the visible color change. But the physical properties of these polymers were not ideal for the investigation of the kinetics or thermodynamics of the mechanically induced transformations of SP mechanophore in bulk polymers. The initial polymer systems studied were not amenable to the kinetic analysis because they were either too soft (elastomeric PMA) or too glassy (PMMA). In PMMA, due to the high T_g, the rate of conversion from MC to SP form could not be investigated at room temperature. And then, polyurethane (PU) with the incorporation of SP mechanophore, which possess the inherent mechanical toughness, elasticity and low glass transition temperature, enables the effects of mechanical force on the SP-MC equilibrium. The equilibrium between the colored MC and colorless SP form can be directly controlled by the mechanical strain. Since the low T_g, equilibrium is reached in experimentally accessible timers at room temperature. Therefore, the kinetics of the mechanically activated SP to MC conversion as well as the thermally activated conversion of MC to SP in a bulk polymer can be studied. When the polymer was uniaxially stretched, a deep purple coloration appeared along the entire gauge section of the sample, demonstrating the stress-induced formation of the open, MC form of the mechanophore. The relatively rapid interconversion between SP and MC forms, coupled with the robust mechanical properties of PU, enables a quantitative analysis of the effects of strain on the equilibrium between the SP and MC forms. The absorbance measurements confirmed the lack of mechanical activation, displaying no MC peak for both the unstretched and stretched spectra. The increasing levels of strain correlate to larger amounts of mechanical force on the mechanophore, which in turn opens an increasing percentage of the mechanophores from SP to MC, resulting in the growing MC peak in the absorbance spectra. The closing kinetics of the mechanophore can be studied by fluorescence imaging. It indicates that the slight increase in the time constant could be a result of slowly relaxing residual strain in the polymer preventing the mechanophore from closing as quickly. In addition, when increasing the strain linearly, and then holding the strain at a constant value, the MC form did not revert back to the thermodynamically preferred SP form, indicating a strain-induced change in the energy landscape of the SP mechanophore system.

7.4 Conclusion

In this chapter we have attempted to provide an overview of the mechanisms of mechanofluorochromic behaviors based on organic/polymer systems and highlight the most significant developments in understanding the origin of mechanofluorochromism. In pursuing this goal, views are mainly focused on the intermolecular effects and intramolecular effects on the mechanofluorochromism, including excimer forming, phase structural transition, J- or H-aggregation and intramolecular coplanarity and so on. It is our wish that this chapter can draw a full view of the mechanisms of mechanofluorochromic behaviors based on organic/polymer system, and help the researchers to understand more deeply the relationship between the molecular structures and the mechano-responsive behaviors.

References

1. Y. Sagara and T. Kato, *Nat. Chem.*, 2009, **1**, 605–610.
2. A. Pucci and G. Ruggeri, *J. Mater. Chem.*, 2011, **21**, 8282–8291.
3. Z. G. Chi, X. Q. Zhang, B. J. Xu, X. Zhou, C. P. Ma, Y. Zhang, S. W. Liu and J. R. Xu, *Chem. Soc. Rev.*, 2012, **41**, 3878–3896.
4. S. P. Anthony, *ChemPhysChem.*, 2012, **77**, 518–531.
5. K. Ariga, T. Mori and J. P. Hill, *Adv. Mater.*, 2012, **24**, 158–176.
6. C. Lowe and C. Weder, *Adv. Mater.*, 2002, **14**, 1625–1629.
7. B. R. Crenshaw and C. Weder, *Chem. Mater.*, 2003, **15**, 4717–4724.
8. J. Kunzelman, M. Kinami, B. R. Crenshaw, J. D. Protasiewicz and C. Weder, *Adv. Mater.*, 2008, **20**, 119–122.
9. Y. Sagara and T. Kato, *Angew. Chem. Int. Ed.*, 2008, **47**, 5175–5178.
10. Y. Sagara, S. Yamane, T. Mutai, K. Araki and T. Kato, *Adv. Funct. Mater.*, 2009, **19**, 1869–1875.
11. M. Teng, X. Jia, X. Chen, Z. Ma and Y. Wei, *Chem. Comm.*, 2011, **47**, 6078–6080.
12. M. Teng, X. Jia, S. Yang, X. Chen and Y. Wei, *Adv. Mater.*, 2012, **24**, 1255–1261.
13. K. Nagura, S. Saito, H. Yusa, H. Yamawaki, H. Fujihisa, H. Stao, Y. Shimoikeda and S. Yamaguchi, *J. Am. Chem. Soc.*, 2013, **135**, 10322–10325.
14. Y. Sagara, T. Mutai, I. Yoshikawa and K. Araki, *J. Am. Chem. Soc.*, 2007, **129**, 1520–1521.
15. M. Sase, S. Yamaguchi, Y. Sagara, I. Yoshikawa, T. Mutai and K. Araki, *J. Mater. Chem.*, 2011, **21**, 8347–8354.
16. S. Yoon, J. W. Chung, J. Gierschner, K. S. Kim, M. Choi, D. Kim and S. Y. Park, *J. Am. Chem. Soc.*, 2010, **132**, 13675–13683.
17. (a) Y. Dong, B. Xu, J. Zhang, X. Tan, L. Wang, J. Chen, H. Lv, S. Wen, B. Li, L. Ye, B. Zhou and W. Tian, *Angew. Chem. Int. Ed.*, 2012, **51**, 10782–10785; (b) Y. Dong, J. Zhang, X. Tan, L. Wang, J. Chen, B. Li, L. Ye, B. Xu, B. Zou and W. Tian, *J. Mater. Chem. C*, 2013, **1**, 7554–7559.
18. Y. Ren, W. H. Kan, V. Thangadurai and T. Baumgartner, *Angew. Chem. Int. Ed.*, 2012, **51**, 3964–3968.
19. J. luo, L. Li, Y. Song and J. Pei, *Chem. Eur. J.*, 2011, **17**, 10515–10519.
20. M. S. Kwon, J. Gierschner, S. Yoon and S. Y. Park, *Adv. Mater.*, 2012, **24**, 5487–5492.

21. A. R. Sheth, J. W. Lubach, E. J. Munson, F. X. Muller and D. J. W. Grant, *J. Am. Chem. Soc.*, 2005, **127**, 6641–6651.
22. J. W. Chung, Y. You, H. S. Huh, B. An, S. Yoon, S. H. Kim, S. W. Lee and S. Y. Park, *J. Am. Chem. Soc.*, 2009, **131**, 8163–8172.
23. X. Q. Zhang, Z. G. Chi, H. Y. Li, B. J. Xu, X. F. Li, W. Zhou, S. W. Liu, Y. Zhang and J. R. Xu, *Chem. Asian J.*, 2011, **6**, 808–811.
24. X. Luo, J. Li, C. Li, L. Heng, Y. Q. Dong, Z. Liu, Z. Bo and B. Z. Tang, *Adv. Mater.*, 2011, **23**, 3261–3265.
25. J. Wang, J. Mei, R. Hu, J. Z. Sun, A. Qin and B. Z. Tang, *J. Am. Chem. Soc.*, 2012, **134**, 9956–9966.
26. X. Luo, W. Zhao, J. Shi, C. Li, Z. Liu, Z. Bo, Y. Q. Dong and B. Z. Tang, *J. Phys. Chem. C*, 2012, **116**, 21967–21972.
27. N. Zhao, Z. Yang, J. W. Y. Lam, H. H. Y. Sung, N. Xie, S. Chen, H. Su, M. Gao, I. D. Williams, K. S. Wong and B. Z. Tang, *Chem. Commun.*, 2012, **48**, 8637–8639.
28. W. Z. Yuan, Y. Tan, Y. Gong, P. Lu, J. W. Y. Lam, X. Y. Shen, C. Feng, H. H. Y. Sung, Y. Lu, I. D. Williams, J. Z. Sun, Y. Zhang and B. Z. Tang, *Adv. Mater.*, 2013, **25**, 2837–2843.
29. Q. Qi, Y. Liu, X. Fang, Y. Zhang, P. Chen, Y. Wang, B. Yang, B. Xu, W. Tian and S. X.-A. Zhang, *RSC Adv.*, 2013, **3**, 7996–8002.
30. F. Chen, J. Zhang and X. Wan, *Chem. Eur. J.*, 2012, **18**, 4558–4567.
31. D. A. Davis, A. Hamilton, J. Yang, L. D. Cremar, D. V. Gough, S. L. Potisek, M. T. Ong, P. V. Braun, T. J. Martinez, S. R. White, J. S. Moore and N. R. Sottos, *Nature*, 2009, **459**, 68–72.
32. C. K. Lee, D. A. Davis, S. R. White, J. S. Moore, N. R. Sottos and P. V. Braun, *J. Am. Chem. Soc.*, 2010, **132**, 16107–16111.

Subject Index